Beginner's Guide to Guns & Shooting

Understanding Firearms:
How They Work & How to Safely Use Them

REVISED EDITION

CLAIR F. REES

E P B M
ECHO POINT BOOKS & MEDIA, LLC

Senior Staff Editor: Harold A. Murtz
Revision Editor: Rick L. Fines
Editorial/Production Assistant: Maria L. Connor
Cover Photography: John Hanusin
Managing Editor: Pamela J. Johnson
Publisher: Sheldon L. Factor

Published by Echo Point Books & Media
Brattleboro, Vermont
www.EchoPointBooks.com

ISBN: 978-1-62654-146-7

Cover design by Adrienne Núñez,
Echo Point Books & Media

Editorial and proofreading assistance by Ian Straus,
Echo Point Books & Media

Printed and bound in the United States of America

Table of Contents

INTRODUCTION

MANY PEOPLE, with the best of intentions and lack of knowledge, simply do not like guns — either for themselves or in the hands of others. If any of us are going to maintain our right to own and shoot guns, we need to understand guns thoroughly before we can expect the uninformed to respect our rights.

Most uninformed people refer to all guns as "weapons." Note that soldiers and police officers carry weapons. The vast majority of guns used for hunting and recreational shooting are not weapons. Guns, yes, but weapons, no. A fine old shotgun or a modern target rifle MAY be used as a weapon, but then a baseball bat may also be misused against another human.

Even some very experienced shooters are quick to take a negative attitude toward guns which are not their particular favorites. For example, many people enjoy collecting and shooting military-type rifles. These guns are not for beginners, but when they are owned for their historical interest and the simple fun of shooting them, they are no more — or less — dangerous than the fine old shotgun mentioned earlier. Yet, I hear shooters who should know better say things like, "No one needs a gun like that ... " Remember, that to people who dislike guns, ALL guns are disliked, whether they are finely engraved and stocked in checkered walnut or a very plain military-type gun. Splitting hairs makes little difference to the anti-gun people. Making such differentiations should not be something shooters do. If we are to expect others to be objective, we should cultivate the same attitudes within our ranks.

We should also give some thought to where the beginning shooter should buy his guns, ammunition and hardware.

In recent years, large discount chain stores have made substantial inroads in the gun business. While it is true that the discounters can often sell for a few dollars less than a local gun shop, the small amount of money saved may not be a real savings. The local gun store people are generally far more knowledgeable and have more time to impart that knowledge than does the clerk at the discount store. The clerk at the discounter may well have been selling saws or shampoo the week before he became the resident "expert" behind the gun counter. Before you decide to save ten bucks on a new gun, think again.

Firearms in and of themselves are harmless. It's when they're in the hands of a careless — or criminally deranged — individual that they become potentially lethal. Properly used, guns — like automobiles, carpenter's saws, lawnmowers and other tools

— are completely safe. To fear any of these useful implements is ridiculous. If improperly used, nearly any tool can be dangerous. But without tools, we would still be living in caves and eating raw what small animals we could catch with our bare hands. Indeed, it's our skillful use of tools that sets us apart from the lower animals, and to avoid tools out of unreasoning fear is irrational.

Firearms are merely inanimate tools or pieces of sporting equipment that *can* be misused, but ordinarily provide only clean, safe recreation in the hands of knowledgeable sportsmen. Because firearms are designed to throw one or more small objects at relatively high speeds, and because these objects, or projectiles *can* cause damage, certain safety rules must be followed when handling guns. Just as you wouldn't push your hand into the whirling blades of a lawnmower, there are things you shouldn't do with a gun. If you learn — and obey — the rules, you'll never have a shooting accident. These "accidents" occur only when a firearm is being misused in some way. Obey the rules, and "accidents" simply can't happen!

Right about here, someone is going to raise this familiar question — "But guns are *designed* to kill. How can they possibly be considered safe?" While no one is sure of the exact date or origin of the first true firearm (historians are still arguing as to which *country* should get the credit for this invention, let alone the name of the individual who first developed the tool), it's probably a good bet that it was used as a means of defense sometime fairly soon after its discovery. As a matter of fact, evidence exists that certain primitive firearms were used to throw arrows rather than bullets and were regarded as somewhat noisy substitutes for the long bow.

But, to say that guns are good only for killing is to ignore the facts. Sportsmen have historically used rifles, handguns and shotguns for pure (and bloodless) recreation, and a large number of shooters today have never fired at a flesh-and-blood target. Many forms of target shooting exist, and competition in these sports extends to the Olympic games. I might add that while formal competitive shooting is a popular — and growing — sport, many people get more enjoyment from a relaxed plinking session with tin cans as the target and no attempt made to keep score.

Fencing is regarded as a perfectly acceptable sport by many of the same people who condemn guns and shooting, and even archery fails to horrify onlookers. Yet both require the same skills and tools once commonly used for hunting and warfare. Javelin throwing falls into a similar category, and even baseball owes its origin to some shaggy-haired Neanderthal who learned to conk small animals (or his neighbor) on the head with a rock.

We could list a number of popular games and sporting events that could be traced to an early battlefield, but that's beside the point. The fact is that marksmanship today — whether it be with a rifle, shotgun or handgun — isn't necessarily a "blood sport." So even if you object to using firearms for hunting or even self-defense, there's no reason you still can't enjoy the recreation firearms can provide.

Although many of us are outdoorsmen who like to hunt and fish, it's not the purpose of this book to defend those pursuits against those who condemn these pastimes. Similarly, this book will not go into the pros and cons of using firearms for self-defense. We simply intend to teach the safe and proper use of rifles, handguns and shotguns, with emphasis toward the purely recreational aspects of learning to shoot.

Shooting is fun. What's more, it's good, clean fun that's neither immoral, illegal (as long as you obey local ordinances), nor fattening. It's a sport both men and women of almost any age can learn to enjoy, and it doesn't necessarily require a lot of expensive equipment. While you'll improve with practice, you can have a lot of fun during your very first trip to the range. The basics are easily learned, and if at first you're too shaky to hit the target, you can always steady the firearm against a fence post or some other kind of rest.

The key to safe shooting fun is learning a few easily memorized safety rules, and then following them until they become habit. These are simple, straightforward rules that are easy to understand and learn, and if you'll only obey them each and every time you pick up a firearm, we can promise you you'll never accidentally shoot yourself or anyone else in the neighborhood. With this knowledge, you'll be able to handle any kind of rifle, handgun or shotgun safely, and you'll soon see how foolish it is to fear guns. It makes as much sense as being afraid of baseball bats, hair dryers or the kitchen sink.

Respect is another matter. No properly trained shooter will point even an empty gun in the general direction of another individual, unless he's a policeman acting in the line of duty. An experienced shooter knows what kind of damage a loaded firearm is capable of producing, and he treats *every* firearm he handles with the same respect he would give a loaded gun out of well-ingrained habit. Even if he's checked and double (even triple) checked to make sure the gun he holds has no ammunition in the magazine or chamber and can't possibly be fired, he never allows the muzzle to point in a direction that would likely cause harm if a bullet *were* to leave the barrel.

That brings up another excellent reason for learning to shoot. Serious firearm accidents almost never happen at the hands of a trained, careful sportsman. It's the untrained child or the inexperienced adult who causes most of the shooting mishaps you read about in the morning paper. Because these people know so little about guns and their proper handling, they — not the firearms — can be considered dangerous. While they may even fear guns a little, they don't know enough about these tools to give them the proper respect. Because of this, one of the first safety rules you'll learn is to keep firearms out of the hands of untrained people. If you take the effort to learn to handle guns safely, you certainly don't want to be the victim of another's ignorance.

Guns are interesting, and the development of firearms over the years makes a fascinating history. There is some evidence that the Chinese used a primitive firearm of bamboo to propel some kind of solid matter through the air late in the 13th century, and documents exist showing that more sophisticated guns throwing either arrows or iron bullets were in use in both England and Italy as early as 1326. Since that time firearms have played a large part in determining the destiny of nations, and if anything, they're more important today than they were 700 years ago.

Even if you're not a historian, it's fun to learn about guns and shooting. Few of man's tools have been so refined over the years, and the modern firearm is much more accurate and easier to shoot than the guns our grandfathers used. At the same time, many old guns are still in service and a great interest is growing around blackpowder-shooting muzzle-loading firearms that were widely used a century or more ago. Low-cost replicas of these old guns are available from an unbelievable number of sources, and most are both fun and safe to use. Nostalgia buffs delight in the pungent cloud of white smoke that hangs in the air after every shot from a blackpowder gun, and the most ardent enthusiasts wear authentically styled clothing made of skins and furs.

Most beginning shooters learn the basics with a 22 rimfire rifle before moving on to handguns, shotguns or big-bore rifles. And some marksmen have so much fun with the little 22 that they don't ever bother with larger guns. A 22 rifle or handgun is all you'll ever need to enjoy a lifetime of informal plinking, and you may never feel the need to graduate to a more powerful (and more costly) firearm. Similarly, shooting clay targets with a shotgun may capture your interest to the point that rifle or handgun shooting remains unimportant. Whatever the case, this book can teach you what you need to learn to safely shoot each type of gun. Reading the whole book will give you a good basic education in shooting, and the knowledge you need to use all three types of firearms in safety.

Shooting is a fun, clean sport that almost anyone can enjoy. With the proper training and the right rules to follow, it's a perfectly safe sport that the whole family can participate in. Our intention in publishing this book is to introduce new and would-be shooters to this fascinating form of recreation and to teach them the basics of firearms in general.

NOTE: *For any terms which may be unfamiliar to you, turn to the Glossary in the back of this book for a full explanation.*

chapter 1

HOW FIREARMS OPERATE

WHAT IS A GUN? A dictionary would define the term as "a mechanical device that throws one or more projectiles, usually through a tube or barrel." That's a pretty broad definition, and there are a number of different devices that fit that general classification. These include rifles, shotguns, handguns, and rifle-shotgun combinations, as well as much larger mortars and artillery pieces.

Most of these guns use a charge of gunpowder or some other highly combustible substance to power the projectile, and these are called firearms. However not all guns are firearms, as steam, compressed air, carbon dioxide, spring pressure and even rubber bands have been used to get the projectile — or bullet — underway. What's more, all guns don't shoot "bullets." Some of the earliest firearms we know about shot arrows instead, and even today some airguns shoot both feathered darts and soft lead pellets ("bullets") interchangeably. Shotguns don't shoot "bullets" either, but throw charges containing hundreds of individual shot pellets each time they're fired. And some specialized guns, used for temporarily tranquilizing wild animals, fire hypodermic syringes.

While there are all kinds of interesting devices around that could be loosely classified as guns,

Explosive-charged arrows were used in conjunction with hand cannons. While this combination was a forerunner of the modern military mortar, it was also dangerous as the shooter really had no idea whether or not the arrow would explode on target or in his hand.

this book will concern itself with only the three basic types used by sportsmen — rifles, handguns and shotguns. While we will be mainly concerned with true firearms, we also will be covering gas and air-powered guns.

This ancient woodcut shows the use of one of the earliest hand-held firearms — the hand cannon. A hot wire or torch was applied to the "touchhole" to get the "gun" to fire.

The cannon lock — merely a hole drilled through the rear of the barrel to allow flame to be passed from a lighted torch to the charge of gunpowder inside — was one of the earliest forms of ignition used in firearms and was still in use in large cannons throughout the Civil War. The cannon shown here is a modern replica.

History

To better understand guns and how they work, we need to know something about how they evolved. How long have guns been around? How did the first guns work? And what changes have been made to firearms over the years? The fact is that modern 20th century guns operate on the same general principles as the firearms used 600 or 700 years ago, and it hasn't been until recent years that any real improvements have been made in the propellant powder burned in guns.

No one knows which early alchemist first discovered that a mixture of sulphur, potassium nitrate (saltpeter) and charcoal could be produced that would explode on contact with flame. Depending on which source you turn to: Gunpowder was discovered in either Europe or China sometime prior to A.D. 850; by a German monk named Berthold Schwartz ("Black Berthold"); or during the 1300s by the Byzantine Greeks; or by the Arabs in "the most ancient of times." Despite these and other widely conflicting opinions — some historians have even claimed that Alexander the Great had seen the use of gunpowder more than

three centuries before Christ appeared on the scene — the most popular opinions among scholars is that gunpowder didn't come into general use until sometime in the 13th century. At least that's when the earliest guns we know of were used, although gunpowder itself was likely employed to frighten enemy soldiers and their horses long before it was used to shoot bullets from primitive firearms.

And the very first guns *were* primitive affairs. Some of the early Chinese "guns" were simply tubes of bamboo reinforced with windings of wire or cord, while in Europe cast metals were used almost from the first. The very first guns probably held a loose-fitting stone or ball loaded ahead of a quantity of powder, which was ignited by simply passing a lighted torch near the muzzle or front end of the gun. This system of ignition left something to be desired (for one thing, you'd probably need a brand-new torch for each firing), and it wasn't long before some bright gunner thought up the cannon lock. All this amounted to was a small hole drilled through the wall of the barrel at the rear, or breech end. After the gun was loaded with powder and shot, some extra powder was trickled

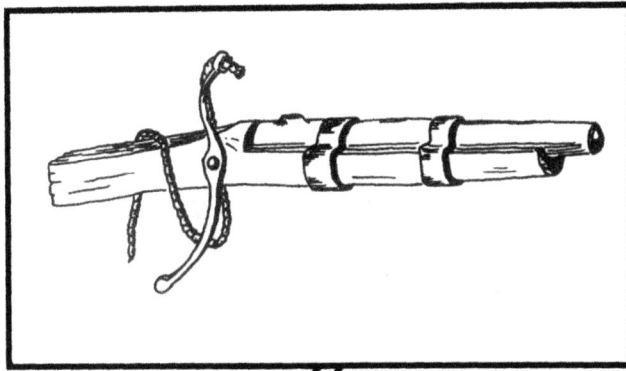

Left — This sketching depicts the design similar to the first matchlock small arms. The serpentine ("s"-shaped arm) holds the match cord. To fire the shooter pulled backwards on the bottom end of the serpentine which brought the burning cord into contact with the powder in the touchhole.

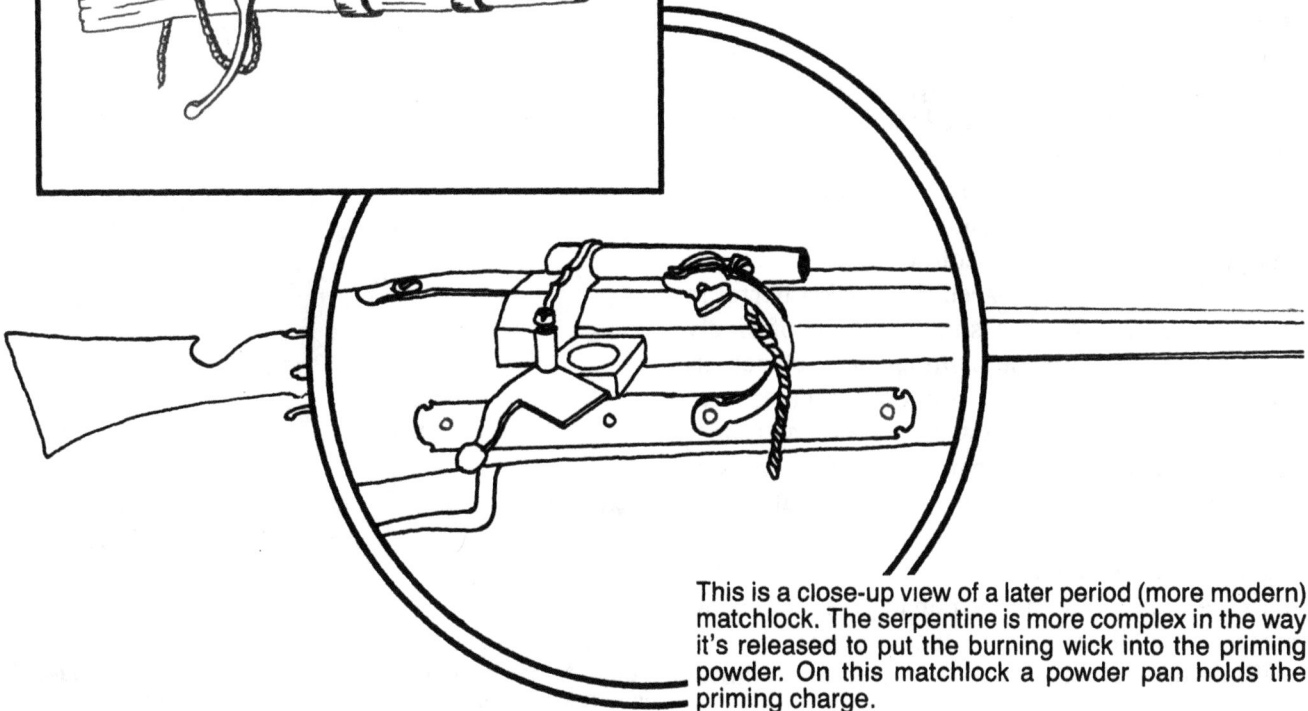

This is a close-up view of a later period (more modern) matchlock. The serpentine is more complex in the way it's released to put the burning wick into the priming powder. On this matchlock a powder pan holds the priming charge.

into this hole. Touching the top of the hole with a torch ignited the powder within the breech, and the gun went off.

This simple ignition system was used almost unchanged in large cannons up through the Civil War. It was uncomplicated, reliable, and — in a properly loaded gun — safe. The guns had to be loaded and cleaned from the muzzle, but at least the gunner could stand well to the rear when he fired his weapon.

The first guns designed to be carried and used by a single man sported stocks that amounted to little more than a simple handle. Later on, short hooks and other projections were added near the muzzle to allow the gunner to brace his weapon against a wall or some other structure; this improvement made it easier to control recoil, as much of this force was then absorbed by the wall the gun was rested against. The addition of these hooks (*haken* in German) caused this adaptation to be known as the *hakenbüchse*.

Because it was very awkward for a single gunner to aim his weapon with one hand while applying a torch or some form of match to the touchhole with the other, a more convenient means of firing was

needed. Around 1450, the matchlock was developed in Europe. The matchlock was merely an S-shaped pivoting arm with a clamp at the top end of the "S" which was fastened to the side of the gun. A piece of slow burning fuse, or match cord was held in the clamp, and when it was time to fire, the gunner simply pulled backward on the bottom end of the S-shaped arm to bring the match into contact with the powder in the touchhole. The matchlock was greatly improved and modified over the years, and more convenient triggers based on the lever principle were installed. Priming pans (covered metal pans holding a small quantity of powder) were added to the simple open touchhole to insure quicker ignition, and other changes were also made before the matchlock was superseded by the wheellock, snaphaunce and flintlock. And in spite of these later improvements, matchlock guns were still being produced in China and Japan as late as 1875, or just over a century ago.

The early matchlock infantry guns, known as muskets, were heavy, clumsy affairs weighing about 16 pounds. They had a wooden stock that was held against the shoulder when firing, and the gun itself was usually steadied in a long, forked rest carried for that purpose. The soldiers or musketeers using these weapons eventually began carrying bandoleers holding a dozen or so containers, with each container holding just enough gunpowder for a single charge. The lead balls used for bullets were carried in a separate pouch.

To load his weapon, the musketeer held it with the muzzle up, poured a charge of powder down the inside of the barrel (called the bore), used his ramrod to tamp a cloth patch over the powder, and then rammed a lead ball down the bore. This was followed with another patch to help keep the charge in place. Then the soldier stuck his forked rest in the ground and placed the musket barrel in it. Finally, he poured a small amount of finely ground powder into the priming pan, closed its cover, clamped a length of smoldering match cord to the cock (or clamp), blew on it to fan the flame, and opened the priming pan cover. Then he was ready to fire. Reloading took about 2 minutes for a practiced musketeer, and he was defenseless during that time. Because of this, 16th century musketeers were paired with pikemen to give them some protection while recharging their weapons.

While the matchlock was used in some countries

1

In the following sequence of drawings you'll see a musketeer firing an ancient matchlock. Here the musketeer is at rest. The loading procedure began when . . .

. . . the musketeer now rammed the ball down the muzzle, making sure it was well seated atop the powder. After returning the ramrod . . .

. . . he primed the flash pan with fine powder from the small priming flask, taking utmost care to keep the burning match at a safe distance. Finally, . . .

4

5

2

. . . he grabbed his powder-filled wooden container from his bandoleer and opened it with his thumb. Next . . .

. . . with the gun fully loaded the musketeer now rested the musket's forend in the forked stick, blew the ashes off the match cord tip and clamped it firmly into the serpentine, and . . .

3

. . . he poured the powder into the barrel, took a metal ball (or bullet) from his pouch and started it into the muzzle with his thumb. With his ramrod . . .

. . . assumed the aiming stance and took careful aim. Our musketeer was now ready to fire.

6

7

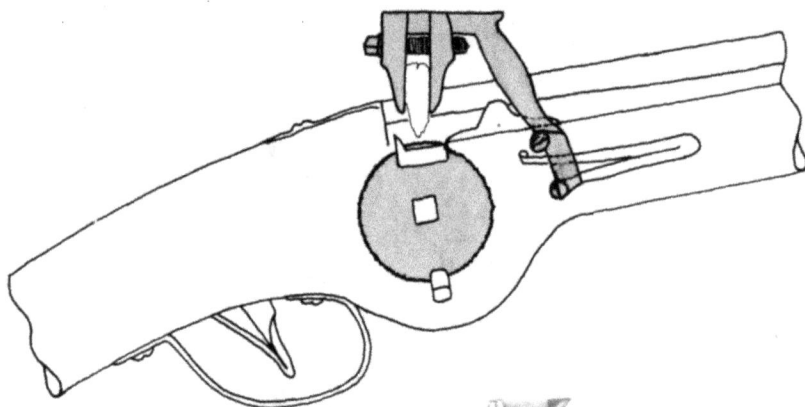

In the process of firearms evolution, the wheellock (shown here) came after the matchlock. When the wheellock is to be fired, the flint-holding cock is positioned against a prewound spinning, serrated wheel. When the trigger is pulled, a shower of sparks are created which ignite the priming powder which in turn ignites the main charge in the barrel.

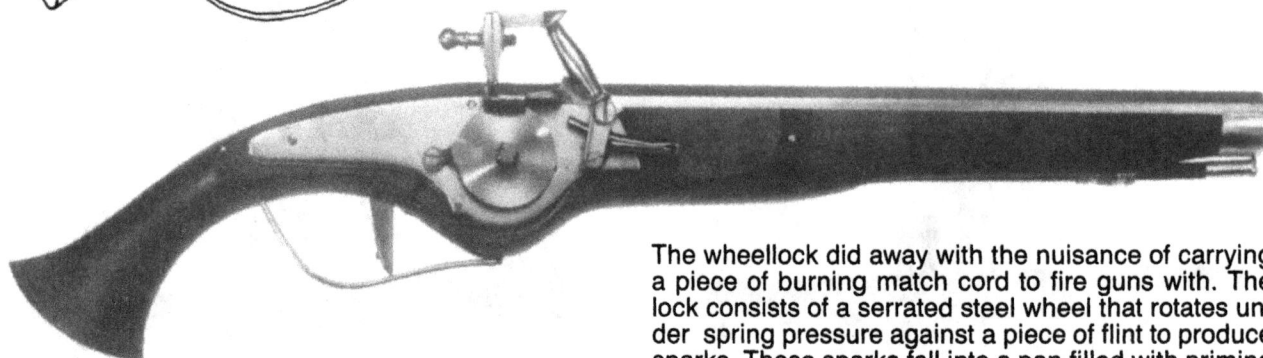

The wheellock did away with the nuisance of carrying a piece of burning match cord to fire guns with. The lock consists of a serrated steel wheel that rotates under spring pressure against a piece of flint to produce sparks. These sparks fall into a pan filled with priming powder, and this in turn ignites the main powder charge. The wheellock shown here is a modern reproduction.

up through the last century, other forms of ignition were soon developed that did away with the need for carrying an awkward length of burning match cord. The first such device to work successfully was the wheellock. The wheellock consisted of a serrated steel wheel that rotated under spring power against a piece of flint. A shower of sparks resulted, and these were directed toward the priming powder held in a pan, causing ignition. The principle is the same as that used in cigarette lighters today.

The loading procedure was much the same as that used for matchlock muskets, except that the burning match cord could be done away with. Instead, the gunner used a key to wind the spring of the lock. Then a pull on the trigger released the wheel, allowing it to spin.

Because the earliest wheellocks were easily broken (as our spring-wound toys are today), they were sometimes equipped with a separate matchlock as a backup system. But as improvements were made, this variation soon disappeared.

Before wheellocks appeared, when ignition was accomplished by manually applying a burning match or torch to the priming powder, mechanical safety mechanisms were not needed. With the introduction of this spring-powered lock, however, accidental firing became possible, and many different kinds of safeties were developed to prevent such accidents. As a matter of fact, many of the rifle and shotgun safeties used today were devel-

oped in primitive form back in the 16th century.

Much attention was also paid to trigger development. Because it required a great deal of pressure on the trigger to release the tightly wound springs in the early wheellock mechanisms, it wasn't long before someone invented what is today known as the set trigger. This was a double trigger device that worked in two separate stages. First, a heavy pull on the first arm of the triggering mechanism all but released the spring-wound wheel. At that point, all that was needed was a relatively light touch on the firing trigger to complete the ignition process. The development of the set trigger greatly improved the accuracy potential of these weapons, and the system is still used in some hunting and target rifles.

While the wheellock was a great improvement in firearms development, wheellock guns were relatively complicated and expensive to make. So when the snaphaunce appeared on the scene not too many years after the wheellock was invented, it was to have a much more lasting effect on guns and gunmaking. The snaphaunce used a spring-loaded arm, or cock, to bring a piece of flint or iron pyrite in sharp contact with a steel anvil. The pyrite was

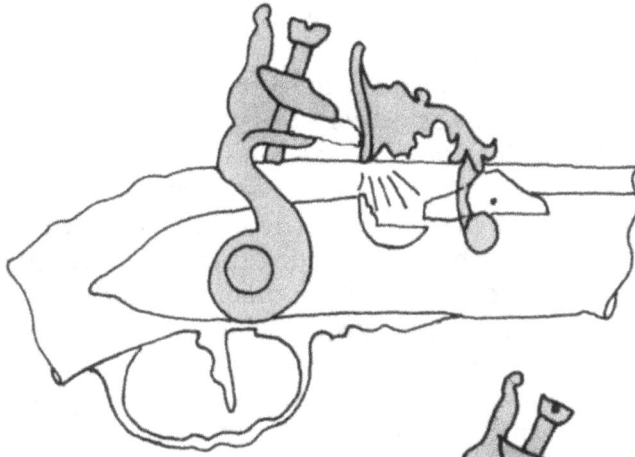

Left — The snaphaunce being fired. The flash pan is now uncovered, the frizzen pulled to the rear allowing the cock (with flint) to strike it, thereby creating the sparks necessary to ignite the priming powder.

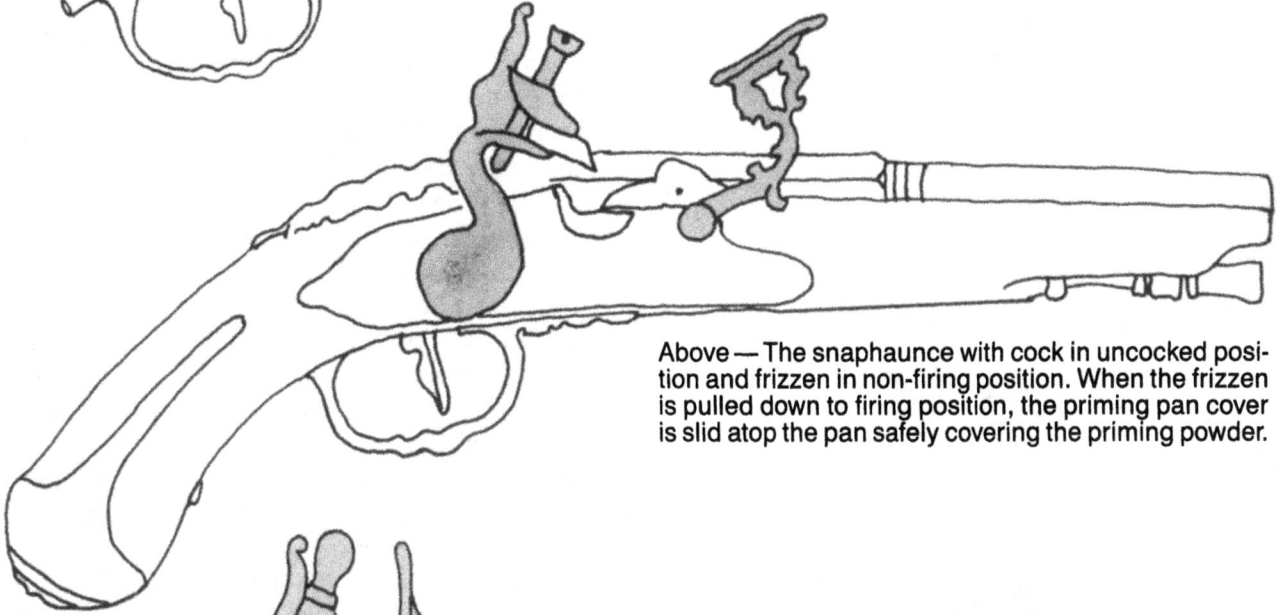

Above — The snaphaunce with cock in uncocked position and frizzen in non-firing position. When the frizzen is pulled down to firing position, the priming pan cover is slid atop the pan safely covering the priming powder.

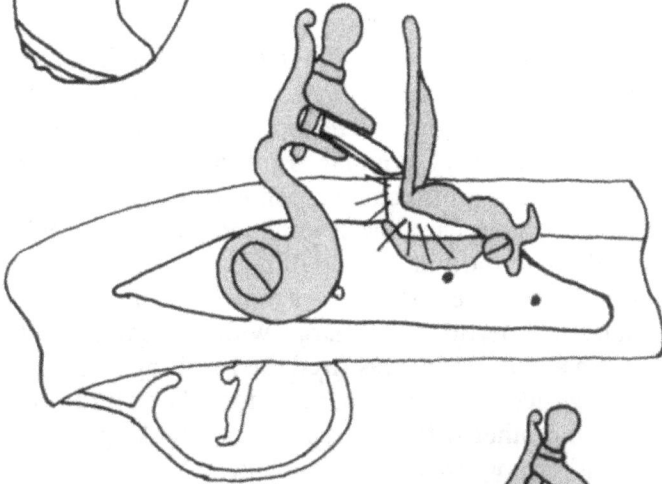

Left — The flintlock being fired. With the bore loaded and the flash pan filled with priming powder, the frizzen has been lowered and the trigger pulled. The flint hits the frizzen, throwing the pan cover open, creates sparks and ignites the priming powder which in turn ignites the main charge in the barrel.

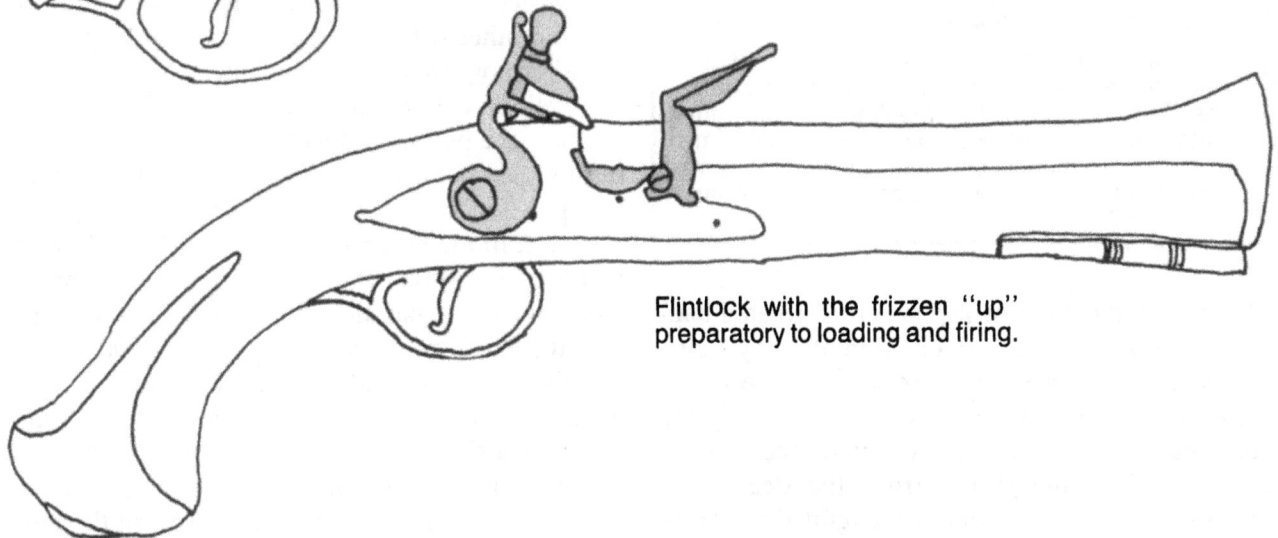

Flintlock with the frizzen "up" preparatory to loading and firing.

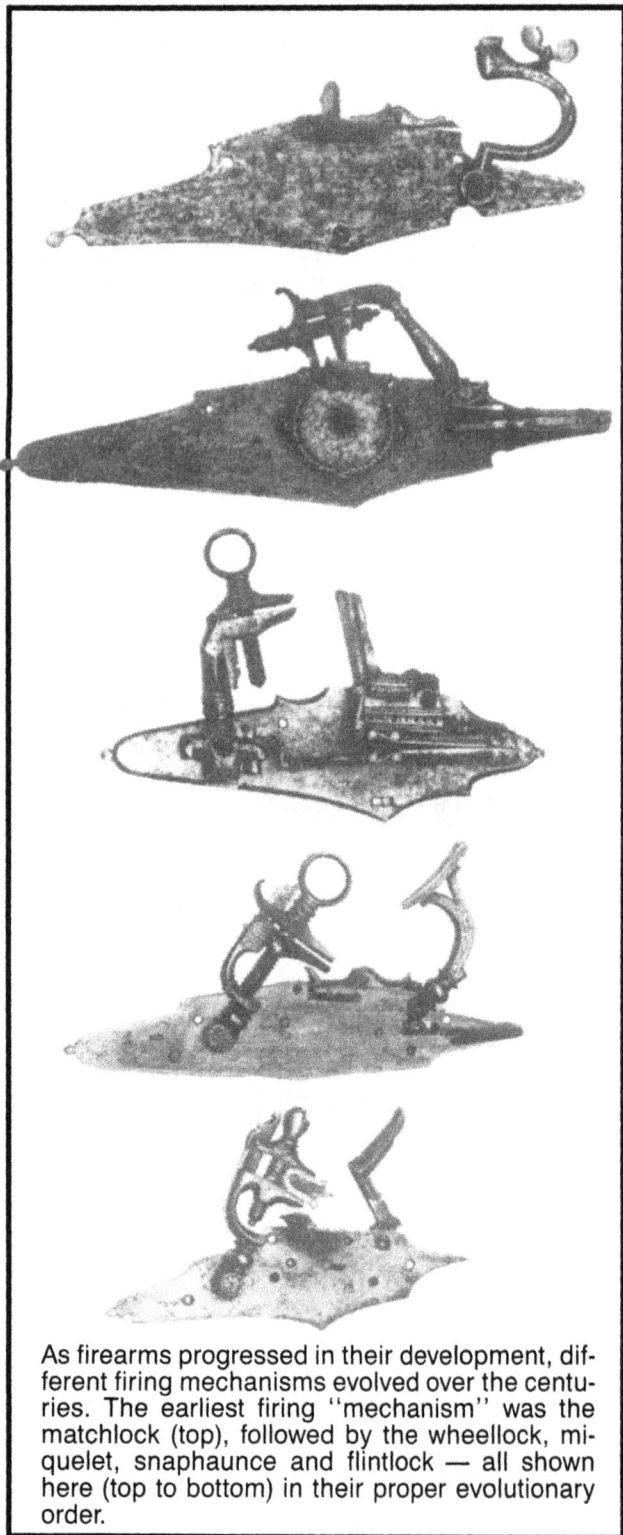

As firearms progressed in their development, different firing mechanisms evolved over the centuries. The earliest firing "mechanism" was the matchlock (top), followed by the wheellock, miquelet, snaphaunce and flintlock — all shown here (top to bottom) in their proper evolutionary order.

ing powder which in turn ignited the main charge in the barrel.

The snaphaunce was the forerunner of the flintlock, which was developed about 100 years later. The flintlock used the same basic firing principle with certain mechanical refinements that made these guns much more reliable and convenient to operate. Rather than striking sparks into an open pan or directly into the touchhole itself, the flintlock used an ingenious device that kept the priming powder intact and safely covered right up to the moment of firing. In the flintlock, the long anvil from which sparks were struck also served as a hinged lid that covered the priming pan. When the spring-powered flint hit the anvil, the anvil was pushed back, causing the lid to open and expose the powder. The sparks produced by the flint striking the anvil ignited the powder, and the gun fired.

In one form or another, the flintlock was used for more than 300 years. It was even carried by some soldiers in the Civil War and is still used today in certain remote areas of the world where modern guns and ammunition are considered an expensive luxury.

Throughout the history of firearms development, various inventors tried different ways to produce a practical repeating gun. Guns with two or more barrels did let the shooter fire more than one shot before reloading was necessary, but early multi-barreled guns were too heavy and cumbersome to be very popular. Firearms with sliding tubes fitted to the rear (breech end) of the barrel, with the tubes carrying several compartments each holding an individual load of gunpowder and ball, were tried. These tubes were moved, or indexed by hand to line up each charge with the barrel, but when a single load was fired the resulting flash often ignited the other charges, too. The flintlock, like the other different types of gunlocks preceding it, simply was not well adapted to rapid fire.

Development of a more sophisticated form of ignition — the percussion lock — wasn't possible until gunmakers turned to the volatile, highly explosive powders made from certain metal fulminates. These fulminates produced a violent explosion when merely struck by a sharp blow. There is some evidence that alchemists knew how to make fulminates of gold and mercury by 1600, but it wasn't until the end of the 18th century that fulminating powder was successfully used in firearms. Some experimenters tried using small amounts of this powerful explosive as the main propellant charge, but only succeeded in destroying the guns

clamped to the cock, which was then rotated back against spring pressure and held in place by a simple catch. Pulling the trigger released the catch, allowing the cock to spring forward and the priming pan cover to slide away exposing the priming powder. When the pyrite struck the steel anvil, sparks were produced, and these ignited the prim-

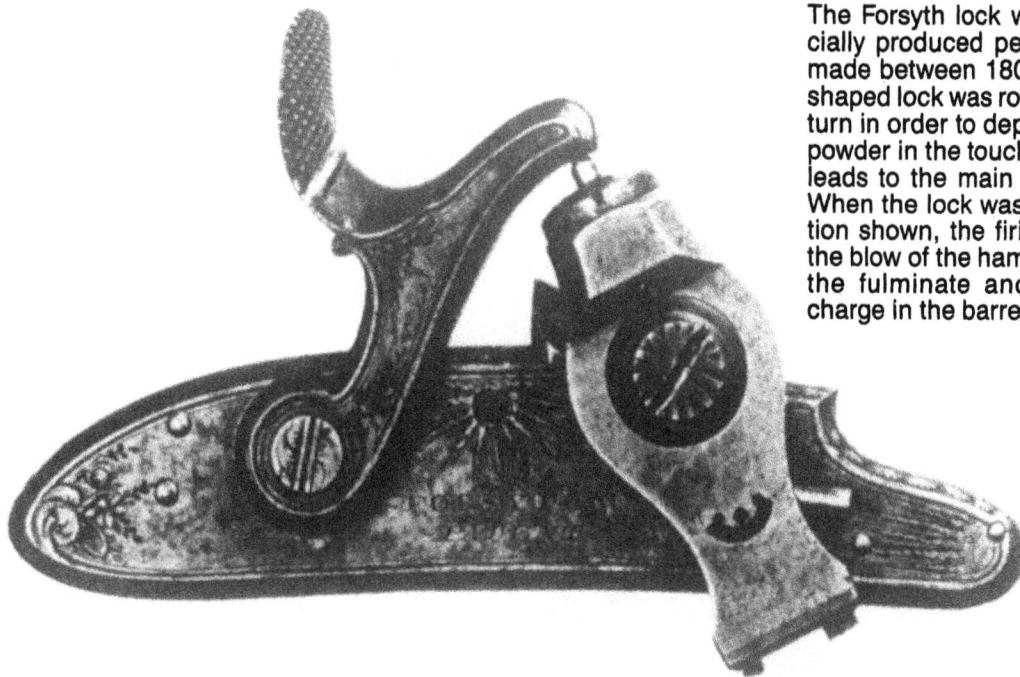

The Forsyth lock was the first commercially produced percussion lock. It was made between 1808 and 1818. The jar-shaped lock was rotated forward one-half turn in order to deposit a bit of fulminate powder in the touchhole in its axis which leads to the main charge in the barrel. When the lock was returned to the position shown, the firing pin was ready for the blow of the hammer which detonated the fulminate and likewise the main charge in the barrel.

this was tried in. Finally, in 1807, a Scottish Presbyterian minister named Alexander Forsyth patented a new gunlock that used an exploding compound (containing oxymuriatic salt) to ignite a regular load of blackpowder. Forsyth's compound was exploded by a blow from a spring-loaded hammer very similar to that used on flintlocks of the day. Since the Forsyth lock eliminated the need for flint, steel and the hinged priming pan, it simplified the chore of shooting and soon became widely popular.

While the Forsyth lock exploded a small amount of loose powder, other inventors soon produced wax-coated balls of the volatile explosive that were faster and easier to handle. Others cemented paper patches around measures of fulminating powder to produce primers much like the paper caps fired in toy pistols today. Continuous strips, or tapes of these caps were used in a number of interesting guns, and several patents were issued for designs using these paper percussion caps. Later, thin copper tubes filled with fulminate powder were used, and finally the percussion cap still used today by blackpowder enthusiasts was developed. Experts disagree on who actually invented the percussion cap, as there were apparently at least three different people working independently toward the same end. Workable caps made their appearance in both the United States and England between 1814 and 1816, and the first American patent for this new priming device was issued in 1822.

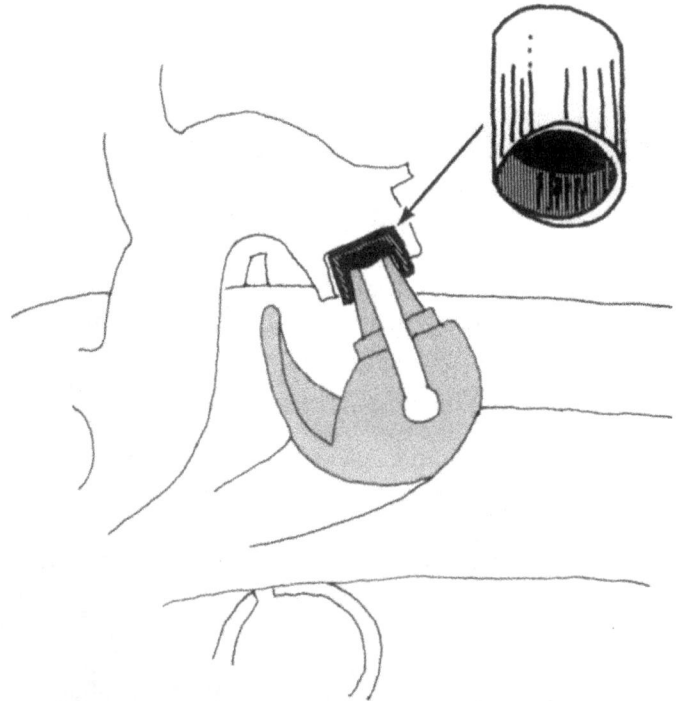

Percussion-type firearms use a "percussion cap" to set off the main charge. Above right is an example of a percussion cap, and the illustration below it shows the cap in position on the nipple about to be detonated by the blow of the hammer.

The percussion cap consists of a small metal (usually copper) cup containing fulminate priming powder in its base. In use, the hollow cup was fitted over a steel nipple which had a hole in its center (this replaced the old touchhole used in earlier guns). When the trigger was pulled, the spring-

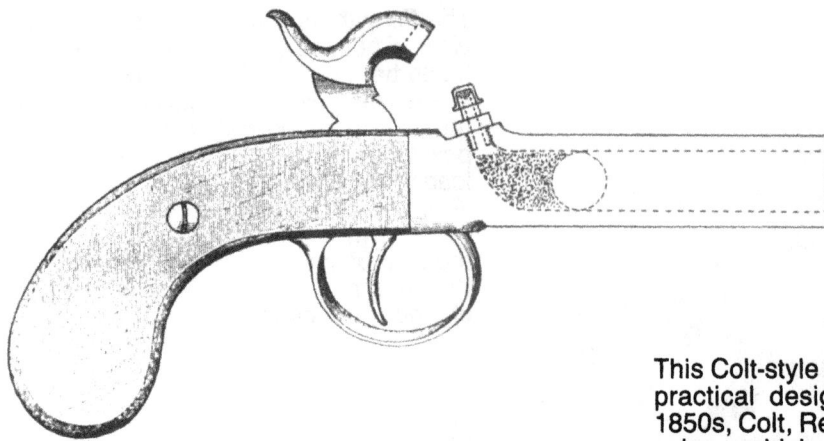

Cutaway view of a typical top-hammer pocket pistol, showing the principle of the percussion cap system. Fulminate in crown of cap over nipple is detonated by blow of cup-nosed hammer; small but powerful jet of fire flashes through nipple directly into powder charge.

This Colt-style percussion revolver represents the first practical design for a repeating handgun. By the 1850s, Colt, Remington and others were producing revolvers which contained most of the mechanical features of modern handguns. As reproductions, they are still being made today.

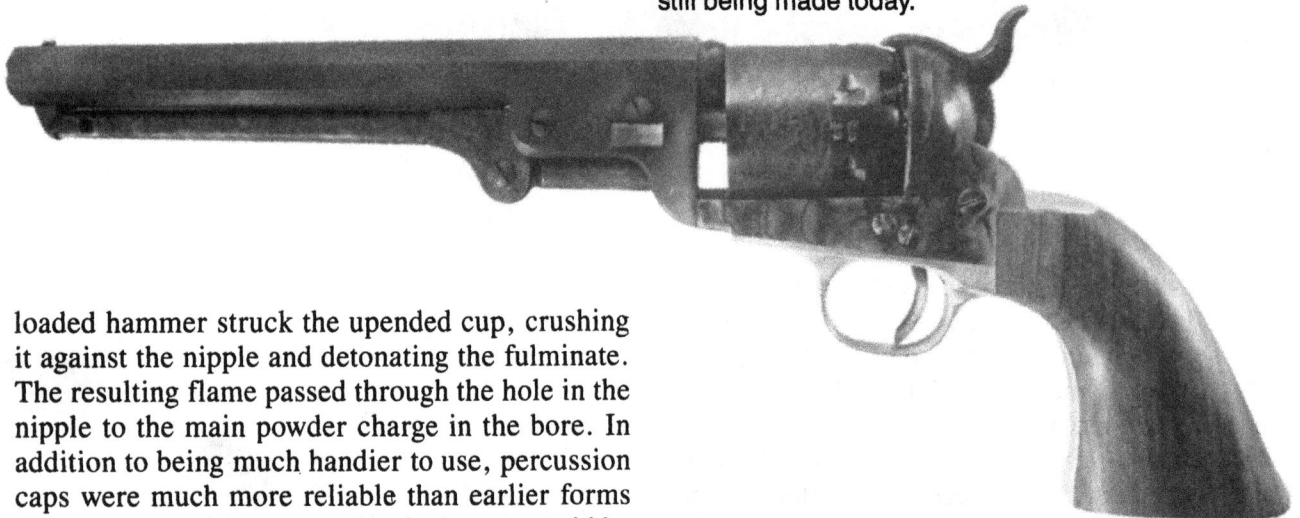

loaded hammer struck the upended cup, crushing it against the nipple and detonating the fulminate. The resulting flame passed through the hole in the nipple to the main powder charge in the bore. In addition to being much handier to use, percussion caps were much more reliable than earlier forms of firearm ignition. Percussion lock guns could be used in weather that would put the older flintlocks and wheellocks out of service, as they didn't rely on dry priming powder carried in a metal pan or open touchhole. Wind couldn't blow the sparks away, either, as sometimes happened with flint-striking guns.

The percussion cap proved so superior that by 1842 it had been adopted as military standard by the United States and several of the European armies. Because the design of the percussion lock was very similar to that used in flintlock guns, it was fairly easy to convert existing flintlocks to percussion cap use. This was one of the factors that led to its rapid adoption, and many military flintlock muskets were so converted. Later, gunmakers began designing smaller, more streamlined locks for the percussion cap, and the first successful repeating firearms were developed. While many attempts had been made over the years to produce a gun that would fire several times before reloading was necessary, it wasn't until after the percussion cap came into use that a workable repeater was invented. Several different types of repeating hand-

guns and long guns were patented in the early part of the 19th century, but the one that seemed to work best was the revolver Samuel Colt patented in 1835.

Colt's revolver was the forerunner of the modern revolver, with five or six individual firing chambers contained in a steel cylinder that revolved around a central axis. The chambers were loaded from the front, a percussion cap was placed on a nipple at the rear of each chamber, and the cylinder was rotated to bring each chamber in line with the barrel, or bore of the gun. This system was first used in both rifles and handguns, but revolving rifles had the annoying habit of showering sparks and small bits of lead onto the arm used to support the rifle. Too, the flash from the chamber being fired would sometimes ignite one or more of the other chambers, and if any fingers were in the way when this happened, it was just too bad. As a result, the revolving rifles first produced by Colt, Browning and some of the other gunmakers of the day were eventually dropped from production. Revolving handguns lacked the latter drawback, as

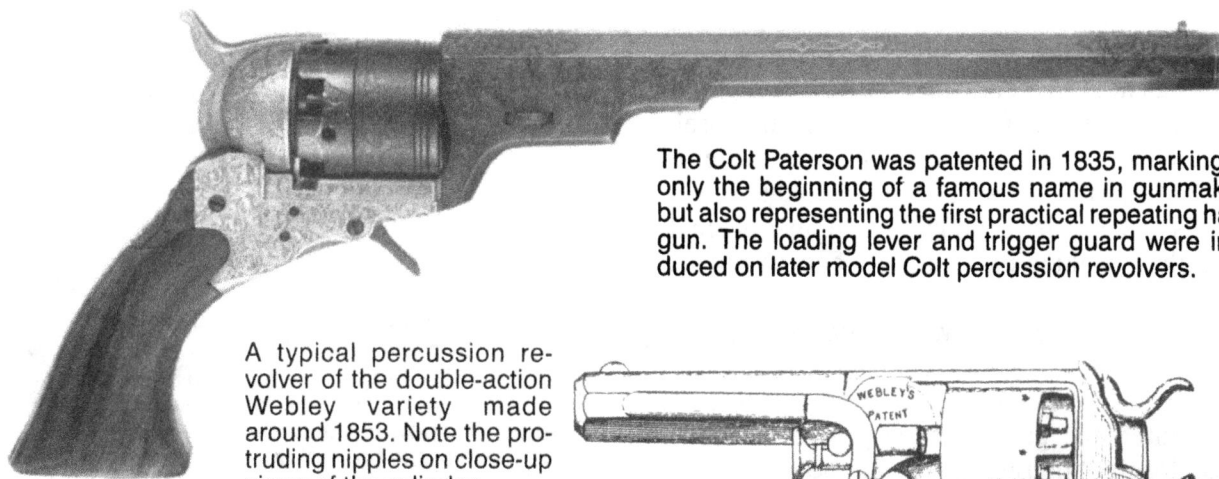

The Colt Paterson was patented in 1835, marking not only the beginning of a famous name in gunmaking, but also representing the first practical repeating handgun. The loading lever and trigger guard were introduced on later model Colt percussion revolvers.

A typical percussion revolver of the double-action Webley variety made around 1853. Note the protruding nipples on close-up views of the cylinder.

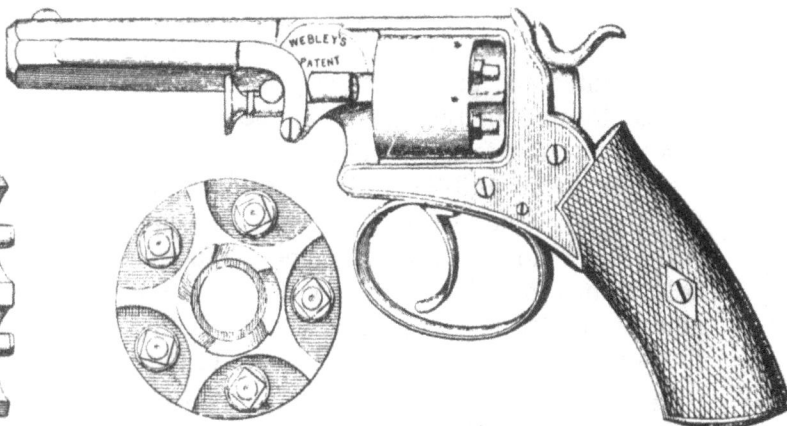

(Below) Paper-wrapped cartridge — powder, ball and wadding all in one. Popular by the 1860s, some of these cartridges used combustible paper so that they could be loaded directly. Others had to be torn open prior to loading.

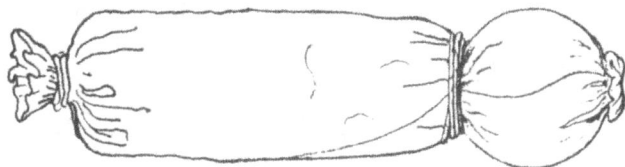

neither of the shooter's hands needed to be held in front of the loaded chambers.

While percussion-fired revolvers made successful repeating handguns, most rifles remained single shot affairs that had to be loaded from the muzzle after every firing. Several new designs soon emerged that allowed the shooter to open the rear, or breech end of the barrel and load powder and ball directly into the bore without the aid of a ramrod. To speed up the loading process even more, cartridges containing a measured powder charge along with the ball, or projectile were developed. The shooter only needed to open the breech, insert one of these paper-wrapped cartridges, close the breech and cap the firing nipple with a percussion cap. Then he was ready to fire.

Cartridges (the word means a case that holds the charge for a gun) were not a new development with breech-loading guns. Paper cartridges were in use several hundred years earlier to shorten the loading time for muzzle-loading firearms. Drawings of cartridges, with instructions on their use, appeared in Leonardo da Vinci's notebooks. The ear-

liest cartridges simply amounted to wooden or leather tubes, each filled with just enough gunpowder for a single load. Musket balls were carried separately, and rammed down the bore after the powder had been dropped in. By the late 1600s, paper cartridges containing both the powder and ball were in wide use. The shooter bit open the rear of the cartridge, poured a small amount of the powder into the priming pan and the rest down the bore, and then rammed the paper-wrapped ball down on top of the charge. The paper wrapping served as a wad to help hold the charge in place.

While the early forms of cartridges did speed up loading times for muzzle-loading guns, these preassembled charges proved much more useful in the new breechloaders. Still, loading a gun required separate steps for placing the cartridge in the firing chamber and inserting a percussion cap on the priming nipple. It wasn't until a metallic cartridge that contained powder, ball and percussion cap in a single unit was developed that a practical fast-firing gun could be invented.

A Swiss inventor named Jean Samuel Pauly patented what is probably the first metallic, self-contained cartridge gun in 1812. The cartridge used in this breechloader was an all-new concept that set the stage for later firearms development, and eventually made possible the modern magazine-fed guns we use today. The Pauly cartridge used a wooden or metal base with a glued paper wrapping to hold the powder and ball. The priming compound was held in a container set into the base of the cartridge. The assembled components closely resembled a modern shotgun shell. As a matter of fact, Pauly cartridges were made to hold both single balls and charges containing a number of small shot pellets, making his guns both convenient and versatile to use.

Pauly's new invention was shown to Napoleon and his generals, but they decided that it wasn't suited for army use. The Pauly cartridge was the first "centerfire" gun cartridge (with the priming device located in the center of the cartridge's base), but other forms of ignition soon followed. Between 1827 and 1829 Johann Nikolas von Dreyse, who once worked for Pauly, invented a bullet that held the priming compound in a small hole formed in its base. The powder was held behind the bullet and its primer, and a needle-like firing pin was driven through the base of the cartridge and the powder to reach the primer. Because of this unique firing arrangement, Dreyse's invention became known as the "needle gun." The Prussian Army adopted the new weapon in 1842, and it performed well in wars against Denmark and Austria in the mid-1860s.

Casimir Lefaucheux, a Frenchman, took still another approach. He invented a self-contained cartridge much like the one used by Pauly, except that the percussion cap or primer was held in the base of the cartridge, with a pin resting against it. This pin projected from the side of the cartridge at right angles, and when the round was loaded into a gun this pin projected through a slot cut in the breech. The cartridge was fired by the hammer hitting the pin, which drove the pin into the percussion cap to cause ignition. The Lefaucheux invention is known as the first pinfire cartridge.

In the mid-1800s, rimfire cartridges became popular and a number of different rifles were produced with rimfire ignition systems. The rimfire cartridge differed from the other cartridges of the day in that the priming compound was distributed all around the protruding rim of the brass or soft metal case. This rim rested against the face of the breech when the cartridge was inserted in the chamber, and the breech face served as an anvil against which the firing pin could crush a portion of the rim and ignite the priming compound.

A Paris gunsmith named Flobert was the first to develop the rimfire principle. He reshaped the percussion cap that was in wide use at the time to

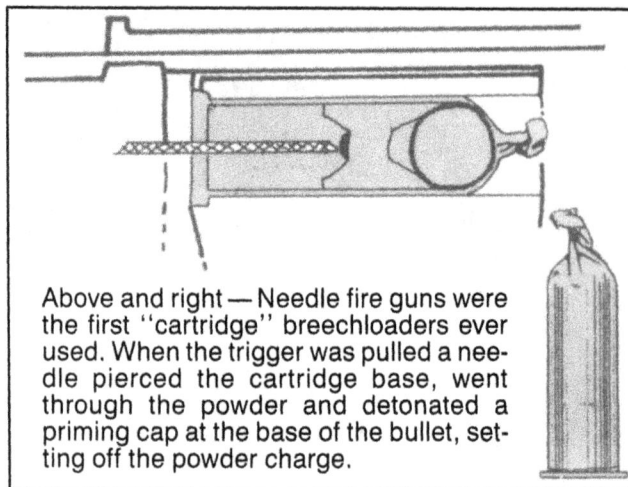

Above and right — Needle fire guns were the first "cartridge" breechloaders ever used. When the trigger was pulled a needle pierced the cartridge base, went through the powder and detonated a priming cap at the base of the bullet, setting off the powder charge.

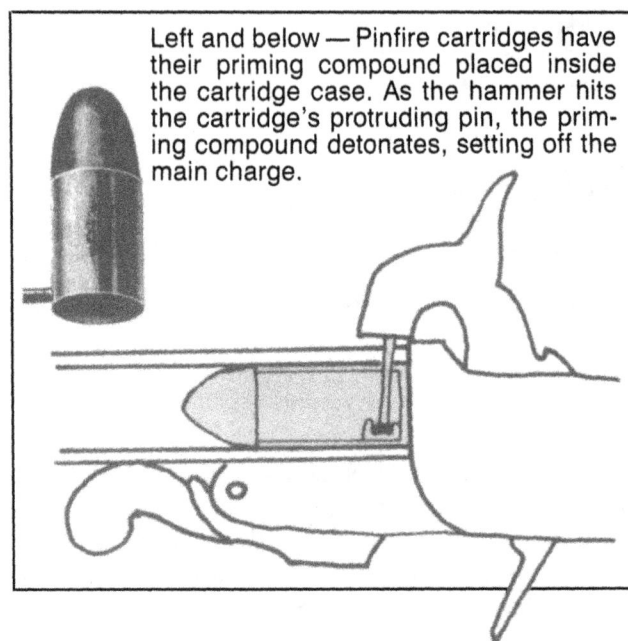

Left and below — Pinfire cartridges have their priming compound placed inside the cartridge case. As the hammer hits the cartridge's protruding pin, the priming compound detonates, setting off the main charge.

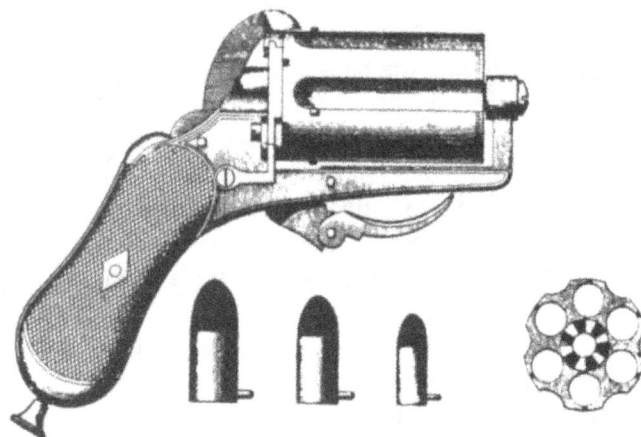

Pinfire pepperbox revolver shown with three cartridges of various calibers. Note cylinder with notches at the top of each firing chamber to accommodate protruding pin of each loaded cartridge.

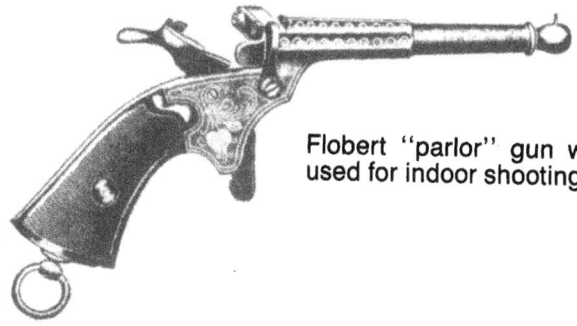

Flobert "parlor" gun was used for indoor shooting.

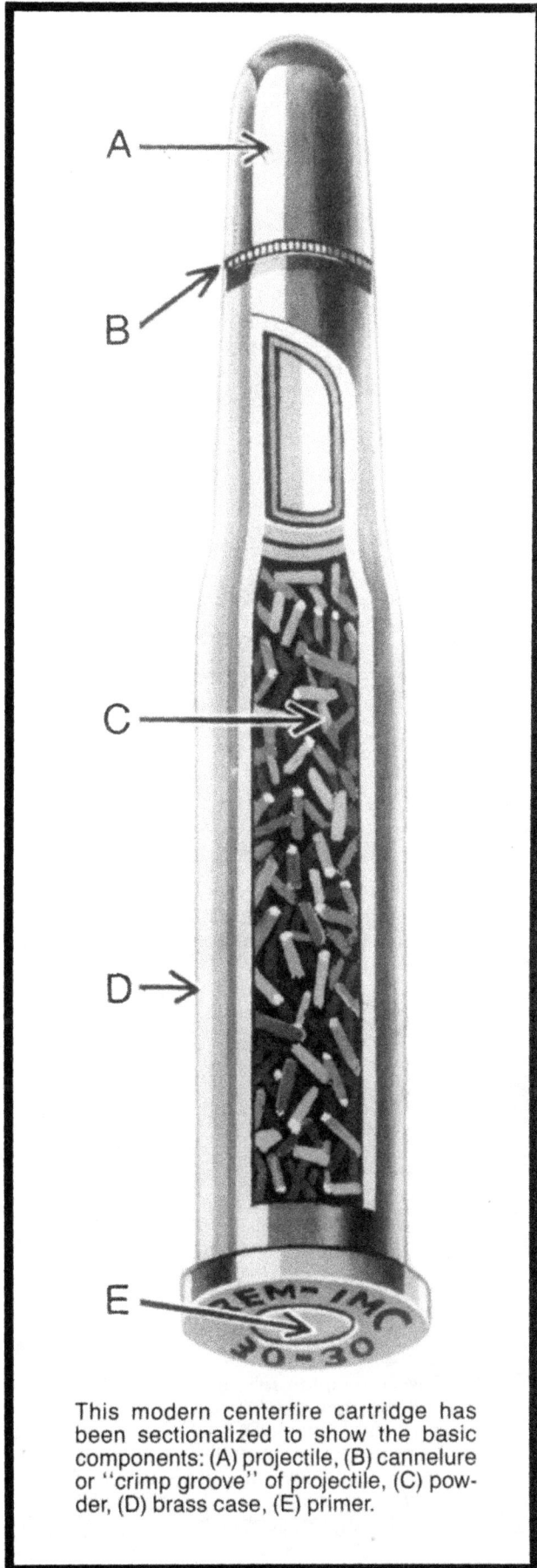

This modern centerfire cartridge has been sectionalized to show the basic components: (A) projectile, (B) cannelure or "crimp groove" of projectile, (C) powder, (D) brass case, (E) primer.

give it a rim, and then loaded the open end of the cap with a small, lead bullet. These cartridges carried no powder charge other than the priming compound itself, but this generated enough power for indoor target shooting. Flobert produced "parlor guns" for indoor, or "salon" shooting from about 1835 to 1848, and this form of sport became very fashionable in some parts of Europe.

Smith & Wesson went on to develop the 22 rimfire cartridge that remains pretty much unchanged today from the original Flobert round. The Smith & Wesson 22 was first produced in 1857, and with it came a low-cost method of producing rimfire brass cases. A number of larger caliber rimfire cartridges were also developed, including the 44 Henry, which later was redesigned and renamed the 44 Winchester.

Most of the breech-loading and repeating rifles used during the Civil War were chambered for rimfire cartridges, and several European rifles were designed to digest this same kind of ammunition. While the early rimfire cartridges made firearms history, the case design proved too weak to handle heavy loads and only small-bore 22-caliber rimfires are still in general use today.

Rimfire cartridges were also easy to ignite accidentally, as the rim holding the priming compound was very exposed and unprotected. To overcome this problem designers worked to develop a new cartridge that would be safer to use and carry. A Frenchman named Pottet is credited with the invention of the first workable centerfire cartridge, which first appeared and was patented in 1857. Pottet used a brass and paper case similar to that used in modern shotgun ammunition. A percussion cap that included a small anvil was set into the center of the base of this cartridge.

Other inventors and gunmakers have improved on the Pottet design, and nearly all shotguns and large bore rifles and handguns in use today use

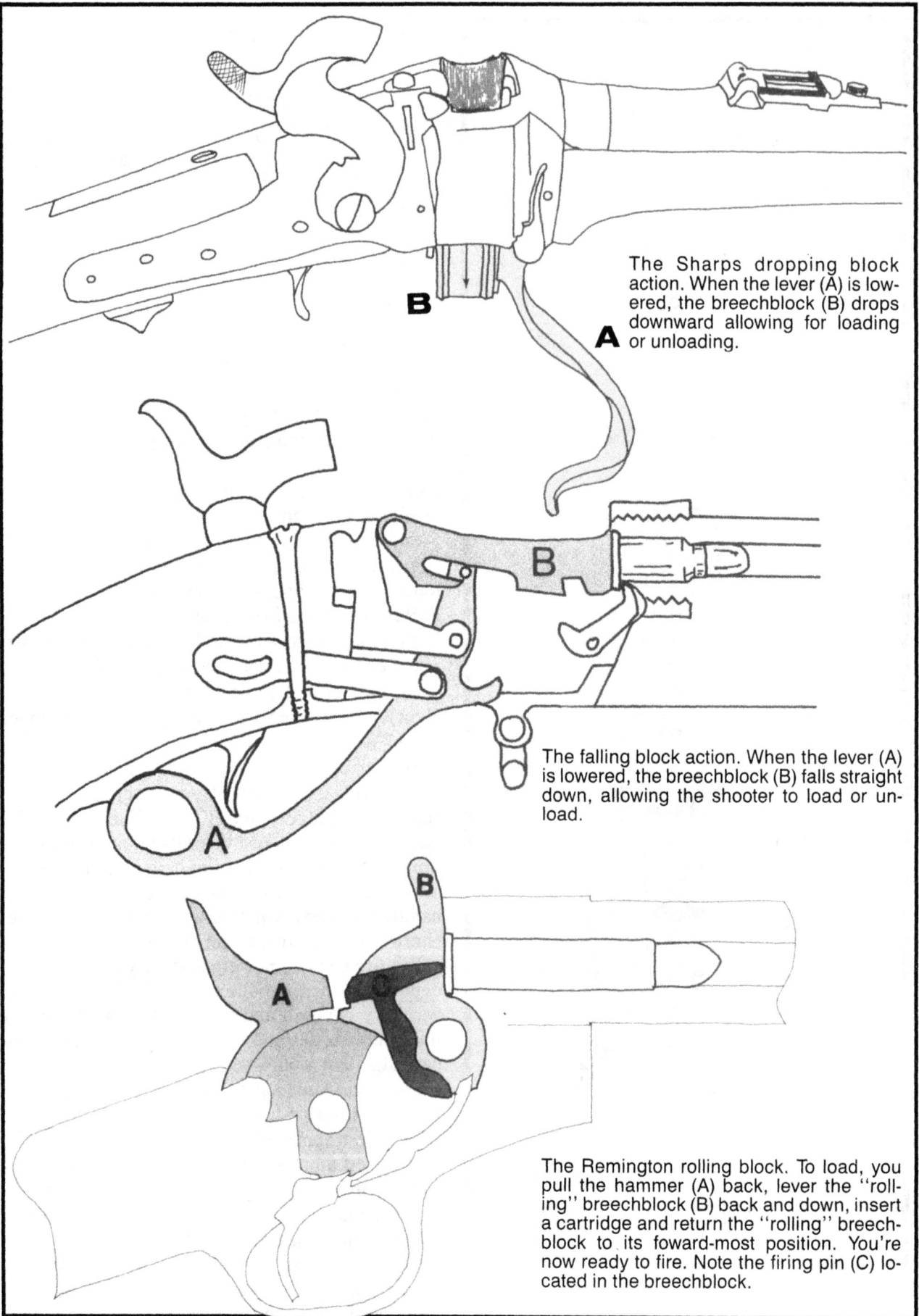

The Sharps dropping block action. When the lever (A) is lowered, the breechblock (B) drops downward allowing for loading or unloading.

B

A

The falling block action. When the lever (A) is lowered, the breechblock (B) falls straight down, allowing the shooter to load or unload.

B

A

B

A

The Remington rolling block. To load, you pull the hammer (A) back, lever the "rolling" breechblock (B) back and down, insert a cartridge and return the "rolling" breechblock to its foward-most position. You're now ready to fire. Note the firing pin (C) located in the breechblock.

centerfire cartridges based on the Frenchman's invention.

With a truly practical cartridge design finally available, firearms development advanced rapidly during the latter part of the 19th century. The Sharps dropping block rifle, originally designed for percussion ignition, was soon adapted to take centerfire cartridges, and other rifles produced by Ballard, Stevens, Winchester and Browning were built around similar actions.

The early dropping block system was a very strong one, and some modern rifles and large military guns still use variations of the original Sharps design. The Sharps rifle and its imitators featured a heavy breechblock that dropped vertically downward through a slot in the base of the receiver to expose the firing chamber. After a live round, or cartridge had been loaded into the chamber by hand, the breechblock was raised to seal the cartridge in the barrel. The sliding breechblock was usually operated by a lever located underneath the receiver, or breech end of the rifle.

There were several other single shot breechloaders developed before the turn of the century, including the falling block design patented by Henry O. Peabody in 1862. Peabody was a Massachusetts gunmaker who used a breechblock hinged at the rear. Pulling down on a lever beneath the receiver unlocked the front of the breechblock, which was then tipped downward at an angle to provide access to the chamber.

The Peabody rifle used an external hammer, which activated a firing pin contained within the falling breechblock. A Swiss gunsmith named Frederich von Martini modified the Peabody by eliminating this hammer (which had to be manually cocked before the gun could be fired) and utilizing an internal striker that was automatically cocked by the same lever that opened the breechblock. A variation known as the Martini-Henry is still being produced in England for small-bore rifle shooters who continue to appreciate the design.

Another well-known single shot design developed at the close of the Civil War is the Remington rolling block. This action is easily distinguished by its two external "hammers," or cocking pieces. These are offset, one behind the other, at the rear of the breech. The front thumb-operated lever actually isn't a "hammer" at all but is a part of the breechblock which pivots around a pin when being opened. After the cartridge has been inserted into the firing chamber, the breechblock is held closed

against the chamber by spring pressure. The second, or rearmost hammer is exactly that and is drawn backward by thumb pressure to cock the gun. Pulling the trigger frees the hammer, which is driven forward by the mainspring to strike the firing pin contained in the breechblock. As the hammer moves forward, a shelf machined into its lower arm rolls behind the breechblock to lock it in place.

Other important single shot designs included the cam lock, which featured a breechblock that pivoted up and forward to allow a cartridge to be loaded into the chamber. This action design was adopted by the U.S. military right after the Civil War because it could be easily adapted to existing stocks of old muzzle-loading weapons. It was used by American fighting forces until replaced in 1892 by the bolt-action Krag-Jorgensen rifle.

While single shot rifles continued to be developed and manufactured long after the metallic, self-contained cartridge was invented, the stage was finally set for a practical repeating rifle that could be fired several times before reloading became necessary. The first successful repeating rifle was the Spencer, which was introduced in 1860 and saw some use among Union forces before the close of the Civil War. The Spencer's breechblock action was operated by a lever, while the cartridges were fed through a tubular magazine contained in the rifle's buttstock. The external hammer had to be cocked by hand.

Another successful repeating rifle using a lever-action design was the Henry. The Henry was patented shortly after the Spencer (both in 1860), and this design became the basis for later Winchester lever-action rifles and carbines in which the cartridges are fed from a tubular magazine located below the barrel.

While American gunmakers concentrated on lever-action repeaters in the latter part of the 19th century, in Europe the turnbolt action was being developed. Peter Paul Mauser, a German inventor, is the man given credit for designing one of the first successful bolt-action rifles. This rifle, the Model 1871, was actually based on the earlier Dreyse needle gun, but featured many significant improvements and used the more modern metallic centerfire cartridge. Frederic Vetterli, in Switzerland, produced a new rimfire bolt-action rifle in 1867, predating the Mauser by a few years. In addition, the Vetterli rifle was a magazine-fed repeater, while the first Mausers were single shots.

The Spencer repeating carbine (and rifle) proved popular with troops during the Civil War. The gun pictured here is the rifle version of that famous 7-shot, tube-fed (from the buttstock) Civil War carbine.

It could be said that the Henry lever-action repeating rifle was the basis for every modern lever-action rifle on the market today. The Henry was patented in 1860

In 1892, the U.S. adopted its first turn-bolt repeater, the 30-40 Krag. The gun pictured is the Norwegian Krag from which the U.S. version was developed.

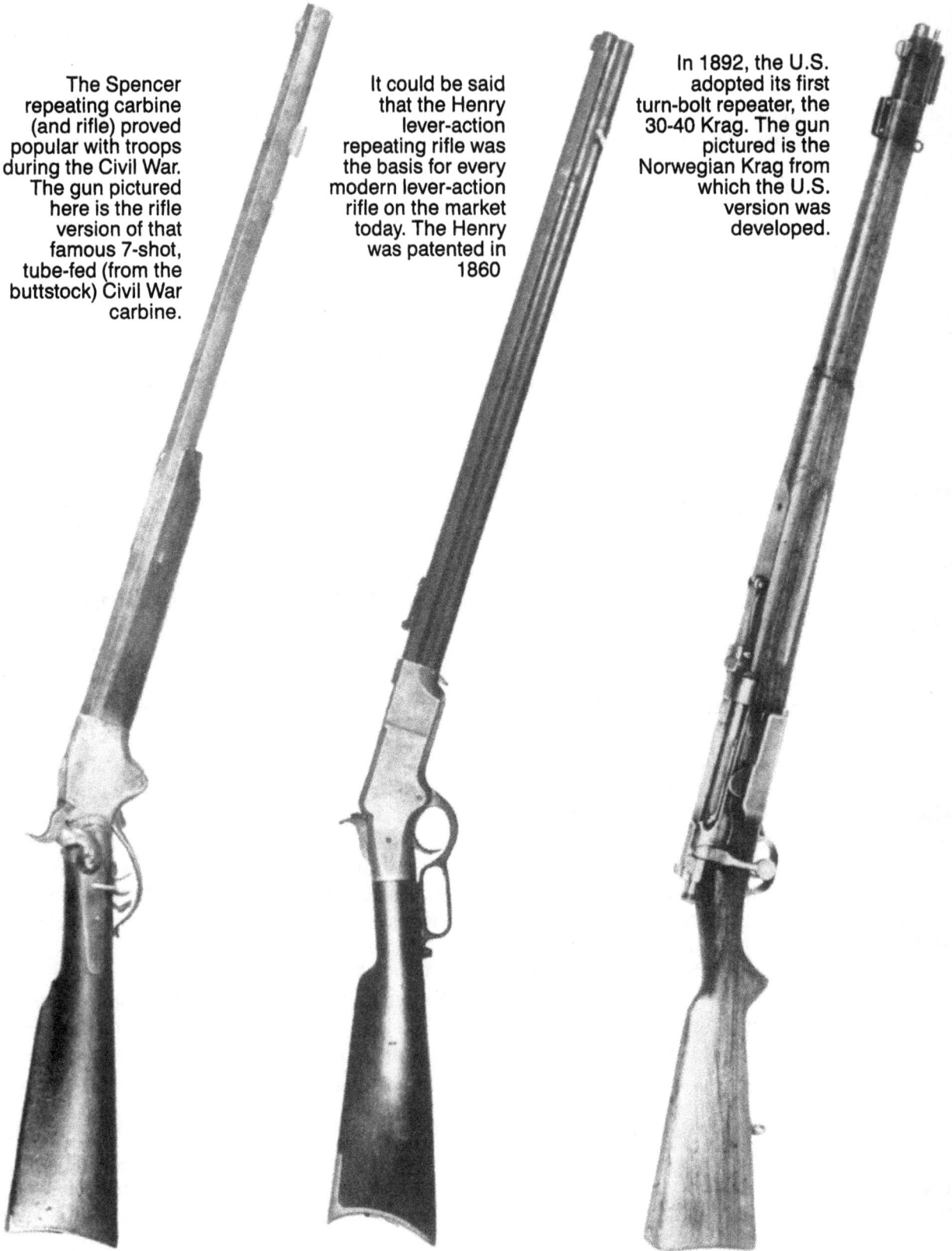

The Mauser and Vetterli turnbolts operated on the same general principle. A sliding breechbolt equipped with a protruding handle was the basis for the action. To open the action, the handle was pulled upward, then back. This unlocked the bolt and brought it rearward. A hooked extractor pulled the fired cartridge case from the chamber as the bolt was slid back, and an ejector threw the case from the action when it cleared the chamber. If the rifle was a single shot, a fresh cartridge was then manually inserted into the chamber and the bolt was closed by pushing forward, then down on the handle, thereby locking the action. Then the rifle was again ready to fire. Magazine-fed repeaters automatically brought a new round into alignment with the chamber each time the bolt was cycled.

The next significant development in firearms technology came when a French chemist named Vieille perfected the first smokeless powder in 1884. The blackpowder propellant used in guns prior to that time hadn't been much improved since the 13th century and had at least one serious drawback from a military point of view. When blackpowder is ignited, a large puff of white smoke is produced. This smoke gave away the position of hidden riflemen to the enemy, and in battle the smoke sometimes became so dense that the shooters could no longer see their targets.

While the new smokeless powder developed by Vieille burned cleanly enough to eliminate those problems, there were other, more important bene-

fits. The new gunpowder was a more efficient propellant that burned in a way that made higher bullet velocities possible without raising pressures of the gases generated by the burning powder to dangerous levels.

By the turn of the century, smokeless powder had largely replaced blackpowder for use in military and sporting arms, and longer, narrower bullets took the place of the larger, slower moving projectiles used in blackpowder guns. And where rifle bullets were once simply molded from common lead, metal jackets were soon added to the new small-bore projectiles to help them stand up to the higher velocities and greater bore friction produced with smokeless powder loads.

The new narrower bullets allowed gunmakers to reduce the bore size which in turn helped make rifles lighter and more compact in size. This, plus the development of truly practical repeating firearms, helped revolutionize the tactics and concepts of warfare. Sportsmen quickly adopted the new, relatively lightweight rifles that were chambered for smokeless powder loads. The 30-30 Winchester, introduced in 1895, was America's first "small-bore" sporting cartridge to be loaded with the new smokeless propellant, and it—along with the Model 1894 Winchester lever rifle that made its debut at the same time—remains in production today as one of our most popular hunting loads.

While most of the world's military forces soon turned to bolt-action repeating rifles as their stand-

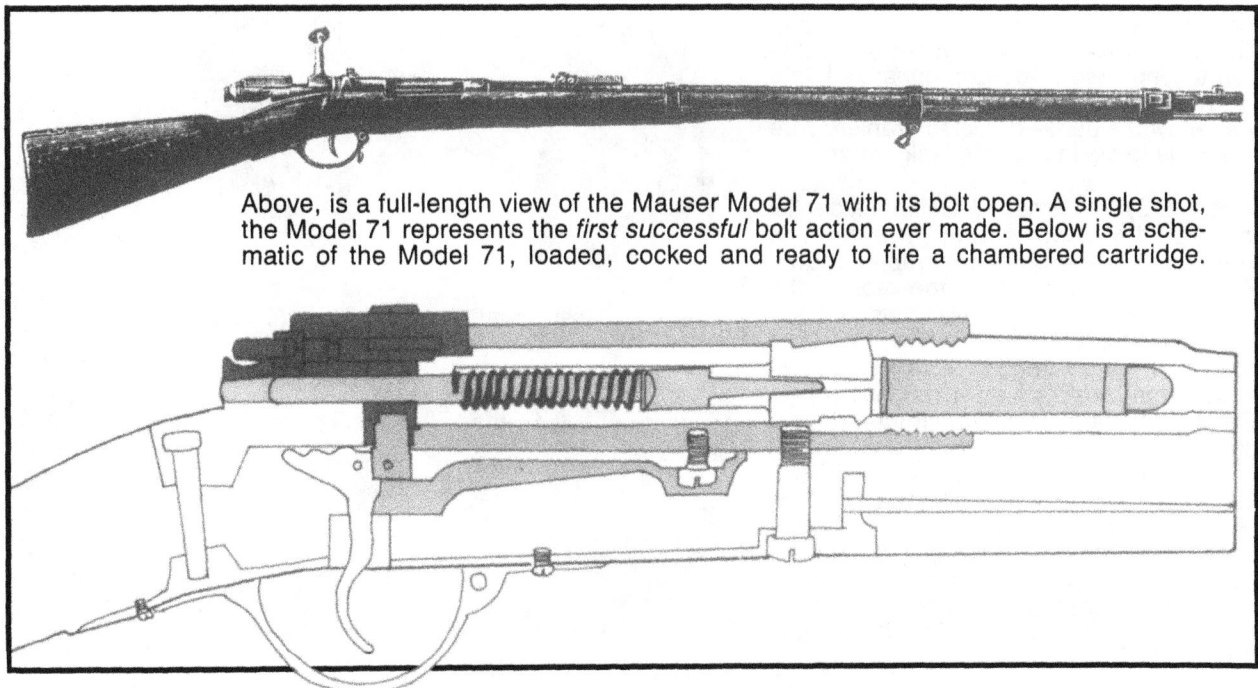

Above, is a full-length view of the Mauser Model 71 with its bolt open. A single shot, the Model 71 represents the *first successful* bolt action ever made. Below is a schematic of the Model 71, loaded, cocked and ready to fire a chambered cartridge.

The A-10 fighter employs a modern 30mm Gatling gun. Though the design is as modern as tomorrow, the machinists who built Gatlings in the 1880s would have no trouble understanding this 1980s version.

(Above) This airman is finishing the cleaning chore on the 30mm anti-tank Gatling. The dummy rounds are used for final function testing before the gun goes back into an A-10 tactical fighter. Live rounds are loaded with a dense depleted uranium core projectile to aid in piercing tank armor.

Patented by Dr. Richard J. Gatling, a dentist in 1862, the "Gatling Gun" was the first successful "mechanical machinegun." When the hand crank at the rear of the breech is turned, the barrels rotate, the gun fires and empty cases are extracted by a system of gears and cams. The "donut" shaped magazine on top of the gun kept the Gatling supplied with fresh ammo, which it could fire at rates approaching 1000 rounds per minute. The modern "Gatling" is a General Electric GAU-8 30mm gun, as mounted in the Air Force A-10 tank buster tactical fighter. The 30mm gun is hydraulically powered and empties ammo at more than 4000 rounds per minute. Other modern "Gatlings" are in use in 7.62mm and 20mm chamberings.

ard armament, an American named Hiram Maxim had produced the first practical semi-automatic rifle as early as 1883. In 1884, he patented a locked-breech recoil-operated design that served as the basis for the modern machine gun. Other inventors like Browning, von Mannlicher, Johnson, and Mauser produced other semi-automatic rifles and fully automatic machine guns in the years to come. These firearms used the recoil force generated by the fired cartridge to cycle the action, but they weren't the first guns capable of a high, sustained rate of fire. That distinction belongs to the firearm patented in 1862 by Dr. Richard Gatling. The Gatling gun was a more primitive firearm that had to be cranked by hand, but it was capable of very rapid fire. In its time, it was used by several governments including the United States. (Although the Union Army refused to consider adopting the gun during the Civil War because its inventor's sympathies were known to lie with the South, a few were used by the Confederacy.) But when recoil-operated automatic guns were developed, the Gatling gun and other hand-cranked firearms became obsolete.

In 1891, Maxim patented a rifle that tapped the expanding gas in the bore to operate the action,

rather than relying on direct recoil force. This invention was the forerunner of most of the automatic and semi-automatic rifles in use today. (An automatic rifle, or machine gun, continues to fire as long as the trigger is held back and the ammunition supply lasts; a semi-automatic arm performs all operating functions—extraction and ejection of the fired case, and loading and chambering of a fresh round—automatically, but requires a separate pull on the trigger for every cartridge fired.)

Handgun and shotgun development pretty much paralleled the advances made in rifle design. No real design changes were possible with the shotgun until after the self-contained cartridge was invented, and then lever-action, bolt-action, slide-action and semi-automatic bird guns appeared on the scene. Break-top shotguns (and rifles), with the barrel attached to the action, or breech, by a pivot pin were popular more than a century ago and similar models are still being manufactured and sold today.

Although there were multi-shot handguns using the revolving cylinder principle in combination with snaphaunce locks that date back to the early 1600s, the Colt revolver patented in 1835 was the first really successful repeating handgun. The Colt single action featured an external hammer that had to be cocked by hand before each chamber could be fired. This gun was originally designed around the percussion cap ignition system, but was easily adapted to the self-contained centerfire and rimfire cartridges when they made their appearance.

Other revolving-cylinder guns were made that were cocked, indexed to bring a loaded chamber in line with the bore, and fired—all by the action of pulling the trigger. These handguns, which didn't require the hammer to be cocked separately by the thumb, are known as double-action revolvers.

Several different methods were devised to eject spent cases from the multi-chambered cylinders for speedy reloading. Early Colt models featured a loading port with a hinged cover at the rear of the frame, and a long rod extending underneath the barrel that could be pressed back to eject fired cases one at a time as each chamber was revolved into line with this port. Fresh cartridges were then loaded into the chambers through the same opening. This system was sure, but slow, and other manufacturers looked for ways to speed things up.

Some designs included revolvers with hinged frames. When the frame was opened, the cylinder was raised clear of the frame and empty cartridges

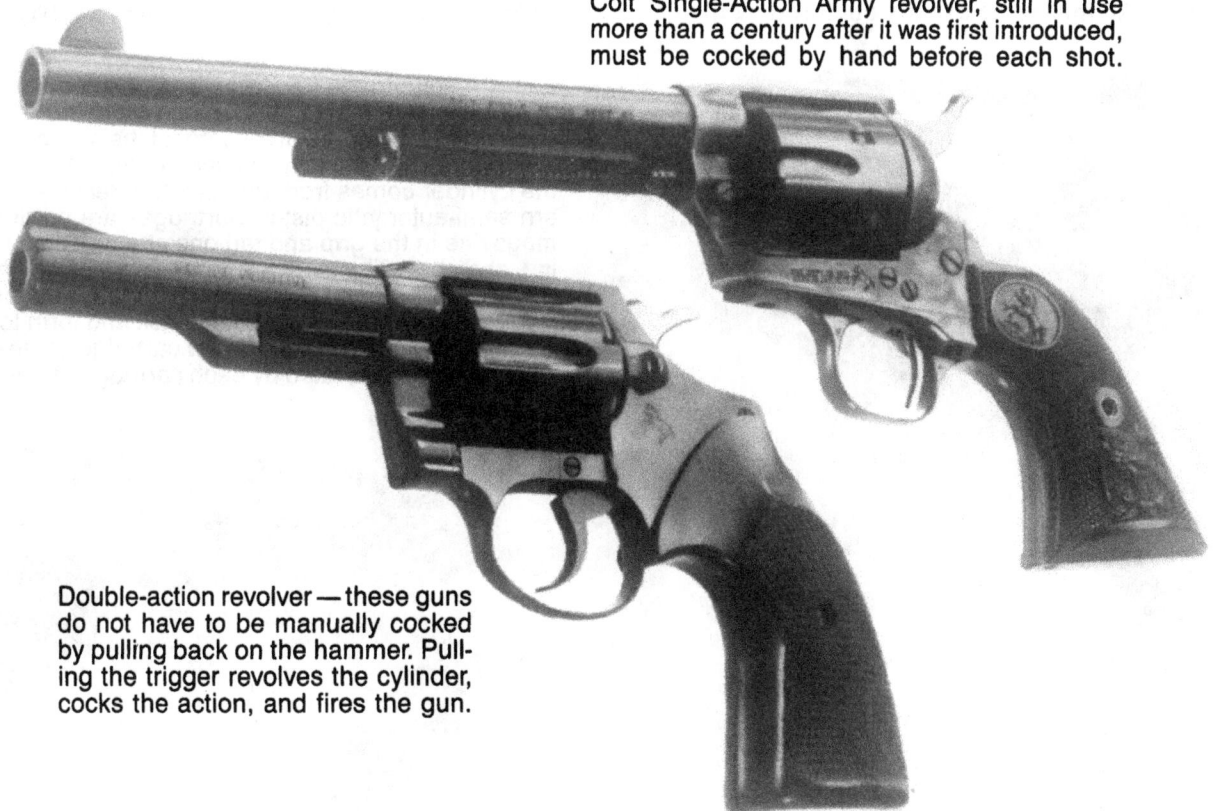

Colt Single-Action Army revolver, still in use more than a century after it was first introduced, must be cocked by hand before each shot.

Double-action revolver — these guns do not have to be manually cocked by pulling back on the hammer. Pulling the trigger revolves the cylinder, cocks the action, and fires the gun.

were ejected at a single stroke. Many are still in use. Most of the older American guns were weaker than solid-frame designs and were chambered for rimfire or low-powered centerfire cartridges. The British Webley hinge-frame design, made until the 1970s, handled more powerful rounds. In the mid-1980s, an American manufacturer attempted to revive the old Webley design. The gun maker failed due to finances—not any inherent weakness in the design.

The most popular of the solid-frame designs is one that features a cylinder that can be swung to one side to expose the chambers. Again, a single stroke of the ejection rod is usually all that's required to kick empty cases clear. Nearly all double-action revolvers being manufactured today are solid-frame guns with swing-out cylinders, while the single-action designs still being sold continue to use the "one-round-at-a-time" ejecting and loading method featured by the first metallic-cartridge Colts.

"Automatic" pistols, usually capable of only semi-automatic operation (a truly automatic pistol would be classified as a small machinegun) have been manufactured since around 1892, and several different designs were introduced in the following decade. These handguns all used some form of reciprocating action that allowed the breechblock to move back and forth. Cartridges were carried in a box-like magazine usually enclosed by the grip, or

handle, and these rounds were fed into the chamber each time the action was cycled. The breechblock was moved back by recoil, and pushed forward by spring pressure.

The most successful early auto pistols made in America came from Johnathan Moses Browning's drawing board. Browning's designs are still being manufactured all over the world. Other early designs, still in wide use, came from Luger and Walther. The 1980s have seen many new auto pistol designs come from nearly every corner of the planet. Most of the new designs combine double-action operation with 9mm chambering.

Firearms development is an ongoing process. New guns, calibers, materials and manufacturing methods continue to be developed. Some of the more interesting developments have involved new ammunition types. In the late 1950s, a company called Gyrojet introduced guns which used a caseless "rocket" cartridge. The designs represented a great idea, but did not work out well in use. Accuracy and power were not great and ammo cost was quite high. The guns were discontinued not long after introduction. The next effort was made by Daisy, known for generations as an airgun maker, in 1968. Daisy introduced the 22 V/L system which used a caseless cartridge with propellant molded to the projectile. The venture lasted only until 1969. More recently, military gun makers have worked to develop the caseless cartridge, as

Left — A solid-frame revolver with a swing-out cylinder that opens to one side for loading/unloading. A single push of the ejector rod at the front of the cylinder ejects all empty cases. Energy to cock the hammer and turn the cylinder comes from the shooter. Below — A modern semi-automatic pistol. Cartridges are carried in a magazine in the grip and fed one at a time into the firing chamber. The basic difference between the revolver and the semi-automatic pistol is that energy to move the breechblock, or slide, back and forth to eject an empty case and to feed a new cartridge comes from recoil energy generated by each cartridge as it is fired.

the military benefits of lighter cartridges are obvious. A workable caseless cartridge is the next logical breakthrough in gun making, just as the percussion cap and the metallic cartridge were in the 19th century.

Improvements in firearms and ammunition are being made all the time. However, sporting rifles, handguns and shotguns represent lifetime investments as most civilian shooters use and prefer traditional firearms. Many sport shooters favor the obsolete cap and ball muzzleloaders to the more modern rifles available, simply because they're fun to shoot and work with. Quality workmanship is also appreciated, and many older firearms now bring premium prices because they were made during a time when skilled handiwork took precedence over mass production.

If properly cared for, a good-quality rifle, handgun or shotgun will give service for several generations. And even though firearms technology continues to advance, guns designed for sporting use become obsolete very slowly. As a matter of fact, many guns appreciate in value over the years and make sound long-term investments. This means it's possible to buy a firearm and enjoy its use for several years, and then sell it for even more than you originally paid for the gun — provided you choose wisely in the first place, and then maintained the gun in top condition.

However, most shooting sportsmen tend to keep the firearms they acquire because of the pleasure they afford. Firearms are an American heritage, and one shooters can take pride in. The long history of firearms development closely parallels the rise in civilization. The technology of an era is often reflected in the sophistication of its weapons and other tools. Firearms are both weapons and tools, and as such serve to mirror history, which is an ongoing process.

Though the military has tried for many decades to perfect caseless ammunition to save space and weight, the concept has also had some civilian exposure. In the late 1960s, Daisy marketed a 22-caliber rifle using a unique caseless round. The Daisy-Heddon 22 resembled an air rifle in that the cartridge propellant was ignited by air compression heat. The rifles and ammo were finely made, but simply did not catch on. By the early 1970s, this latest commercial example of caseless ammunition was dropped.

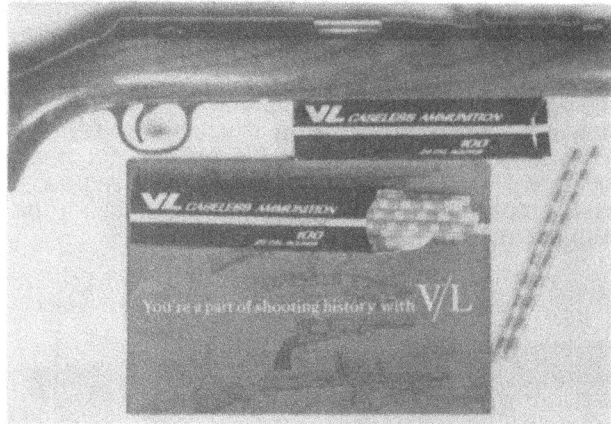

RIM FIRE CARTRIDGE

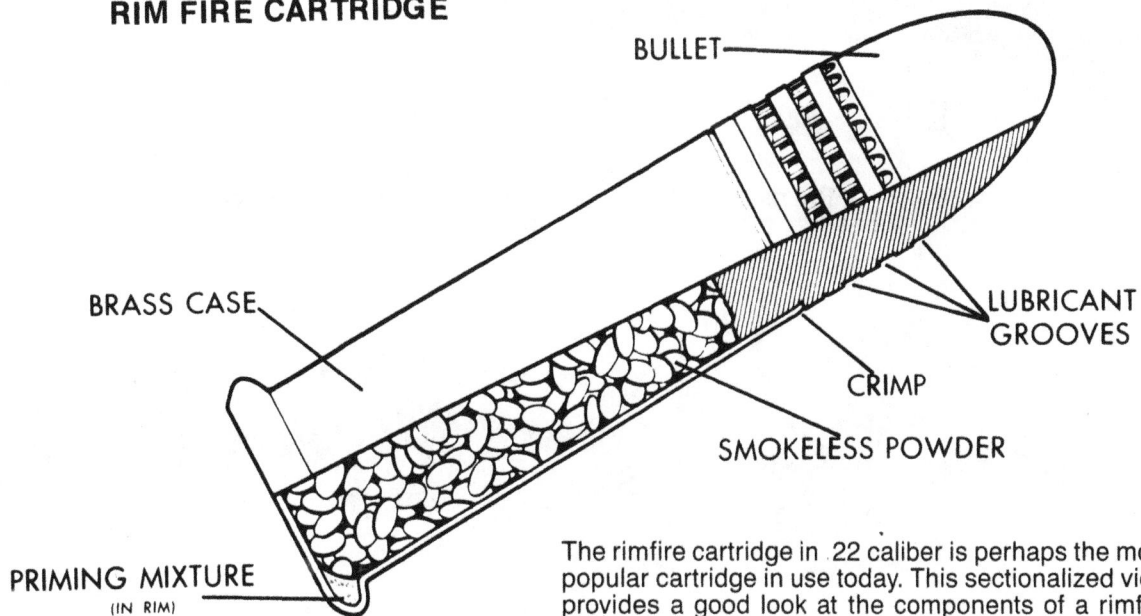

BULLET

LUBRICANT GROOVES

CRIMP

SMOKELESS POWDER

BRASS CASE

PRIMING MIXTURE
(IN RIM)

The rimfire cartridge in .22 caliber is perhaps the most popular cartridge in use today. This sectionalized view provides a good look at the components of a rimfire round.

Unlike the centerfire cartridge, the rimfire round has its priming compound in the outer areas of the rim itself. When the firing pin (A) impacts on the rim of the cartridge (B), the priming compound detonates, sending a burning flash into the case (C) igniting the powder. As the powder burns and creates gases (D), the projectile (E) is forced out of the case, down the bore (F) and on its way to the target.

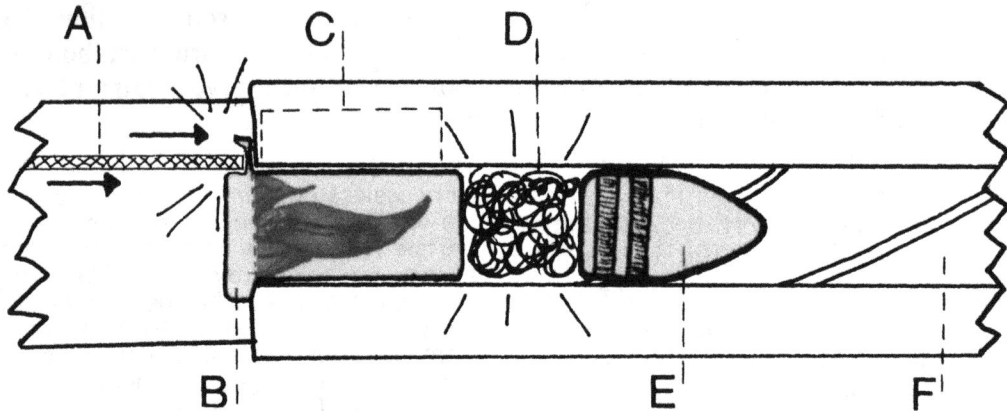

(Below left) Here's how a centerfire cartridge works. The firing pin (A) strikes the primer (B), detonating the fulminate compound inside. The resulting flash passes through a small hole in the base of the case (C) to ignite the main powder charge, which burns very rapidly to produce large volumes of expanding gas (D). It's this gas that drives the bullet (E) down the bore (F) and out the end of the barrel. The brass walls of the cartridge case expand momentarily to form a tight seal in the chamber, but then contract as the pressure drops.

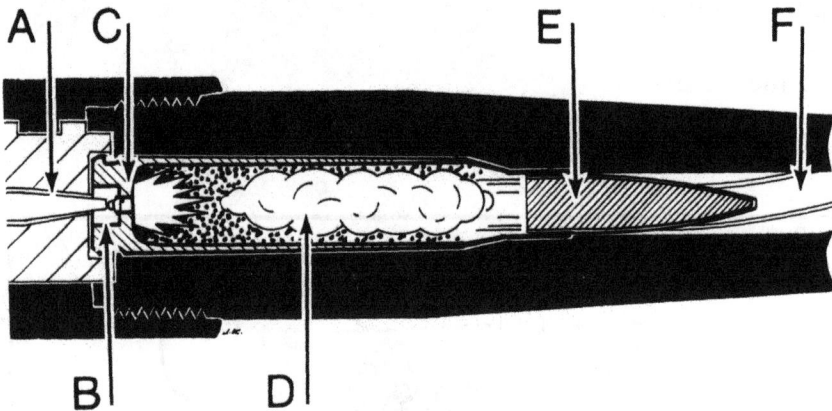

(Above) Cartridges for modern rifles and handguns come in two different varieties: rimfire (at left) and centerfire (right). Centerfire cartridge cases can be reused; rimfire cases cannot be reloaded.

Below — The modern rifle or handgun cartridge consists of four components: projectile (bullet), powder, case and primer.

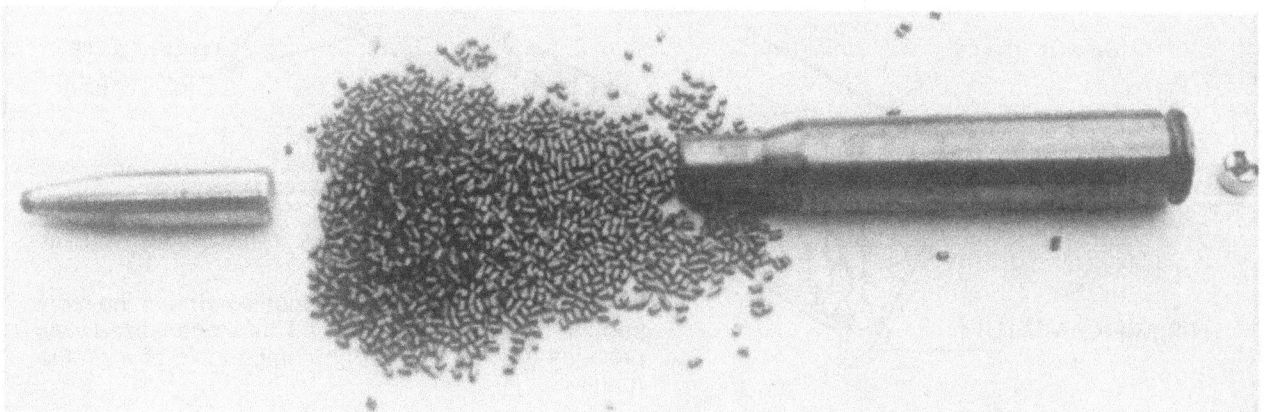

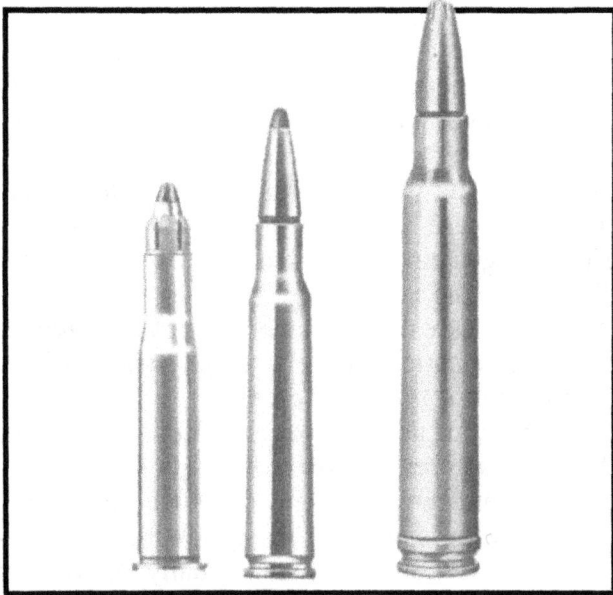

Modern metallic centerfire cartridges are found in three basic case/rim styles. Left, the earliest style is the "rimmed" cartridge; center, the "rimless" type; on the right a "belted" cartridge.

How Modern Cartridges Work

The modern firearm cartridges used in rifles, handguns and shotguns all operate on a similar principle. Basically, a cartridge consists of a case — which may be made of brass or some other soft metal, a combination of metal and plastic, or brass and layers of waxed paper — a charge of propellant powder, and a primer to ignite the main charge. Finally, there is some kind of projectile to be propelled from the gun. This is usually a single bullet, or ball, in cartridges designed for use in rifles and handguns (although this isn't always true). Similarly, cartridges intended for shotgun use normally

contain several hundred tiny pellets of lead or soft metal alloy. Again, this rule isn't absolute, as there are shotgun loads that throw a single lead ball, or slug for deer hunting use, just as there are shot-loaded cartridges that can be used in rifles or handguns.

Almost without exception, modern sporting arms cartridges use one of two types of priming mechanisms — rimfire or centerfire. A rimfire primer is an integral part of the case itself, with the volatile priming compound contained in the protruding case rim. It is ignited by crushing the rim between the firing pin and the face of the chamber in which the cartridge is loaded. Centerfire primers usually consist of a small, metal cup (normally made of brass) containing the explosive priming compound. Because a portion of the primer must be crushed in order to have this compound detonate, a metal anvil is included to give the firing pin something to work against. Unlike rimfire priming, centerfire primers are not an integral part of the cartridge case, but fit tightly within a centrally located hole in the base of the case (hence the term "centerfire"). These primers are held in place by friction only, so a snug fit is a must.

The cartridge case itself serves to hold the other components together, but also acts as a gas seal within the chamber when the cartridge is fired. To give a proper seal, the case must be able to expand quickly under pressure, fitting tightly against the walls of the firing chamber. Then it must be elastic enough to spring back into an approximation of its former shape so that it can be easily removed from the chamber afterward.

Cartridge cases come in many sizes and shapes,

Modern shotgun cartridges (shotshells) are made up of primer, case, powder, wad and plastic shot protector (a single unit here) and shot.

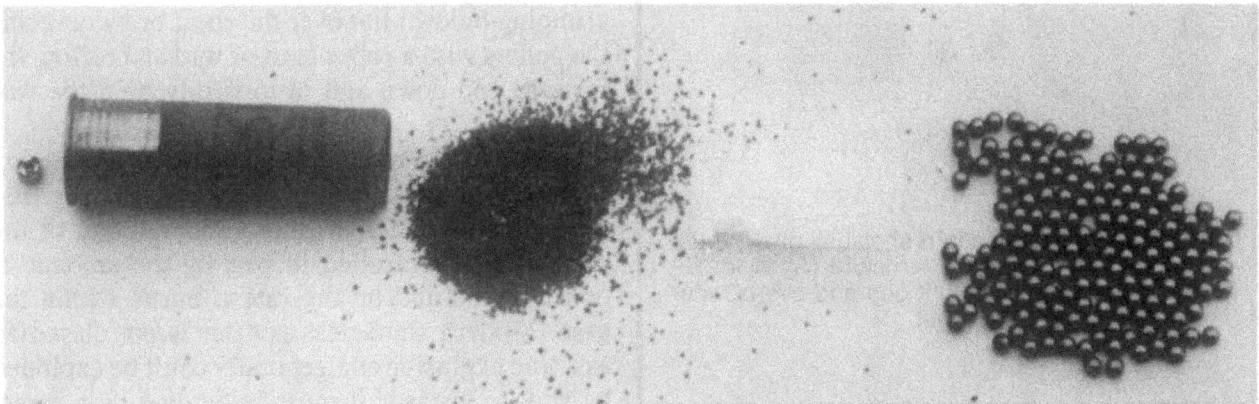

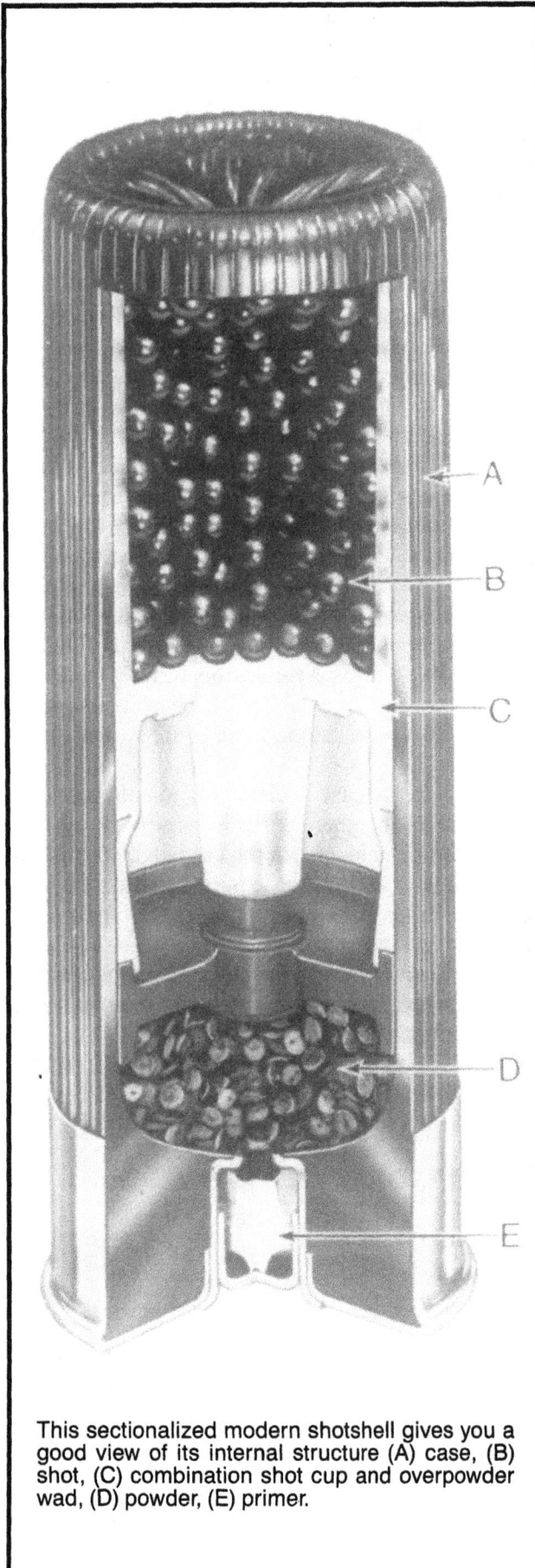

This sectionalized modern shotshell gives you a good view of its internal structure (A) case, (B) shot, (C) combination shot cup and overpowder wad, (D) powder, (E) primer.

but all have certain things in common. Except for rimfire cartridges, each case features a centrally located primer hole in its bottom end, or base. All cases have some sort of rim surrounding the base to give the gun's extractor something to hold on to so that the case can be removed from the chamber after firing. This rim can protrude past the outside diameter of the case body (such cases are known as "rimmed" cases), or it can be formed by making a ringed indentation in the case just ahead of the base (this is a "rimless" case).

The base, or head of the cartridge is stamped by the manufacturer to show caliber (in the case of a rifle or handgun cartridge) or gauge (shotguns), and the maker's name or initials. Military ammunition will show caliber and usually the date of manufacture.

The body of the cartridge case can be straight, although many modern rifle cartridges narrow down toward the front. Some high-intensity cartridges also have a narrow ring that protrudes ahead of the base — these are known as "belted" cases. Metallic cartridge cases used in rifle and handgun ammunition are generally referred to as "cases," while containers used in shotgun ammunition are called shotshells, shells, or hulls. Rifle and handgun cases are commonly made of brass, while shotshells usually feature a brass head or base, with the rest of the shell body composed of paper or plastic.

Rifle and handgun projectiles are called bullets, while the multiple pellets contained in a shotgun shell are known as the "shot charge," or simply "shot." Bullets are made simply of lead or lead alloy, or are formed of lead with a soft metal jacket. Bullets are held in the cartridge case by a friction fit or crimp, and usually project beyond the mouth of the case. Shotshells completely enclose their charge of shot pellets, which are held in place by folding the forward end of the shell and crimping it down flat over the shot, or by covering the pellets with a paper card or wad and rolling the forward end down and in to firmly hold the wad in place.

Several different kinds of gunpowder are used in shotshells or metallic rifle and handgun cases, and these burn at different rates. The velocity of the projectile is controlled in part by the amount of powder used and by the rate it burns within the case. Modern smokeless powder is not classified as a true explosive and generally can't be exploded by shock. Instead, it must be ignited by a direct

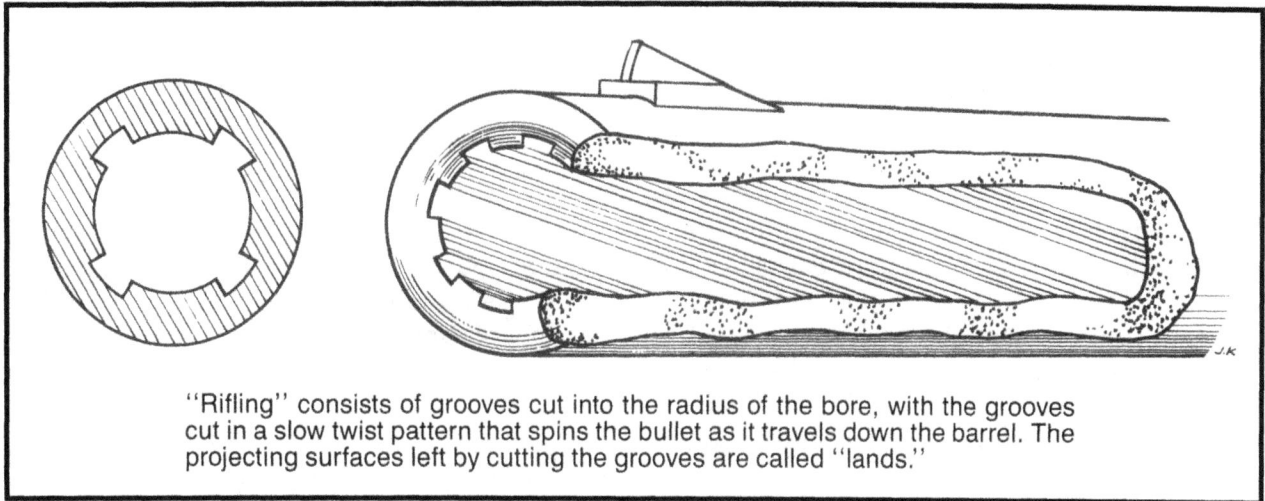

"Rifling" consists of grooves cut into the radius of the bore, with the grooves cut in a slow twist pattern that spins the bullet as it travels down the barrel. The projecting surfaces left by cutting the grooves are called "lands."

spark or flame, which is provided by the primer. Once ignited, the powder burns very rapidly and this, in turn, generates expanding gas. It's this expanding gas that produces the pressure needed to drive the projectile from the gun at high speed. The pressure generated by the expanding gases sometimes exceeds 50,000 pounds per square inch — but only momentarily. No sporting firearm could withstand such pressures for more than a small fraction of a second, but this pressure drops off the moment the bullet leaves the barrel.

That's the principle used to make modern firearm cartridges operate. When the gun's firing pin moves forward to strike the primer, the wall of the primer is crushed and the shock of this action detonates the sensitive priming material. The resulting flash of flame passes through a hole in the cartridge base to ignite the powder inside, and this burns very rapidly, producing large volumes of gas and creating high pressures almost instantaneously. The walls of the cartridge case expand under this pressure to form a tight seal with the walls of the firing chamber, and the bullet (or shot charge) is forced out of the cartridge and through the bore. Once the bullet leaves the gun, pressures inside drop and the cartridge walls retract.

Rifled Barrels vs. Smoothbores

While rifle, handgun and shotgun cartridges operate on the same general principle to get their projectiles underway, these projectiles behave in widely varying manners once they leave their cartridges and start traveling down the inside of the gun's barrel (which is called the bore).

The earliest firearms all had smooth, reasonably straight bores and fired a variety of projectiles ranging from small stones to fletched darts and arrows. Sometime during the 16th century, it was discovered that a series of spiral grooves cut into the surface of the bore could be used to improve accuracy when a soft lead projectile was fired. These grooves gripped the surface of the ball or bullet, causing it to spin as it passed through the bore. This spinning motion helped to stabilize the bullet in flight, and prevented elongated projectiles from tumbling in the air.

Such a system of spiral grooves became known as "rifling," and firearms operated from the shoulder position that had rifling grooves cut into the bore became known as "rifles." Long-barreled shoulder guns with smooth, unrifled bores were called muskets. Even though rifles could be fired at longer ranges and with greater accuracy than the smoothbore muskets could, muskets remained the standard arm of most military forces long after rifling was introduced, mainly because they were faster and easier to load.

Rifled barrels are almost universally used in military firearms today, and you can be sure that all sporting arms known as rifles have a system of grooves cut into their bores. Handguns also have rifled barrels. In fact almost every firearm intended to throw a single projectile each time it's fired takes advantage of the stabilizing spin offered by a rifled bore.

While rifling works extremely well with the elongated cylindrical bullets used today, it isn't well suited to guns designed to throw loose charges of shot. A typical shotgun load contains hundreds of tiny pellets, and these are easily deformed if forced through a rifled bore. Too, these small, soft lead spheres tend to flatten and fill in the rifling

31

grooves, and are difficult to remove when it's time to clean the gun.

When a load of pellets fired from a shotgun shell leaves the gun barrel, it begins to spread out. If it spreads too far and too fast, the pattern of pellets rapidly becomes too thin to be effective on game or clay targets. Shotguns are designed to throw a load of pellets so that the pattern expands in a predictable, controlled manner and produces an evenly distributed shot cluster. The rate at which the shot pattern expands is regulated by slightly constricting the size of the bore near the front end, or muzzle of the barrel (this constriction is known as "choke").

Shotshells fired in rifled barrels tend to produce wide, erratic shot patterns and simply don't per-

form efficiently. So shotgun barrels are smooth inside. Single lead balls, called "slugs," can be used in shotguns, but are only effective at limited range.

Rifles, Shotguns & Handguns

While the basic differences between the three types of sporting firearms—rifles, shotguns and handguns—are easily understood, there are different types of each in use, and sometimes it's not all that simple to tell one from another by a quick glance.

It's important to know the difference between the several kinds of firearms available, and be aware of what each can and cannot do. Let's take a short look at each type.

Rifles are best fired from a stable shooting position and are designed to hit targets at extended ranges. They may be used in conjunction with precision sighting aids like the scope sight shown here. The scope extends the useful range of the rifle and improves the shooter's ability to hit targets in the poor light conditions found at early morning and late evening. Shooters with impaired vision, who could not otherwise shoot well, may find their abilities enhanced by the use of a scope.

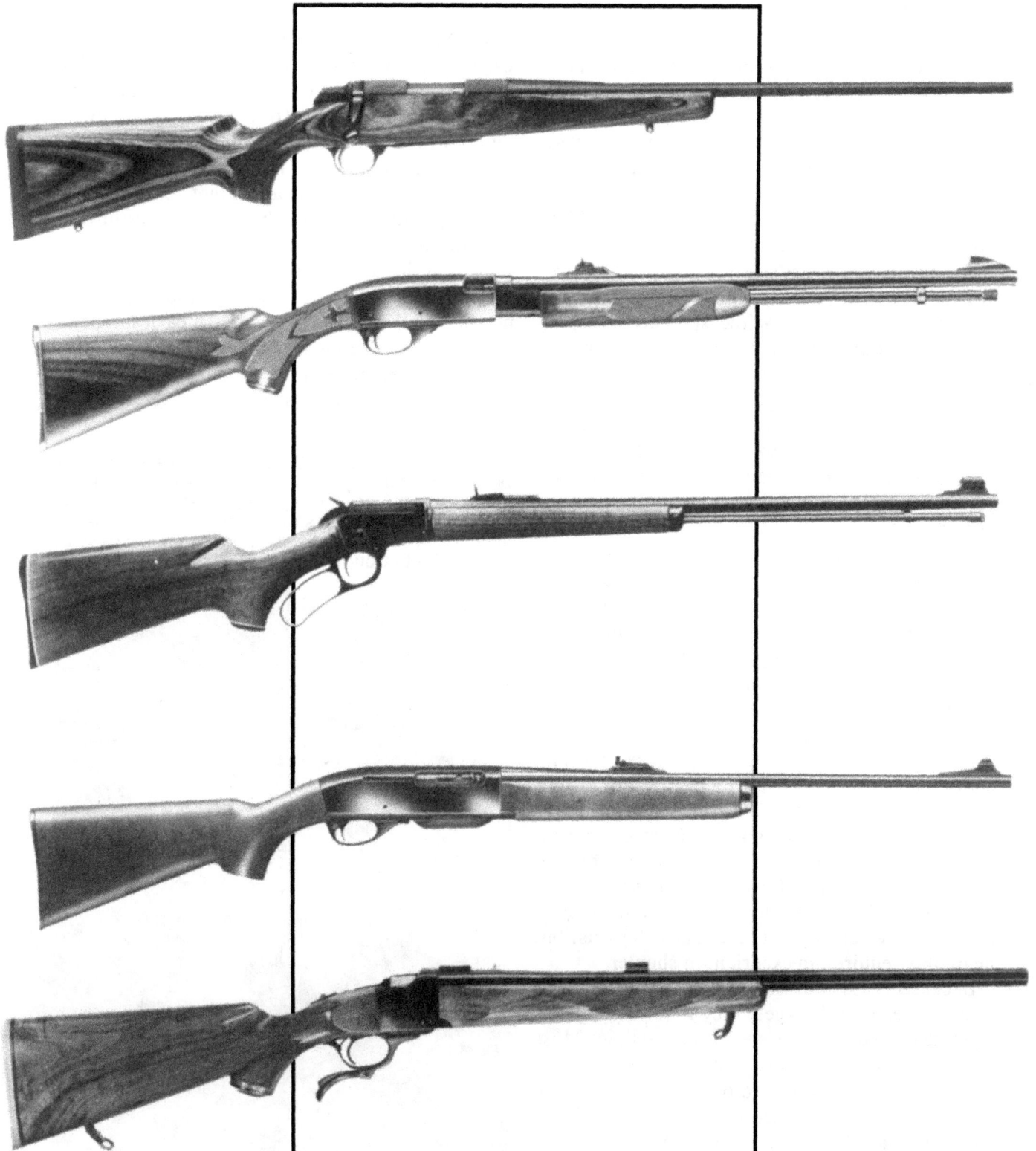

These rifles represent the action types found in modern firearms. From top to bottom: Browning bolt-action centerfire, Remington pump-action rimfire, Marlin lever-action rimfire, Remington centerfire autoloader and Ruger single shot centerfire.

RIFLES

As I've already pointed out, rifles are firearms with a spiral rifling pattern cut in their bores. This pattern is formed by "lands" and "grooves." The grooves are just that—grooves cut into the surface of the bore. The lands are the raised surfaces between the grooves. Because handgun bores are also rifled, we need to tighten the definition a bit to exclude this class of firearm.

For our purposes, let's define the rifle as a firearm with a rifled barrel, fitted with a stock designed to be held against the shoulder. Because they have longer barrels, and are larger and heavier than handguns, most rifles require the use of both hands for their operation.

There are several different types of rifles available to sportsmen, as well as specialized types (like fully automatic assault rifles and machineguns) that are usually reserved for military or police use only.

Rifles can be classified by action type—bolt, lever, single shot, pump or autoloader—and by the type of use they are designed for. Thus there are small game rifles, big game rifles, varmint rifles, and rifles intended for target work only. A separate arbitrary classification is also made for rifles with barrels shorter than 22 inches—these are called carbines.

Rifles are designed for precision shooting, and some are accurate enough to hit a 2- or 3-foot bullseye at 1000 yards. However, these are fairly specialized rifles, and it takes spectacular rifle control and marksmanship to hit any kind of target at that extreme range. Most good quality sporting rifles are capable of placing a bullet within an inch or two of the actual point of aim at 100 yards, but again this requires an experienced shooter.

Because rifles are used to hit relatively small targets at extended ranges, they require precision sighting devices to make such feats possible. Most sporting rifles come equipped with a set of iron sights that consist of a bead or post mounted near the muzzle and a notched blade set farther back along the barrel. To align these sights, the front bead (or post) is either superimposed over the target or placed directly under it, and then the rifle is maneuvered until the bead appears to rest in the rear sight notch when viewed by the shooter. When the target, front sight and rear sight are all in alignment, the rifle is fired.

Another type of sight—the receiver sight (also known as an aperture or "peep" sight)—also uses a front-mounted bead or post, but in this case the rear sight consists of a circular aperture mounted farther back, on top of the rear part of the action (or receiver). In use, the front bead is again superimposed on the target, but this time the shooter merely looks through the hole, or peep, of the rear sight to see the front sight and target. The eye automatically centers the front bead in the aperture, so it doesn't have to try to keep three different objects in focus at the same time.

Many sporting rifles have telescopic sights mounted on them, and these are the easiest type of sights to use. The most popular telescopic sights magnify the target from three to nine times the size the eye would normally see it. Looking through a

Specialized rifles are available which are capable of extreme accuracy. This Olympic-grade rifle is equipped with precision sights and many other special features. Yet, the basics of shooting this rifle are the same as those involved with shooting an inexpensive plinker. (Photo Courtesy of NRA)

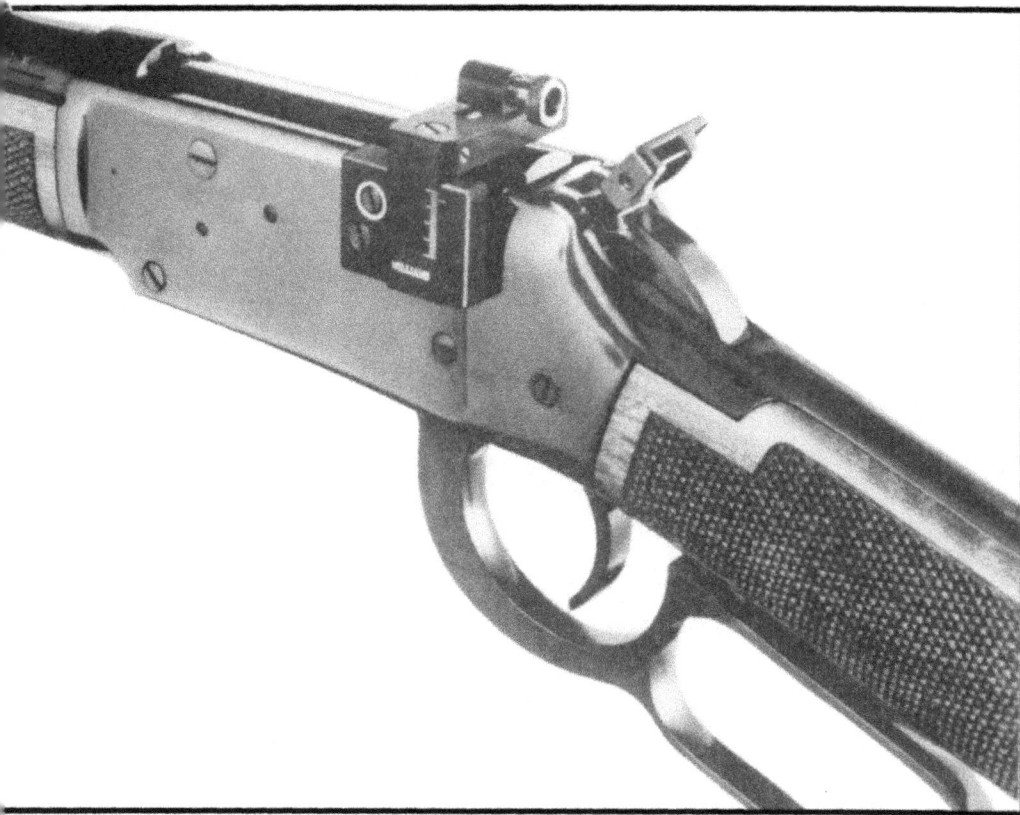

The receiver sight provides the hunter or target shooter with more precise adjustments than the conventional open sight. Shown here on a Winchester Model 94, this Williams receiver sight is adjustable with a screwdriver for windage and elevation. The aperture — the knurled disk atop the sight — may be removed when shooting in poor light.

Since the 1940s, the scope sight has gained tremendously in popularity. This modern Redfield Widefield is variable from 2 to 7 power. Various degrees of magnification are selected by turning that collar at the rear of the scope with numbers on it representing the magnifications. Adjustments for windage and elevation are beneath the serrated caps near the middle of the scope tube. The oblong shape of the lens provides a greater field of view than conventional round tubes.

Right — For years the peep sight has been popular with hunters and target shooters alike. This tang-type peep sight (behind the hammer) is mounted on a Winchester Model 94.

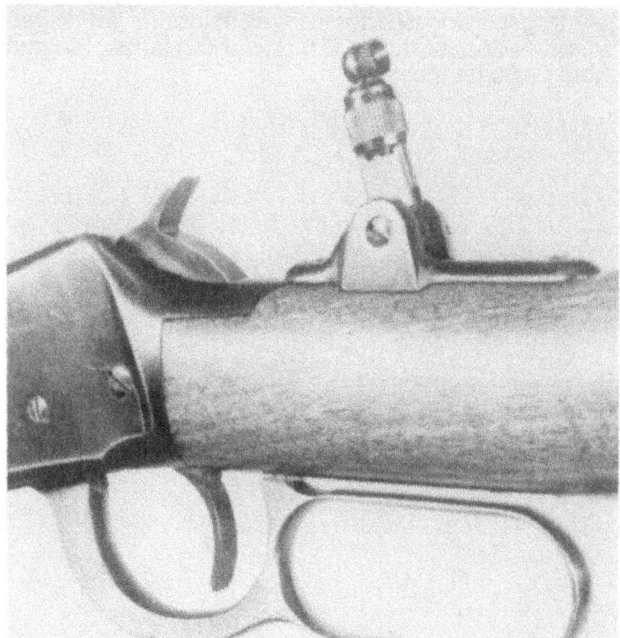

Shotguns are usually fired at a moving target, like the clay target being shattered here by author (indicated by arrow at upper right). To hit a moving target, the gun should also be moving.

typical rifle scope, you see the image of a crosshair superimposed over the field-of-view. Merely place the cross hairs so that they intersect over the target, and pull the trigger.

SHOTGUNS

Like rifles, shotguns are also designed to be fired from the shoulder, and they, too, feature a long stock to fit against this part of the anatomy. What's more, there are double barrel, slide-action, bolt-action, single shot and autoloading shotguns just as there are rifles built around these same kind of actions. These similarities create possible confusion for the beginning shooter, who may have some problem in telling the two types of firearms apart.

Even though rifles and shotguns resemble each other, there are easy ways to tell them apart. For instance, shotguns usually have much fatter barrels, with larger holes in the end (bore size). Another giveaway is that while rifles almost always have either separate front and rear sights or some kind of telescopic sight mounted over the barrel, shotguns usually have only a single, small bead attached to the muzzle end (sometimes a second bead that's even smaller will be fastened mid-way along the barrel). Many shotguns also have a

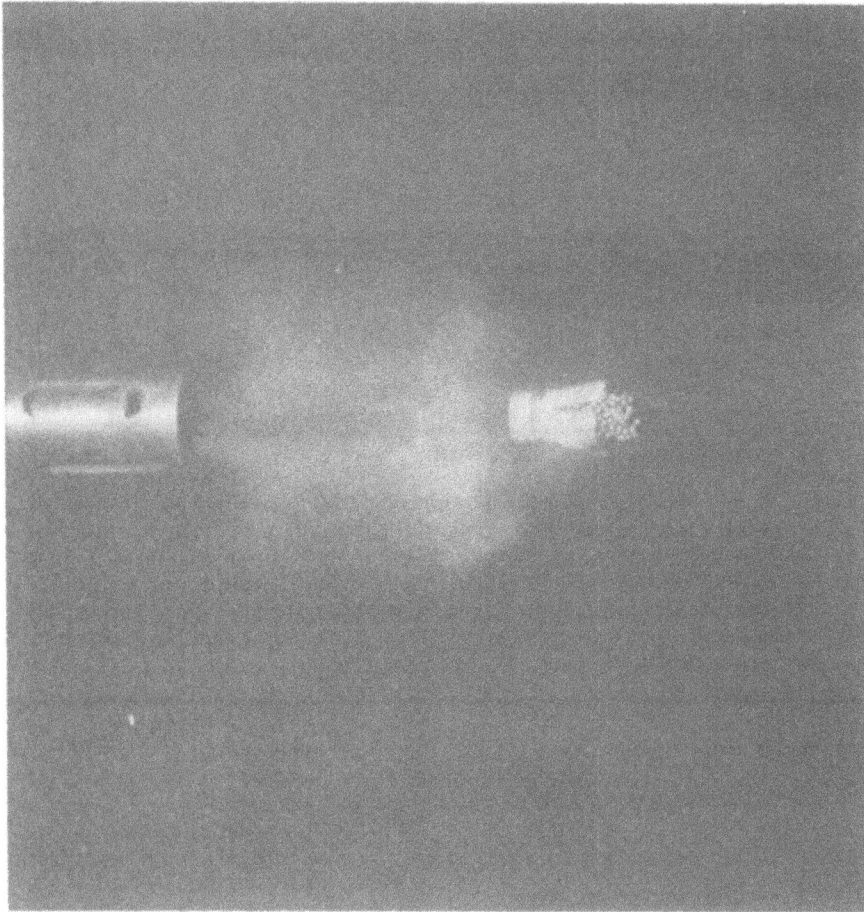

These high-speed photos from Winchester show what happens to a shotshell charge as it leaves the muzzle. In the photo showing the shot charge closest to the muzzle, note that the plastic wad has not yet fully opened and the shot column has not yet opened up. The next photo shows the wad spreading and ready to drop away from the shot column. This sequence also illustrates the fact that it is easy to miss with a shotgun at close range, before the shot column spreads out.

Pump shotguns like the Remington 870 (top) have long been an American favorite. The ventilated barrel rib aids in quick sighting. The autoloader, as represented by this Browning Auto-5, has been available since early in this century. Double-barrel shotguns like this double-trigger Stevens are also popular. Less expensive doubles generally use two triggers; more expensive doubles often use a single trigger to fire both barrels.

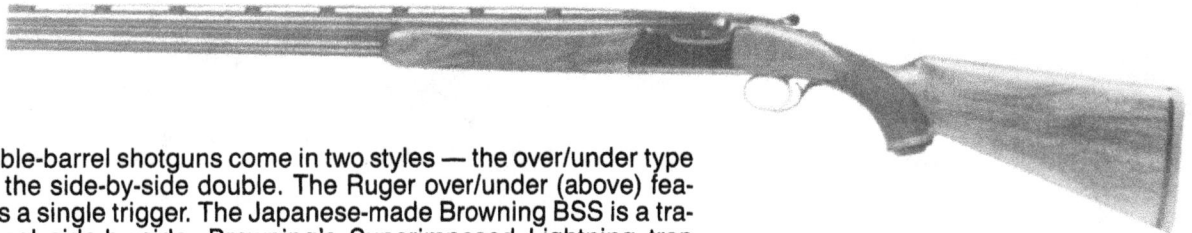

Double-barrel shotguns come in two styles — the over/under type and the side-by-side double. The Ruger over/under (above) features a single trigger. The Japanese-made Browning BSS is a traditional side-by-side. Browning's Superimposed Lightning trap model is another classic over/under double.

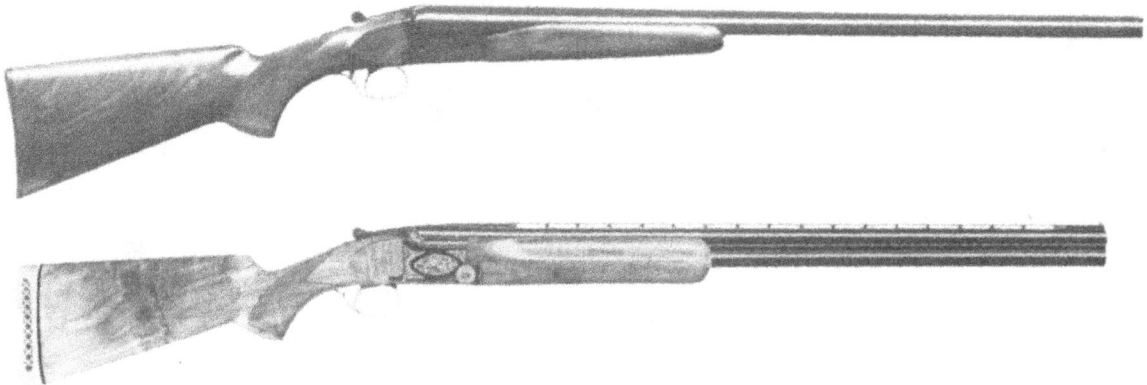

raised platform, or rib mounted on top of the barrel and running full length along it.

Shotguns are designed to throw several hundred individual pellets at the target at a time, and have smooth, rather than rifled bores. Because the internal pressures created by firing a shotgun shell are much lower than those generated by a high-intensity rifle cartridge, the wall of a shotgun barrel is usually much thinner than that found in a rifle.

Unlike rifles, shotguns are pointed, rather than precisely aimed at the target—and since the target is usually a moving one, the gun should also be in motion when fired. Riflemen try to shoot at a stationary mark, and do their best to eliminate even the tiniest bit of motion when squeezing the trigger.

Some shotguns have two barrels mounted together, with the twin tubes either side-by-side or fastened one over the other. There are a very few rifles also made this way, and some rifle-shotgun

Handguns, like this Ruger GP-100 double-action 357 Magnum revolver are easily identified by their small overall size and the fact that they are designed to be fired with one hand. The GP-100 is a modern double-action revolver. A pull on the trigger indexes the cylinder, cocks the action and releases the hammer to fire the gun. It can also be fired single action, however.

combinations (with one rifled barrel fastened to a smooth bore shotgun tube), but most double-barreled guns you will see in the United States will be shotguns.

HANDGUNS

Handguns are easily identifiable by their small overall size and the fact that they have a small handle, or grip, designed to fit in one hand. Most handguns have barrels running from 2 inches to a usual maximum of 8 inches in length, although there are some exceptions to this rule.

Handguns fall into three basic categories, or action types—revolver, semi-automatic, or single shot. Revolvers are easily identified by the revolving cylinder mounted in the center of the gun's frame—this will contain five or six firing chambers that are brought into alignment with the bore one at a time as the firing mechanism is operated. Revolvers generally have external hammers, normally with some sort of grooved or serrated gripping surface to allow easy cocking by the shooter's thumb. Revolvers that must be thumb cocked before firing are called "single-action" guns, while those that can be fired by simply pulling the trigger are "double-action" revolvers.

The auto pistol differs from the revolver in that recoil energy is used to cycle the action, feed a fresh cartridge and cock the hammer for the next shot. Despite their small size, some pocket autos can produce surprising accuracy in the hands of an experienced shooter.

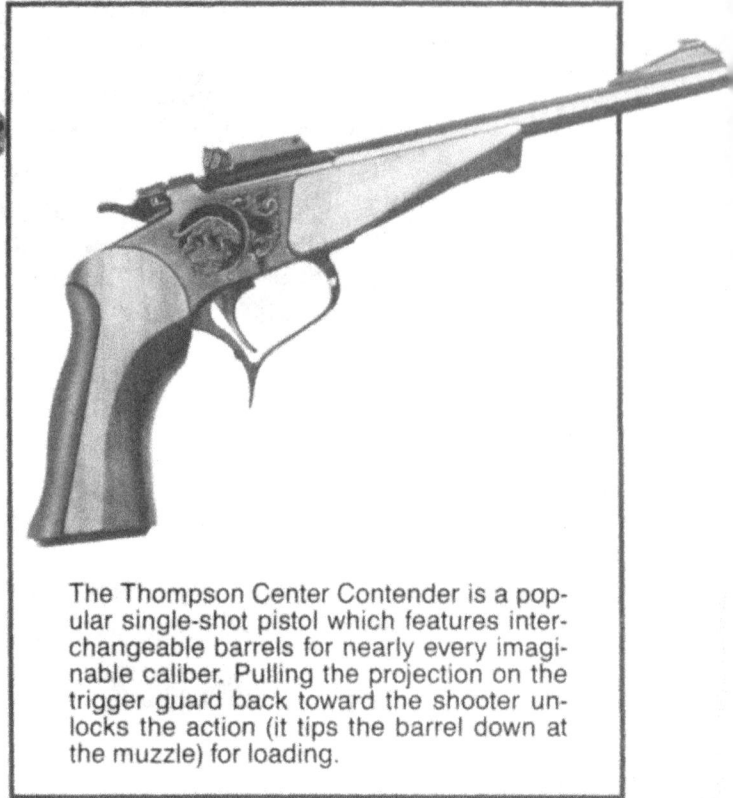

The Thompson Center Contender is a popular single-shot pistol which features interchangeable barrels for nearly every imaginable caliber. Pulling the projection on the trigger guard back toward the shooter unlocks the action (it tips the barrel down at the muzzle) for loading.

Semi-automatic handguns, often called "selfloaders," "automatics" or "auto pistols," feature a reciprocating slide that travels back and forth along the gun's frame. Cartridges are contained in a removable box magazine, called a clip, usually located within the gun's handle, or grip. Some auto pistols have an exposed hammer that can be cocked by thumb pressure, although such hammers will be much smaller than those found on revolvers. Other auto pistols are "hammerless," which only means that this part of the firing mechanism is hidden within the slide. These pistols are normally cocked by pulling back on the slide and then releasing it. This action also strips a loaded cartridge from the magazine and moves it forward into the chamber, where it is ready for firing. Some pistols (with inertia-type firing pins) can be carried with the hammer lowered against a loaded cartridge as the firing pin does not touch the primer. It's fired by simply pulling on the trigger. Like revolvers having the same capability, these guns are known as "double-action" pistols. Auto pistols that must be cocked before firing is possible are called "single-action" guns.

Single-shot handguns have no magazine or revolving cylinder to hold extra cartridges. Instead, the cartridge is loaded directly into the firing chamber by hand, and the action must be opened and reloaded each time the gun is fired.

Ruger's 357 Magnum Blackhawk has been produced since the 1950s. The Ruger single-actions are thoroughly modern adaptations of the 19th century single-action revolver. Fully-adjustable sights and coil springs throughout bring the Ruger designs firmly into the 20th century.

(Below) Ruger's new P85 is an American entry in the double-action 9mm auto pistol market. Since the late 1970s, most European manufacturers have introduced new double-action 9mm autos for military and police use. Most of these "duty" autos feature fixed sights and large magazine capacity. The Ruger holds 15 rounds in the magazine and one in the chamber. The advantage over a typical six-shot revolver is obvious for police use.

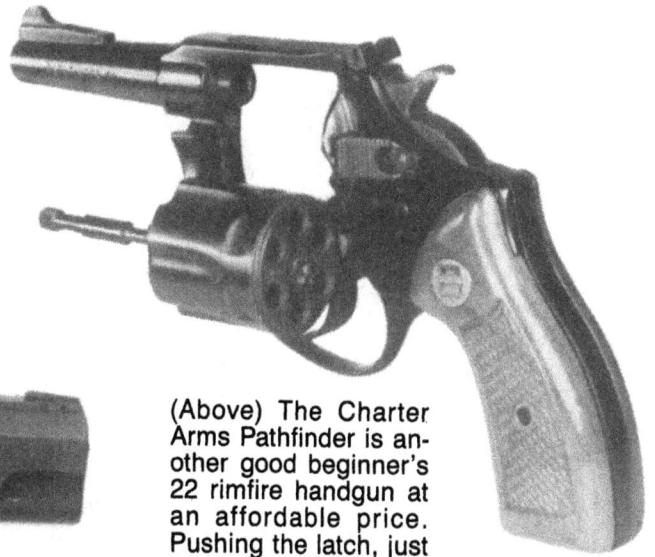

(Above) The Charter Arms Pathfinder is another good beginner's 22 rimfire handgun at an affordable price. Pushing the latch, just to the left of the hammer, releases the cylinder. Pushing the ejector rod at the front of the cylinder dumps all six empty cartridge cases at once.

(Right) AMT's stainless Lightning 22 is as good a beginner's rimfire autoloader as can be had. The Lightning offers very good balance and feel and fully-adjustable sights along with very simple, durable construction.

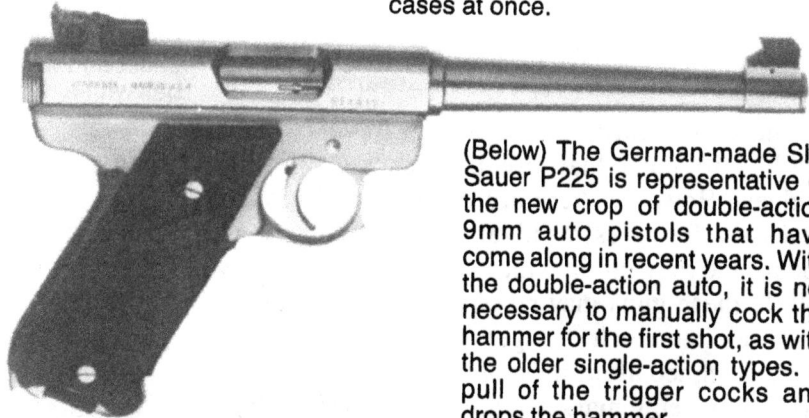

(Below) The German-made SIG Sauer P225 is representative of the new crop of double-action 9mm auto pistols that have come along in recent years. With the double-action auto, it is not necessary to manually cock the hammer for the first shot, as with the older single-action types. A pull of the trigger cocks and drops the hammer.

This Ruger P85 9mm autoloader illustrates the major components of a typical, modern auto pistol. Recoil energy unlocks the action and opens the slide. The coiled spring provides energy to pick up a cartridge from the magazine and return the slide to firing position.

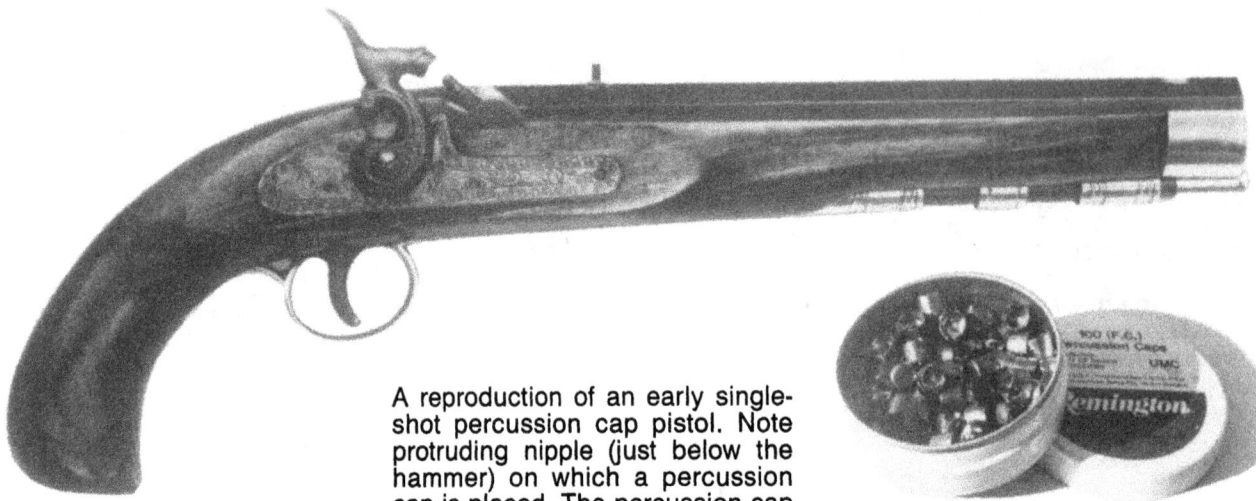

A reproduction of an early single-shot percussion cap pistol. Note protruding nipple (just below the hammer) on which a percussion cap is placed. The percussion cap is a small metal cup containing fulminate priming compound which detonates when the cap is struck by the hammer.

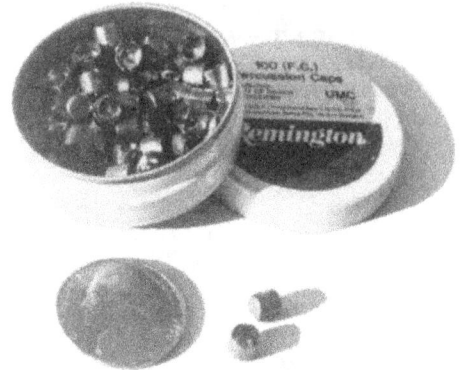

Many modern sport shooters favor obsolete cap and ball muzzleloaders, simply because they're fun to shoot and work with. Here a gunner rams home a lead ball with a ramrod.

Modern Blackpowder Firearms

The classifications used above are most often applied to guns that fire modern metallic cartridges. However there are a number of modern blackpowder rifles, handguns and shotguns in use today, and they fall into their own special category. While these guns may be of modern manufacture and are often much stronger than the originals they were designed after, most of the blackpowder firearms now being marketed are more or less faithful copies of early percussion or flintlock guns dating back 100 years or more. *Under no circumstances should you ever use smokeless powder in guns specifically designed for blackpowder.* Blackpowder rifles and shotguns are usually single- or double-barreled affairs that must be loaded from the muzzle in the traditional manner, while most of the blackpowder handguns in use today are percussion-fired revolvers.

Summary

Firearms of all types can provide hours of enjoyment and challenge for both the novice and experienced shooter. The key to understanding any firearm, is, in two words, "*patience* and *safety*." *Patience* on the part of the student as well as the instructor; plus, *safety* at all times on the part of everyone who has occasion to own, handle or be in the presence of a sporting firearm. If you choose to follow this approach, I can simply say, "Welcome to a sport that's truly an American heritage!"

chapter 2

FIREARMS SAFETY

SAFETY, SAFETY AND SAFETY are the three most critical aspects of gun ownership. Without a thorough understanding of safety, no other information is important. The shooter — beginner or expert — must be conscious of safety when handling any gun, at all times. If the shooter is not certain of how to handle a particular gun, he or she should decline to handle it until instructed. Don't guess. People who guess end up subjects of news stories with headlines like, "Man Thought Gun Wasn't Loaded."

While most of the basics are similar, there are important nuances to be learned in the safe handling of the many types of guns in the field, on the range and at home.

It might seem, at first thought, that the rules would be the same for all guns. The new shooter must consider that rifles, handguns, shotguns, and blackpowder guns all shoot — but all have differing handling characteristics and varying degrees of safety risk. Knowledge is the way to avoid accidents. An experienced pilot would not consider taking off in an unfamiliar aircraft type without a thorough checkout. The shooter should feel the same way about new guns and safety questions; don't be embarrassed to ask questions.

There are 10 basic rules cited in gun safety courses. I'm going to change the number and order a bit based on my personal ideas of importance:

THE COMMANDMENTS OF GUN SAFETY

1. *Open the action AS SOON as you pick up ANY gun and inspect for live ammunition.*

This is not only good safety practice, but good etiquette. While opening the action shows respect for safety and those around the shooter, it's surprising how many experienced shooters fail to observe this cardinal rule. Even if another shooter opened — and looked — in the action before handing you the gun, stop. Open it and look again. If you aren't sure how, ask. Let's think about how this sort of thing might differ with various types of guns. With a single-shot rifle, or a single-barrel or double-barrel shotgun, a simple opening of the action and a glance will verify if the gun is or is not loaded. Other firearms types are a bit different. A rimfire rifle with a tube magazine comes to mind, along with most pump-action shotguns and lever-action rifles. In these guns, it's possible to open the action, glance at the empty chamber, close the action and chamber an unseen live round from the tube magazine. The ammo isn't as visible as it is in a box magazine or a single-shot.

2. *Treat EVERY gun you handle as if it were loaded.*

Even though you carefully checked for the presence of ammunition and verified that the gun is unloaded, treat the gun as though it were ready to fire. Dozens of people are killed or injured every year by so-called "unloaded" guns handled carelessly. Don't be one of them.

3. *ALWAYS keep the muzzle pointed in a safe direction.*

Make certain that the muzzle is pointed away from people, inhabited buildings, animals — and yourself. In a range setting, make sure the gun's muzzle is pointed downrange (toward the target). In the field, straight up is generally the safest policy. Some people suggest that carrying a gun with the muzzle downward is a safe practice. In the event of an accidental discharge, the muzzle-down carry can cause a dangerous ricochet or a hole in someone's foot. No matter what the circumstances or the type of gun involved, MAKE CERTAIN that the muzzle is pointed in a safe direction.

4. *Be sure the barrel is clear of obstructions before firing.*

A plugged barrel is seldom a problem on the target range, but can easily happen in the field. Dirt, mud or snow in the barrel can cause serious damage to the gun and injury to the shooter. Make sure you are using the correct ammunition for the gun you are firing. Cleaning patches have a way of being forgotten. With the gun unloaded, check for obstructions before shooting. If the barrel of the gun being inspected can not be viewed from the breech end, open the action of the empty gun and place a scrap of white paper in front of the breech bolt. The paper will reflect enough light so the bore may be inspected from the muzzle.

5. *Unload the gun when not in use.*

That means before you leave the firing range or shooting area. It is very easy to forget that a gun is loaded. A gun you "thought" was unloaded is an accident waiting to happen. Transporting a loaded gun is also illegal in many areas.

6. *Be sure of your target and backstop.*

Before you pull the trigger, you should know exactly what you're shooting at — and where the bul-

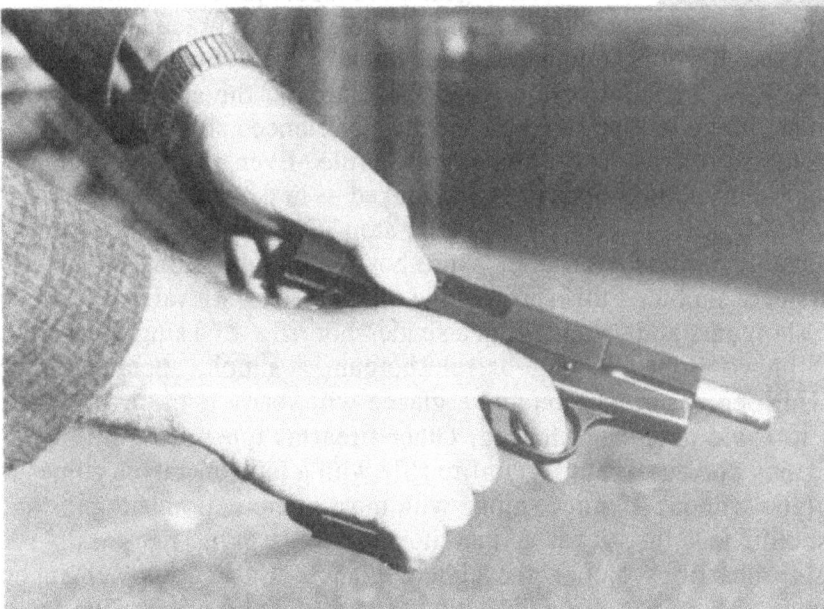

Above and left — Always open the action and check to make sure the gun you're holding is unloaded. This shooter should *not* have his finger in the trigger guard until he's on the firing line and ready to shoot. To inspect the chamber of an auto pistol, the magazine is first removed and then the slide is drawn back.

Watch that muzzle. This is a view no one should have to put up with — even if the shooter is "only kidding around." Pointing a gun at another individual is serious business, and anyone who does so intentionally has no business with a gun in his hand.

let will go if you miss. You should be absolutely certain that no one — seen or unseen — will be in danger when you fire. When you're in the countryside for a plinking or target shooting session, take care to always shoot into a high, soft bank or hillside. This backstop should be tall enough and wide enough to stop even stray bullets accidentally fired while the rifle is pointing downrange. If you're hunting, make sure you don't fire in the direction of buildings or other people. Even a rimfire 22 bullet will carry more than a mile, and it's dangerous until it falls to the earth and stops. Most centerfire rifle bullets travel even farther.

If you fire at the wrong target during a shooting match, you'll lose points from your score. If you shoot at the wrong kind of target while hunting or even plinking, the results could be far more serious. Never shoot at glass or at a hard, unyielding surface. Some glass bottles are tough enough to cause a 22 rimfire bullet to ricochet wildly, and broken glass is a dangerous nuisance in the field. Leaving this kind of litter gives other shooting sportsmen a bad reputation and often results in

land being posted "off limits" for future use.

Because bullets can ricochet, always make sure that no one else is downrange (anywhere in front of your muzzle) when you're shooting. Even if someone is out of your direct line of fire, he could fall victim to an unexpected ricochet or an accidentally fired round. Any shooting accident can be prevented, and observing this rule faithfully will help *you* prevent accidents.

In addition to being sure you know what you're shooting at, you should be equally certain that doing so won't cause loss of property or damage. Indiscriminate shooting — even at "safe" targets — sometimes amounts to vandalism, and this is something every real sportsman avoids at all costs.

7. *Never point a gun at anything you don't intend to shoot.*

The key word here is NEVER.

8. *Never shoot at a hard, flat surface or water.*

Bullets frequently ricochet, or bounce, from such surfaces.

9. *Store guns and ammunition separately and beyond reach of children.*

We might also add that secure storage keeps guns out of the hands of thieves. Some people keep a loaded gun in their home for protection. Using a gun for personal protection is not an activity for the beginner.

10. *Don't mix alcohol and gun powder — no drinking before or during a shooting session.*

Most of us are aware of the problems caused by drinking and driving. Drinking and shooting is an equally poor practice.

11. *Never climb a tree or a fence or jump a ditch while carrying a loaded gun.*

Disobeying this rule can save a minute or two — and cause a lengthy hospital stay for the shooter or a bystander. Proper procedure is to open the action, remove the cartridge or shell from the chamber and hand the gun to a companion while you negotiate the obstacle. If a companion isn't handy, unload the gun, open the action and lay it flat on the ground several yards away from where you want to be. Once you do your climbing or jumping, retrieve the gun. Avoid leaning the gun against fence posts or trees — the gun may fall and be damaged, or fire if it's loaded.

While the basics apply to ALL guns, there are a number of very important differences to remember when handling various types of guns. It is important to understand the differences between safe handling of rifles, handguns, shotguns and blackpowder guns.

Blackpowder Guns

Shooting reproduction blackpowder guns has become a very popular shooting sport, but requires considerable care to enjoy safely.

Note that blackpowder is an explosive. Blackpowder is also very easy to ignite accidentally. If the powder is not in a confined space, it will burn very quickly with an intense, smokey flame. In a confined space, the result of accidental ignition will be the same as what happens in a gun barrel — an explosive and very dangerous "Bang."

While it should be obvious that smoking is a no-no while handling bulk blackpowder, many shooters persist in smoking while shooting. Don't be one of them.

Blackpowder is generally sold in 1-pound metal cans. Handle the can carefully — don't shake it unnecessarily. When shooting, do not leave an open powder can near the firing point. Smoldering wadding or loose grains of powder have been known to find their way into powder cans with explosive results. Unless actually filling a powder measure, keep the can closed and in a safe, cool place.

When shooting a blackpowder revolver, remember to smear grease over the loaded chambers. The object of this exercise is to prevent the flash from causing more than one chamber to fire at once.

Some shooters say that blackpowder guns cannot be overloaded. Don't believe it. Blackpowder guns *can* be damaged by overloading. Though it is difficult to overload a revolver, simply because of the limited volume of the chambers, a rifle, shotgun or single-shot pistol can be dangerously overloaded. If you aren't sure how much powder to use, check the literature that came with your gun or ask an experienced blackpowder shooter.

Checking to see if a blackpowder revolver is loaded is reasonably simple, as a visual check of the cylinder will quickly tell the tale. A rifle, shotgun or single-shot isn't nearly as easy to check, as there is no convenient way to look. If uncertain as to whether a blackpowder gun is loaded, use the ramrod or a wooden dowel. Slide the dowel down the barrel to determine whether or not it will reach at least to the nipple (or pan on a flintlock). If the dowel misses reaching the nipple by an inch or two, you may be reasonably certain that the gun is

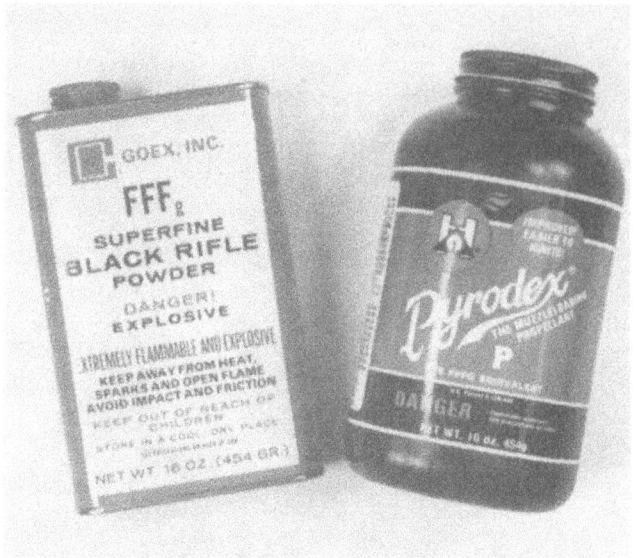

Many new shooters, in associating blackpowder with antiques, forget that blackpowder is a powerful, dangerous explosive. Reloaders handle bulk smokeless powder in a controlled environment, while muzzleloaders handle loose blackpowder in the field where it is easier to make mistakes. Pyrodex, a modern, cleaner burning blackpowder substitute carries the same cautions. BE CAREFUL!

loaded. If you didn't load the gun, you really don't know what's in there. Don't try to unload it by shooting. Special tools called "worms" are available from gun shops which screw into the ball or bullet to allow it to be pulled out of the bore.

Blackpowder guns often leave small amounts of smoldering powder or patch material in the bore between shots. Make sure there is nothing hot in the bore before you pour in the next charge.

Blackpowder guns tend to get dirty much quicker than modern, smokeless powder guns. One common problem in a dirty gun is that the ball becomes very difficult to ram down the barrel. The proper way to avoid that problem is to swab the bore with a blackpowder solvent when bullet seating becomes difficult. Some shooters get a bit impatient and resort to literally hammering the ramrod to seat the ball, with the sad result that the ball becomes firmly stuck part-way down the bore. Extracting the ball can be a real bear, as the ball won't want to come out any easier than it tried to go in. Attempting to shoot the stuck ball out can be a very dangerous practice, as more than a few blackpowder guns have been known to come apart when that sort of thing is attempted. Remember to keep things clean and be patient.

Percussion caps are not especially dangerous to handle, but they should be handled with care. The

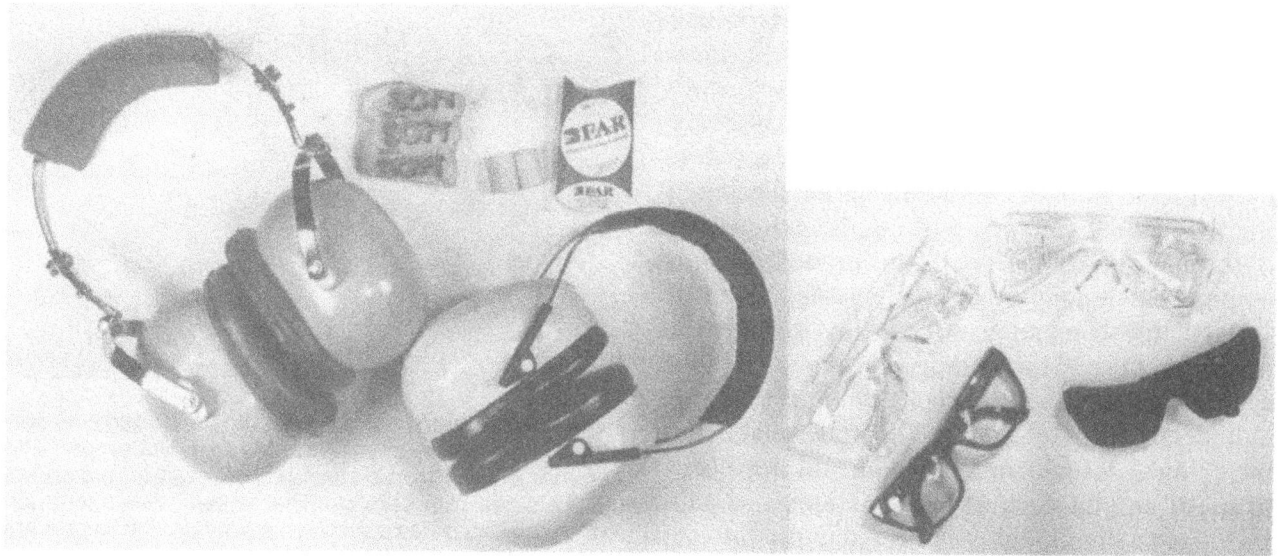

same cautions which apply to handling fixed ammunition suffice.

Never forget that you are handling a bulk explosive when shooting blackpowder. Treat it with respect.

Handguns

All the general rules apply, plus a number of special cautions, simply because handguns are small and easy to mishandle. Most handgun accidents take place because people lose track of where the muzzle is pointing. It is very easy to combine violation of the basics with failure to remember where the gun is pointing, with the predictable result being an accident. Note also that many handgun accidents result in injury to the shooter while companions or bystanders are more often injured in accidents involving rifles or shotguns.

Single-action revolvers, double-action revolvers, auto pistols and single shots all require slightly different care and cautions. Whatever type you shoot, get to know the gun well before buying the first round of ammunition. Particularly in the case of auto pistols, learn how the various buttons and switches work *before* the first trip to the range.

SINGLE-ACTION REVOLVERS

Single-action revolvers must be cocked by hand for each shot. Though this type of handgun was little more than a collector's gun in the early 1950s, modern single actions are now among the most popular handguns for the sport shooter and hunter, as well as the beginner.

Left — Ear protection is critical to prevent progressive hearing loss. Types of ear protectors range from disposable earplugs which cost only a few cents to the more elaborate muffs, which might cost up to $30. There are still more complex electronically augmented headsets which run about $250. These headsets permit the shooter to clearly hear normal conversation, but block out shooting noises.
Right — Eye protection comes in many styles, ranging from very low prices. Note that glasses must have impact-resistant lenses to provide real protection.

Older (pre-1974) single actions usually have three distinct notches, or "clicks," as the hammer comes back. One notch is intended to be engaged to load the gun and permits the cylinder to be turned freely by hand. The other notch, called the half-cock position, is often inadvisedly used as a safety. In the half-cock position, the firing pin does not touch the primer of the cartridge under the hammer. Many accidents have taken place when the older single actions were dropped, or when a "false" half-cock takes place. The false half-cock happens when the half-cock position is not fully engaged. A pull on the trigger can then cause the hammer to drop hard enough to fire the gun. The safe way to carry the older single action is with only five chambers loaded — with the hammer fully lowered on the empty chamber (see below). The possibility of accident is serious enough that newer Ruger single actions have been designed to eliminate the half-cock notch entirely.

Most of us don't like to admit it, but there are people who derive some sort of childish pleasure by playing cowboy quick-draw with single actions and live ammo. These would-be Matt Dillons have been known to shoot TV sets as well as their own feet. Quick-draw with live ammo is a questionable activity at best and certainly not for the beginner. Those who disagree should stick with cap pistols.

Unloading a single-action revolver may be accomplished in two ways. The first is by depressing the cylinder pin lock, withdrawing the pin, opening the loading gate and pushing the cylinder out of the gun, toward the right. This procedure is just cumbersome enough that most shooters avoid it. Instead, the hammer may be pulled back to the loading notch and the loading gate opened. Inspect each chamber through the gate while rotating the cylinder by hand. Count the chambers, and turn the cylinder through more than one full revolution to make certain each chamber is empty. Again, this check should be performed with the gun pointed in a completely safe direction, and you should keep your fingers *away* from the trigger guard. If your single-action revolver has any cartridges left in the chambers, they should be removed by pushing on the ejector rod (located underneath the barrel) as the chambers come into alignment with the open loading gate. Remove *all* cartridge cases from their chambers, even if they've been fired (you can see the indentation made by the firing pin in the face of the primer). Misfires sometimes occur, and a punched primer is no guarantee that the cartridge has, in fact, been fired. The gun isn't safe until *all* the chambers are *empty*.

For safe carrying in a holster, most single-action revolvers should *always* be loaded with only five rounds (if the cylinder has a six-shot capacity) to leave one firing chamber empty. This chamber should then be lined up with the bore by rotating the cylinder by hand (you'll need to place the hammer on half-cock to do this), and then the hammer should be fully lowered over the empty chamber. The empty chamber is necessary when holstering these old-style six-guns, because most feature fixed firing pins that protrude into the chamber when the hammer is down. If the firing pin was allowed to rest on a live primer, a sharp blow on the hammer could cause the gun to fire.

Ruger's most recent single-action design is an exception to this rule, as it features a rebounding firing pin and a hammer block safety that prevents the face of the hammer from contacting the rear of the firing pin unless the hammer is first cocked and the trigger pulled. These guns can safely be loaded with all six rounds and carried with the hammer down over a live chamber.

Most single-action revolvers should be loaded with cartridges in only five chambers, leaving one empty. The hammer should then be lowered over the empty chamber. This new model Ruger Blackhawk is an exception to this rule, as it has a modern safety device to prevent accidental firing if a blow should strike the hammer.

DOUBLE-ACTION REVOLVERS

With a modern double-action revolver, all you need to do to render the gun entirely safe is to swing its cylinder out of its frame. This is accomplished by pushing the cylinder release button or slide on the left side of the frame, using the thumb, and then, using the other hand, move the cylinder (gently) out of the frame to the left. With the cylinder craned to one side, the gun can't possibly be fired. To carry the double-action revolver safely, swing out the cylinder and grasp the gun by hooking the fingers and thumb of one hand through the opening the cylinder vacated. Hold the gun by the top of the frame, butt forward, with the barrel pointing rearward. In this condition there's no way an accident can happen. If you want to carry safety a step further, you can remove the shells from their chambers by tipping them out into your palm.

If you are on the shooting range or in the field and want to carry your double-action revolver loaded in some kind of holster, the safe condition is with the hammer all the way down over a loaded chamber. (Lower the hammer *gently* while pointing the gun safely at the ground, if the hammer happens to be cocked.) Most revolvers — double-action or otherwise — have no additional safeties to contend with, and they need none. *Caution*: Even though this carrying condition is safe, the gun can still be fired by a long double-action pull on the trigger.

AUTO PISTOLS

Note that various auto pistols differ in many

48

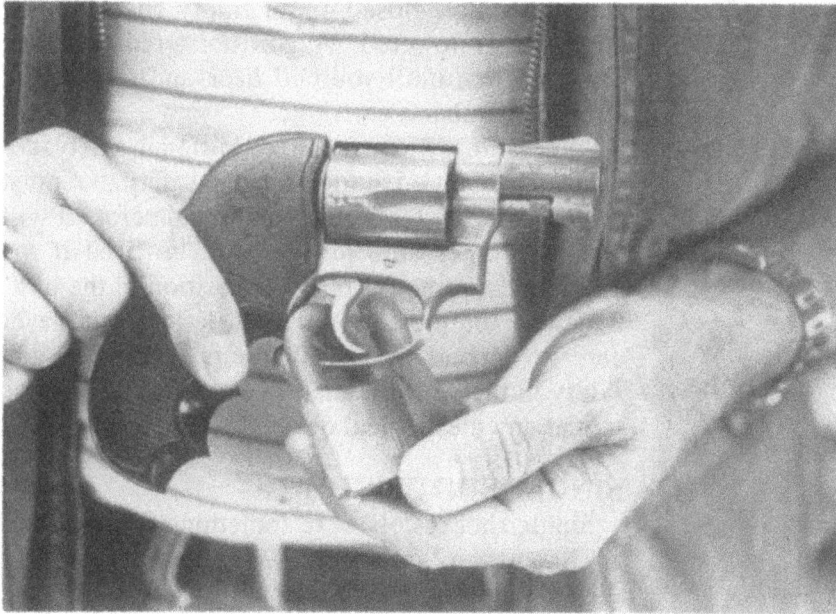

Here is a simple trick used by people who want to keep a loaded gun handy but still keep risk at a minimum. Locking the padlock into the trigger guard, behind the trigger to block rearward movement, makes the gun incapable of being fired until the lock is removed. A piece of rubber tubing protects the gun from scratches. Wearing the padlock key on a neck chain means the gun can be put in shooting order very quickly.

ways. For example, the Austrian Glock has no separate manual safety. The Colt Government Model uses a manual safety which will lock the hammer in the full-cock position. Many Beretta and Walther models use a safety which drops the hammer when the safety is engaged. Some auto pistols incorporate a magazine safety which prevents the pistol from being fired with the magazine removed. Other auto pistols do not incorporate this feature. The point to be made here is that the new shooter *must* learn exactly how his pistol works before the first range session.

Auto pistols, or semi-automatics, are a little more complicated to clear and inspect. The first thing to do in rendering an auto pistol harmless is to remove its magazine and put the magazine in your pocket or insert it in your waistband. Keep your finger *off* the trigger and then use the free hand to pull the slide back far enough to allow the chamber to be inspected. If a live round remains in the action, gently pull the slide all the way back to eject the cartridge. If there's a slide locking lever (usually located on the left side of the frame at about mid-point along its length), the slide can be locked open.

When you're ready to reload, all you need to do is replace the magazine and (if the slide doesn't automatically ride forward at this point) pull the slide slightly to the rear and release it, or depress the slide locking lever. The slide will strip a cartridge from the magazine and chamber it as the action moves closed. Once you move the safety lever to the off safe, or "fire" position, the gun will be ready to shoot.

Some auto pistols can be carried in a holster in three different states of readiness: 1. With a loaded magazine in place and the hammer lowered over an *empty* chamber; 2. With both the magazine and the firing chamber loaded, and the hammer lowered — this time over a live cartridge and (in the case of some single- and most double-action automatics) with the safety on; 3. Or with both magazine and chamber loaded, and the hammer "cocked and locked" — that is, the hammer drawn back to the full-cock position, and the safety lever moved to the "on safe" position. When at the range or afield, the *only* way a beginner should carry an autoloader is with the hammer down on an empty chamber. Never put a loaded magazine in an autoloader until you reach your destination and observe the safety precautions.

The only handgunners who have any need for the last condition of readiness outlined above are law enforcement officers or other individuals who legally carry a handgun for the protection of life and property. The pistol can be readied for immediate firing by simply thumbing off the safety — *but this "ready-to-shoot" condition shouldn't be used by anyone who is not thoroughly familiar with his gun and well grounded in its use. Definitely not for the beginner.*

The second method, carrying the auto pistol fully loaded and the hammer lowered over a live cartridge in the chamber, is a little safer as modern

This small boy is learning to check the chamber of his rifle for live ammo, no matter who hands him the rifle. His father isn't insulted and should not be, as this practice is among the cardinal rules of gun safety.

auto pistols are designed with inertia firing pins that are too short to reach from the face of the hammer to the firing chamber — even a sharp blow on the back of the hammer won't make the gun fire accidentally. A "single-action" auto (an automatic pistol that must have its hammer cocked to be fired) carried in this manner must have the hammer manually thumb-cocked before it's ready to shoot. Newer "double-action" auto designs are capable of being fired in the hammer-down position (by a long, double-action pull on the trigger). In either case, avoid using the "half-cock" hammer notch (if your gun has one) — an accidental blow on a hammer that's at half- or quarter-cock *could* result in an unexpected discharge.

The safest of all "loaded" pistol carrying conditions, and the one I recommend, is the first one mentioned — with a full magazine in place, but with the action closed on an empty chamber and the hammer lowered. A gun holstered this way can't be fired until you pull back and release the slide to chamber a cartridge.

Some auto pistols feature additional safety devices. The Colt 45 auto has a grip safety that must be depressed by the web of your hand before it will fire, and other auto pistols can't be fired if the magazine is removed from its position in the butt. Be aware of these auxiliary safeties, but never come to depend on them. Remember, the best gun safety in the world is merely a mechanical device — and any mechanical device can fail.

SINGLE SHOT PISTOLS

Single shot pistols were something of a relic until the newer guns achieved popularity in the 1970s. The single shot is not only a versatile gun, but is ideal for the beginner to use to achieve mastery of shooting skills before moving to more complicated guns. While all the safety cautions must be observed, the single shot is easier for the beginner to manage than a revolver or auto pistol as there are fewer parts and gadgets to understand. Some single shots also have interchangeable barrels that allow the new shooter to experience many calibers at far less expense than buying a different gun for each caliber.

Some single shots must be manually cocked before firing, much like a single-action revolver. The important thing to remember here is *not* to thumb that hammer back until you're ready to shoot.

In a range setting, always keep the action OPEN and the chamber EMPTY until on the firing line, ready to shoot.

The most popular single shot incorporates an automatic safety device. Others use a cross-bolt type. Make certain you are familiar with your single shot before shooting. Don't wait until you get to the range to figure things out.

Rifles and Shotguns

Since rifles and shotguns handle in similar fashion, safety discussions may be combined, with cautions unique to one or the other noted.

Note that ammunition for many rifles tends to look alike. For example, mixing 8mm Mauser ammunition with 30-06 has happened many times. Depending on the gun involved, the result of dropping an 8mm cartridge into a 30-06 rifle can range from severe stress to destruction. In the case of shotguns, many 12-gauge guns have been de-

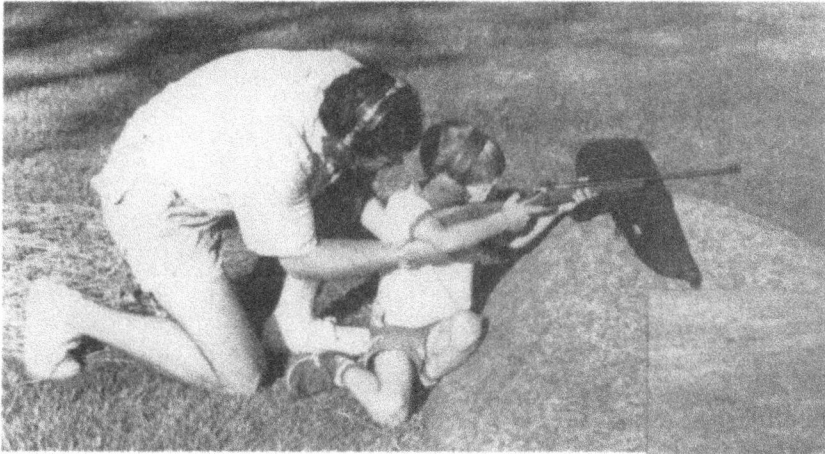

Left and below — After the basics are understood, it is time to go with live ammo in a carefully supervised setting. Note that this little boy is wearing eye protection, as well as hearing protection. The rifle case is being used as an improvised rest to help this new shooter get off to a good start.

stroyed — and shooters injured — when a 16-gauge shell was inadvertently loaded. In this case, the smaller shell slides forward into the barrel, but still allows the larger shell to chamber behind it. If the gun is fired in this condition, plan on buying a new gun after your hospital stay. Make sure you know the correct ammunition for your gun. *Don't* mix it at home or at the range.

Until about 1920, many shotguns were made with Damascus, or twist-steel barrels. These guns were intended to be fired with BLACKPOWDER ONLY. Even though modern smokeless shotshells will chamber, they must never be fired in a Damascus-barrel shotgun. Many experienced shooters — who should know better — suggest that low-base modern shotshells can be fired in the Damascus guns. *Don't believe it! Don't do it!* While most Damascus guns are now retired to the collector category, millions were made and many are still around. They may usually be recognized by the intricate pattern evident in the metal of the barrel(s), but not always.

Beyond the basics, the important thing to learn about rifles and shotguns is proper operation of the safety devices.

Until quite recently, nearly all lever-action rifles used a half-cock hammer safety similar to that found on single-action revolvers. The same false half-cock can happen, with potentially dangerous results. Best practice, even in the hunting field, is to carry these rifles with the chamber empty. Newer lever-action rifles use a positive cross-bolt safety. While this is a safer system, sound practice is still to carry with a dry chamber and to lever a cartridge into the chamber only when ready to shoot.

Most pump and autoloading shotguns and some

rifles use a cross-bolt safety, which consists of a small button near the trigger. The problem here is that it is very, very easy to forget which way is "safe" and which way is "fire." It is also very difficult to feel the safety position while wearing gloves.

Some shotguns and a few rifles, including most over/under and double-barrel shotguns, use a tang safety. The tang safety is a small switch located atop the grip, just to the rear of the action. This device is not only small and easy to forget, but also very easy to accidentally move to the "fire" position.

Many centerfire bolt-action rifles use a "wing" safety on the bolt sleeve. This is a very good arrangement as the safety is easy to see and requires deliberate effort to operate. Other bolt actions use a smaller sliding safety located near the bolt handle. These safeties certainly work well and facilitate easy scope mounting, but are not as safe for the beginner.

Mechanical safety devices, in general, require some additional explanation. Regardless of the

If you think a handgun is a toy, take a good look at the back of this mail-order catalog. Play it safe — *all the time.*

type involved, some safeties positively block the firing pin from contact with a cartridge. On many other guns, note that the mechanical safety may only block the trigger. That means that a hard bump may still cause the gun to fire, even with the safety engaged. While it is important to understand how mechanical safeties work, never forget that the only totally reliable safety is between your ears. Don't rely on a mechanical device — use your head to practice safe gun handling.

Any sort of horseplay is out of place where firearms are involved. Handling guns is a serious business in which you don't get a second chance to repeat *some* mistakes. Always assume that *all* guns are loaded — and treat them that way even if you are certain they are not loaded. Watch that muzzle. Be sure you know where your bullet will go if you miss your target. Be certain you *know* what your target is. More than a few careless types were certain that they were aiming at a deer, only to find out that their target was really a cow, a tractor — or another hunter. Be *sure* before you shoot, not sorry later.

Safety and courtesy demand that you clear any gun — check for live ammunition — before handing it to another person. Don't be offended if the other person repeats the operation. Doing so is an accepted part of safe shooting. The muzzle of any gun should *always* be pointing in a safe direction. If someone steps in front of the muzzle, immediately point the gun elsewhere. That action should be automatic and reflexive. If you must think about safe gun handling, you haven't yet mastered it.

In match competition or on a supervised range, there are more formal rules to be followed. The rangemaster will be able to answer your questions, but they will cover these basic points:

1. *Don't load your gun (or insert a loaded magazine) until you're on the firing line and actually ready to shoot. If you're firing with a group, don't load until the appropriate command is given by the rangemaster.*
2. *Keep the action open anytime you're not actually firing. Make sure the action remains open and that the chamber and magazine are empty before you leave the firing line.*
3. *Keep the muzzle pointed downrange at all times when you're on the firing line. When behind this line, keep the gun unloaded and the muzzle pointing in a safe direction (straight up is usually a good choice).*
4. *Keep your finger outside the trigger guard until you're actually ready to fire.*
5. *Don't let your attention wander when you're holding a loaded gun. Never turn around at the firing line with a gun in your hand.*

Safety is a full-time proposition whenever you have a gun in your hand. Used with knowledge, skill and reasonable precautions, a gun is as safe as any other tool or piece of sporting equipment. But in careless hands it can be deadly. Safety *must* be practiced religiously and continuously whenever you handle *any* firearm. Safety is one habit you never want to break.

chapter 3

RIFLE, HANDGUN, OR SHOTGUN?

THERE ARE THREE basic types of firearms available to shooters — rifle, handgun and shotgun. While all three can be used for target shooting, hunting or even self-defense, each type has its own strengths and weaknesses, and each demands a separate set of shooting skills.

Before you start learning to shoot, you need to decide which kind of firearm you have the most interest in. Not that you can't learn to use all three — you certainly can. Many shooters become excellent shots with rifle, shotgun and handgun and have no trouble at all switching from one type of gun to another. But it pays to choose one particular gun in the beginning and stick with that firearm until you've mastered it before moving on to another type of shooting.

Although rifles, handguns and shotguns all work on the same general principle — the shooter visually lines up the barrel of the gun with the target, and then pulls the trigger to start the projectile on its way — each type of firearm requires a distinctly different shooting technique. Because of this, switching from shotgunning to rifle shooting (or vice versa) during the learning process can cause serious confusion — at least at first. So my advice is to choose one particular type of gun and stick with it until you've mastered the basics of safety, operation and markmanship. Of these, safety is by far the most important — and the safety skills you'll learn with a rifle will also apply to shotgun and handgun shooting.

In deciding which firearm to begin with, there are several factors to consider. First, are you most interested in learning to hit moving or stationary targets? If plinking at tin cans or putting holes in paper targets is your immediate goal or if big game hunting is your long-range goal, then either a rifle or handgun would fill the bill. But if you want to hunt pheasants or ducks, or join the crowd at the local trapshooting club, a shotgun is the logical choice. Where you live will also influence your choice — it's difficult to legally own or even acquire a handgun in some large cities, and if your home isn't within easy driving distance of a big-bore (large caliber) rifle range or some sparsely populated countryside, you won't get much use out of a magnum hunting rifle.

With those thoughts in mind, let's take a closer look at the three types of sporting firearms to see which best fits your own needs.

You've got to crawl before you walk! As a result, we strongly suggest you master the basics with a 22 rimfire, *before* you graduate to a centerfire.

Rifles

Rifles are designed for precision, long-range marksmanship, while handguns, though more convenient to carry, are used over much shorter ranges. Shotguns are short-range guns, intended primarily for hitting fast-moving targets out to an extreme range of about 50 yards.

The rifle is perhaps the easiest kind of firearm to learn to shoot. The target is usually stationary, and you can lean the rifle's forend against a tree or fencepost, or even use a commercial benchrest of sandbags to help steady your aim. Aligning the sights is a simple mechanical operation, and magnifying optical sights are available to make things even easier.

Rimfire rifles — 22s — are great fun to shoot. While it is still possible to buy a new 22 for about $100, it is also possible to spend more than ten times that amount. In recent years, a number of very high quality — and very pricey 22s have hit the market. The one great thing about the 22 is that ammo *is* inexpensive, setting the shooter back a tiny fraction of the price of centerfire ammunition.

While larger centerfires are necessary for hunting deer and all other large game, the 22 is the rifle to start with. The 22 lacks the muzzleblast and recoil of the larger guns. It is just plain fun to shoot — few shooters ever tire of knocking over cans with a 22.

Recreational shooters who want nothing more from the sport than the fun afforded by informal "plinking" and target shooting can stop right here. The 22 rimfire is far and away the most popular

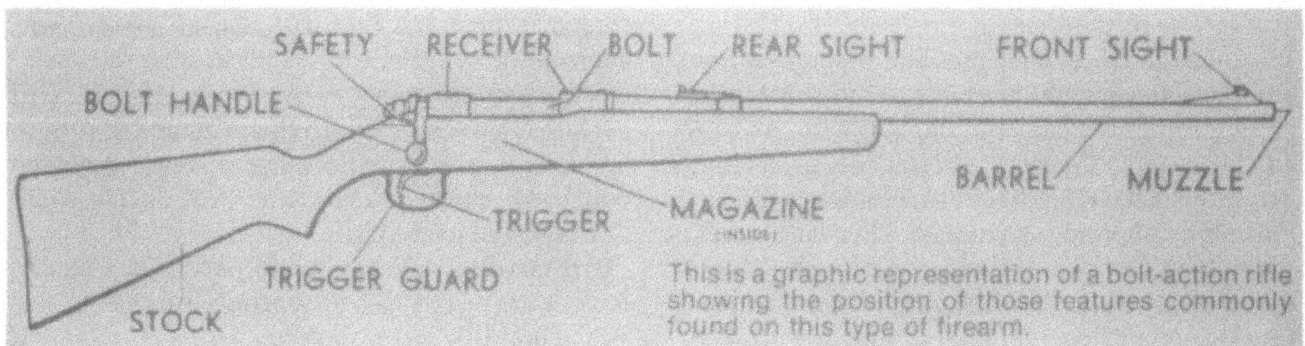

This is a graphic representation of a bolt-action rifle showing the position of those features commonly found on this type of firearm.

54

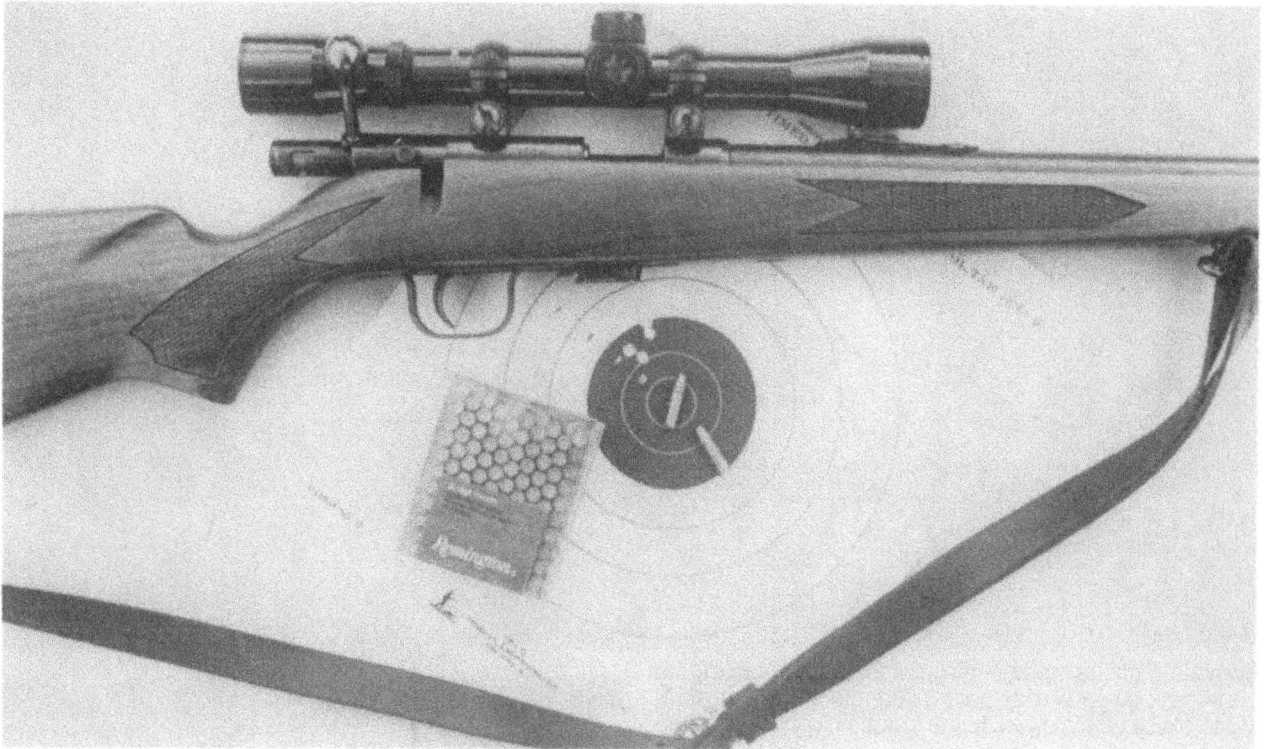

Rifles are designed for precision marksmanship, and a 22 rifle like this one is the easiest type of firearm for a beginner to learn to use. Magnifying scope sight is used for precise aiming.

caliber on the market and is capable of providing a lifetime of shooting pleasure. It's also a fine small game cartridge out to 75 yards or so; however, you must remember that the 22 is still dangerous at a distance of one mile or more.

For those who don't want to go to the effort and expense of moving up to a more powerful arm, a 22 rifle can be made to serve double duty as a home defense weapon. It may not be the ideal choice for self-defense against intruders and housebreakers, but it will do the job. Although it lacks the bark and bite of the big-bore calibers, the diminutive 22 can be very deadly and *should be treated with respect*.

If hunting deer or larger game is your bag, or if you'd like to take on prairie dogs and similar var-

Once you've mastered the basics and moved on to centerfire rifles, your interests may take you in the direction of big game hunting. The exhilaration of successfully taking that first trophy elk or deer can't be matched.

Above — This bench rest target shooter is checking out his scope, target and shooting gear just before attempting to place all his shots in the smallest group possible.

Right — For those who believe that shooting is strictly a sport for the pickup truck and good ol' boy folks, note that shooting has been an Olympic sport since the inception of the modern Olympic games. The gold these shooters haul home requires at least as much skill as figure skating. The 22 rimfire rifles used in Olympic competition are a very long way from your plinker — some of them can cost nearly as much as a pickup truck.

mints at extended ranges, you'll eventually need a rifle that shoots centerfire ammunition. If this is your aim, you still need to learn the basics with a 22 — and once you've got the basics down pat, it's relatively easy to transfer those skills to big-bore rifle shooting.

Even if you don't plan on becoming a hunter, there are a number of shooting sports that require centerfire rifles. Benchrest target shooters use a variety of hot centerfire calibers to do their thing with, and many of the more prestigious national and international rifle matches are held with centerfires. Silhouette shooting, a popular sport that came here from south of the border, with contestants trying to knock over steel life-size silhouettes of birds and animals, is a shooting event for both rimfires and centerfires.

As I've said, the logical starting point for a beginning rifleman is with a 22 rimfire. But don't rule out the possibility of moving up to a centerfire rifle later on — rifle shooting grows on you, and in time most rimfire shooters get the urge to graduate to something larger and more powerful.

56

Handguns

If a 22 rifle is the easiest kind of firearm to learn to shoot accurately, a big-bore handgun ranks as the most difficult. A rimfire handgun is a bit easier to master, but any revolver or auto pistol requires a lot of practice before you're able to shoot it well. With a handgun, there's no shoulder stock to offer firm support, and it takes practiced muscle control to hold the gun steady enough to hit the bullseye consistently.

Still, a beginner can learn to hit empty beverage cans at reasonable range the very first time out, provided he takes advantage of sandbags or some other kind of rest. And once learned, handgun skills can provide a high degree of satisfaction and many hours of enjoyment.

From a practical standpoint, the handgun offers certain advantages over "long" firearms. A pistol or revolver is much more compact than a rifle or shotgun and can be conveniently carried in a holster to leave both hands free when the gun is not in use. This makes the handgun a favorite of hikers and campers who enjoy taking time out for an informal plinking session, and a small handgun can also be used to put meat in the pot.

Many varmint hunters prefer handguns to rifles, partly because they're less bulky — but mainly because using a handgun makes their sport more challenging and therefore more enjoyable. Similarly, some hunters have used handguns to successfully bag even large, dangerous game like the big, northern bears. These individuals have all been exceptional marksmen, however, and wisely have been backed up by a guide carrying a large-caliber rifle.

In many states, it's legal to hunt deer and other animals with a handgun, and a number of trophy bucks are taken every year by this method. At the same time, no real sportsman will attempt to collect his venison with a handgun until he's become really proficient in its use. Big game handgunning is for experts only.

Handguns are widely sold for self-defense. Note that personal protection with a handgun is in the same category as big game hunting — an activity NOT for beginners. Besides skill questions, there are more legal implications connected to carrying and shooting handguns than with any other gun type. Before you buy a handgun with self-defense — or any purpose in mind — you will need to check on the laws that apply in your area. In a few areas, handgun possession is totally illegal. Other areas require a permit to purchase. Virtually *all* states require a police permit to carry a concealed handgun. Most police agencies are reluctant to issue the permit unless the applicant can prove a real need — not just a desire — to carry a gun. Some

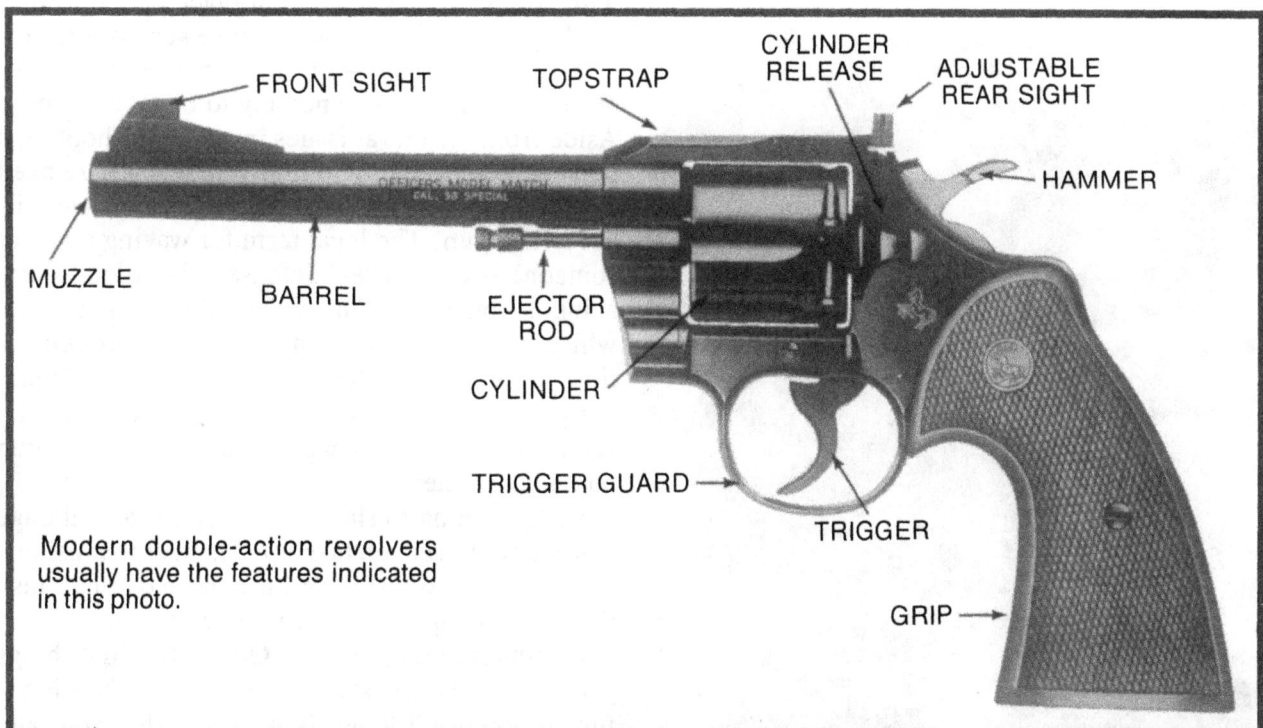

FRONT SIGHT   TOPSTRAP   CYLINDER RELEASE   ADJUSTABLE REAR SIGHT

HAMMER

MUZZLE   BARREL   EJECTOR ROD

CYLINDER

TRIGGER GUARD

TRIGGER

Modern double-action revolvers usually have the features indicated in this photo.

GRIP

Above — A 22 rimfire handgun is ideal for plinking and shooting at targets. Ammunition costs little, and recoil and noise are held to a minimum. Below — More compact than a rifle or shotgun, a handgun can be carried on the belt in a holster to leave your hands free when not in use.

permits also limit the times and places where the gun may be carried. Penalties for violation can be severe. Unless you like sleeping away from home with strange new friends in a place called "jail," don't guess about gun laws. Make sure you *know*.

Using a handgun — or any gun — as a self-defense weapon is not something to be taken lightly. Aside from the moral issues involved in shooting a person, you are likely to find yourself in dire need of a lawyer if you shoot or even threaten an assailant with a gun. The legal term for waving a gun at someone — even in self-defense — is called "menacing." You could find yourself in situations in which having the gun is more trouble than protection. In any event, few people manage to obtain permits for concealed carry in most urban areas. As a home-defense weapon, another kind of gun could be a better choice.

Still, learning to shoot a handgun is a challenge that many sportsmen take up. The fact that using a revolver or auto pistol well is a more difficult task than shooting a rimfire rifle doesn't make handgunning less pleasant. One of the first things you learn when you start to shoot is that shooting is fun. If you don't let early misses bother you, and

Silhouette shooting is a reasonably new and very popular shooting sport imported from below our southern border. Best described as highly organized plinking, the goal of the silhouette shooter is to hit heavy steel cut-outs of animals and birds. Conventional handguns may be used, like the Smith & Wesson revolver being used by the handgunner on the shorter, less formal range. The shooter backed up by the spotting scope is using a Thompson/Center Contender single-shot. The Contender is very popular as its interchangeable barrels are available in rifle calibers; these heavy numbers are quite effective in tumbling the iron animals. The thump, ''bang'' and dust generated in this sport appeal to many who find punching holes in paper less thrilling than hammering steel plates.

Accuracy with any handgun comes as a result of continued practice. Once you've proved your skill "on paper" try moving on to more animate "fun" targets such as tin cans.

you practice holding the gun firmly while taking careful aim — and then gently squeezing the trigger — I can promise you'll do well with a handgun. It may take a bit more time than learning to shoot with a rifle, but that won't make it any less fun.

As with the rifle, the beginning pistoleer should confine his first efforts to shooting rimfire handguns. A gunner shouldn't even consider firing a centerfire revolver or pistol until he's become good enough with a 22 handgun to consistently hit what he's shooting at at 25 yards or so — and *without* the help of an artificial rest.

Again, you may have no desire to graduate to big-bore gunning. The little 22 is ideal for plinking or shooting at targets, and rimfire ammunition is considerably less costly than centerfire pistol cartridges. Recoil and noise are held to a minimum making rimfire handguns much more pleasant to shoot than their larger brothers. The majority of handguns sold in this country are chambered for the 22 Long Rifle rimfire, and this popularity is well deserved.

At the other end of the scale, the really big centerfire handguns — the Magnum 41s and 44s — are noisy, hard-kicking and expensive to shoot. But even these potent loads can be mastered. My youngest son started shooting a 44 Magnum when he was only 8 years old, and he never thought of it as anything but all kinds of fun! He didn't hit much

with it at first, but he certainly wasn't afraid to shoot it. If an 8-year-old can learn to handle the recoil of a big 44 without complaining, *anyone* should be able to.

Between the 22 rimfires and the big magnums lie a number of medium-powered handgun cartridges that anyone can handle after learning the basics of control, safety and marksmanship with a rimfire pistol or revolver. Some of the more popular cartridges in this range include the 32, the 380 ACP, the 9mm Parabellum and the ever-popular 38 Special. These rounds are noisier than the 22s and recoil a bit harder, but lack the intimidating bellow and wrist-popping kick of the large magnums.

Most people turn to handgun shooting after first becoming proficient with rifles, and this procedure has a lot to recommend it. Many of the mechanical skills needed to use both kinds of firearms — sight alignment, trigger squeeze, etc. — are more easily learned with a rifle, and most riflemen have little difficulty transferring those skills to handgun use.

From a safety standpoint, the rifle is also a better beginner's choice. While the safety rules are basically the same with both firearms, it's easier for a beginner to accidentally shoot himself in the leg with a short-barreled handgun if those rules are forgotten or ignored. If all the rules are carefully observed, rifle and handgun are equally safe.

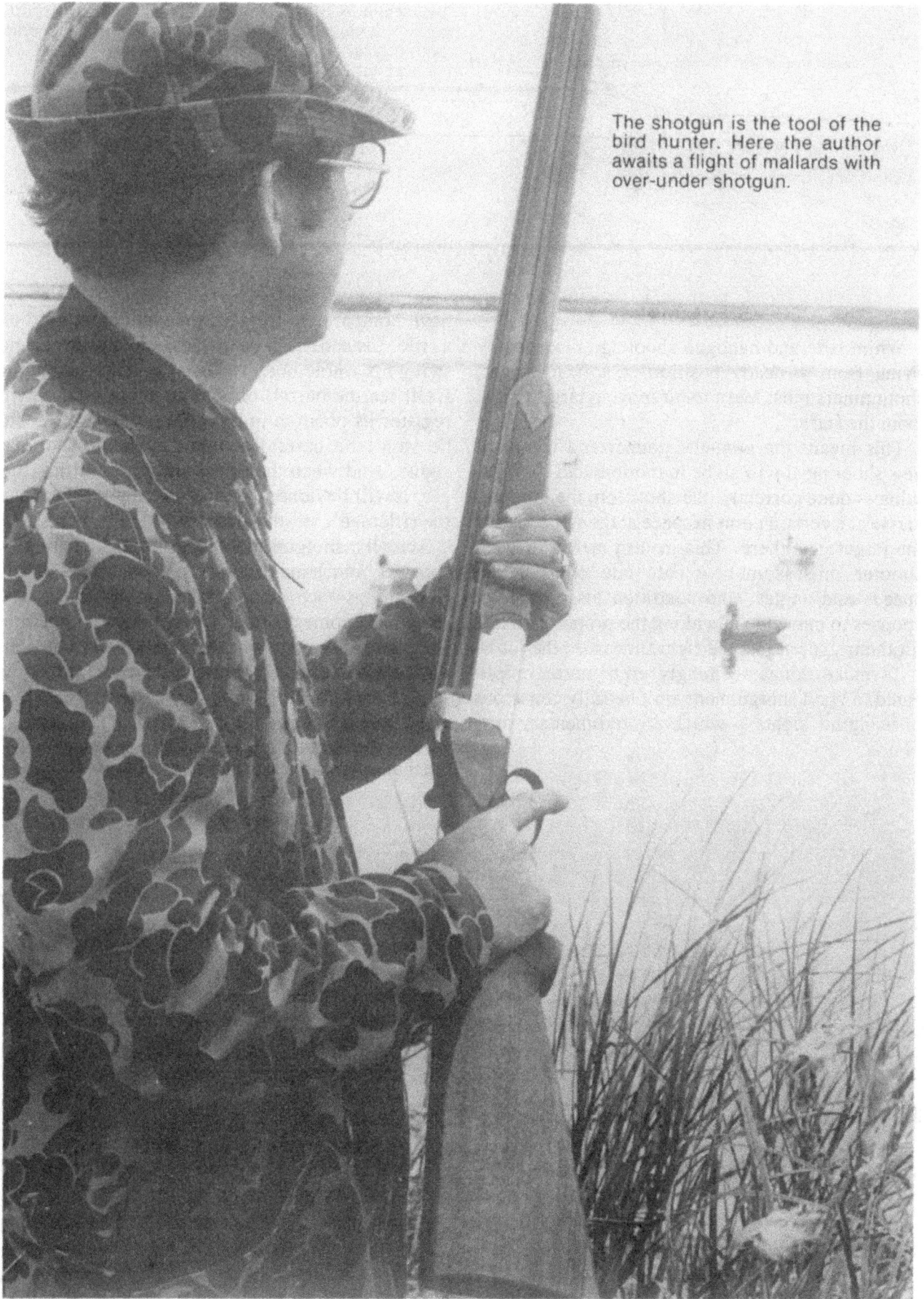

The shotgun is the tool of the bird hunter. Here the author awaits a flight of mallards with over-under shotgun.

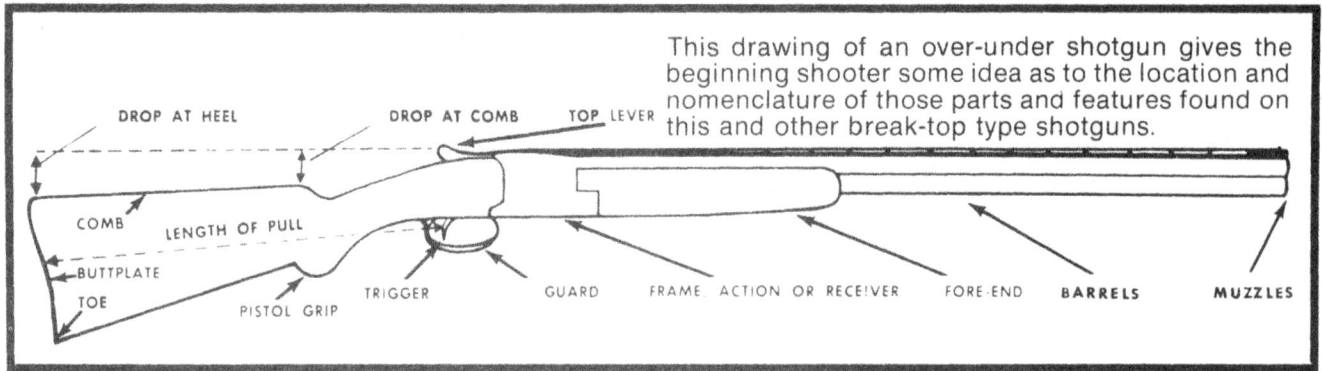

DROP AT HEEL   DROP AT COMB   TOP LEVER

COMB   LENGTH OF PULL

BUTTPLATE
TOE

PISTOL GRIP   TRIGGER   GUARD   FRAME, ACTION OR RECEIVER   FORE-END   **BARRELS**   **MUZZLES**

Shotguns

While rifle and handgun shooting is learned by firing from a steady position at a fixed target, shotgunners must learn to hit moving targets right from the start.

This means the gun, the gunner and the mark he's shooting at will all be in motion, and if everything is done correctly, the shot from the gun will arrive at a certain point in space at the same instant the target gets there. This in turn means that the shooter must somehow calculate comparative speeds and angles, and condition his or her responses to move the gun along the proper path and hit the trigger at just the right time to do the job.

To make things seemingly even more complicated, a good shotgunner won't be fully conscious of using his sights — which are rudimentary anyhow, compared to the sighting equipment used on a rifle. Instead, he'll keep his eye on the target and swing his whole body to keep the gun tracking it. He'll see the barrel, of course, and his mind will register its position in relation to the target — but he won't be carefully lining up front and rear sights. And when the time comes to pull the trigger, it will be done with a quick snap, rather than the rifleman's steady squeeze.

Actually, shotgun shooting isn't as difficult as it sounds. Your hands and body instinctively do the right things once you've picked up the knack, and that knack comes to most new gunners with a little practice. As a matter of fact some fledgling scattergunners seem to be blessed with a high degree

Upland game, such as the rising pheasant pictured here, offers the wingshooter a challenging, airborn target that's the start of a gourmet meal!

of natural coordination, and these lucky individuals are able to hit clay targets from a hand trap with satisfying regularity right from the start.

If shooting a rifle can be called a science, shotgunning could be called an art. Widely differing skills are needed, and good riflemen don't automatically become good shotgunners—or vice versa.

The shotgun is the tool of the upland hunter or waterfowler. Taking a bird on the wing or rolling a running rabbit is the thing a hunting scattergun does best, and the majority of shotgunners are hunters. However, many shooters "hunt" only clay targets thrown from a mechanical trap, and Skeet or trap clubs (two different forms of the sport) can be found near practically any fair-sized city. By joining one of these clubs, you can compete on local, state or national levels. And because shooters are "handicapped" according to their experience and skill, you can start competing seriously almost right from the first.

While shotguns are usually used to fire loose loads of small lead or steel pellets, they are sometimes used with larger projectiles. In certain areas, deer can be hunted only with a shotgun loaded with either buckshot (pellets of relatively large diameter, as compared to birdshot) or single, large slugs. When used with slugs, shotguns are aimed like rifles. Some shotguns designed for hunting deer are equipped with rifle-type sights or scopes. Many hunters buy shotguns with interchangeable barrels so that by buying only an extra slug barrel, they have the utility of two guns at lower price.

Home Defense

While it is unpleasant to dwell on the non-sporting uses of firearms in an introductory book on shooting, home defense is becoming of increasing concern to law-abiding people most everywhere. Some authorities believe a shotgun is the best choice for home defense, but there are some misconceptions involved. The first involves the contention that it is not necessary to aim precisely; the inference being that the shot pattern is so large that a near miss is as good as a hit. Wrong! Note that most home defense shooting is done at very close range, before the shot column has time to open up. When fired at typical across-the-room range, the shot column is little larger than the bore of the gun. Yes, you can miss with a shotgun. Some writers also discuss "psychological" deterence involved with pointing a shotgun at an intruder. This writer is not a psychologist, and few burglars are

practicing psychologists. The burglar might not understand that he is supposed to be intimidated. The point to be made is that pointing a gun at another human being should NEVER be done unless you are fully prepared to pull the trigger — and face potentially enormous consequences.

Which to Choose

Which kind of firearm most appeals to you? Rifle, handgun or shotgun — all have their uses, and all three can provide simple recreation, defense, or meat in the pot.

If you want to hunt deer or other large game, chances are the rifle will be your best choice. Learn the basics with an inexpensive rimfire (which can give a lifetime of plinking pleasure even after you've "graduated"), then move up to a more potent centerfire to do your hunting with.

If target shooting or informal plinking has special appeal, either a rimfire rifle or handgun will do a good job. Ammunition is cheap, and neither noise nor recoil reach objectionable levels. Twenty-twos are great for hunting small game.

While there are a number of shooting competitions you can enter with the right 22 rifle or handgun, once you get hooked on rimfire target shooting the odds are good that you'll eventually move up to centerfire matches, too. Handgun competitors, in particular, tend to accumulate a variety of both rimfire and centerfire target guns as they get serious about their sport.

Handgunners aren't limited to paper punching, either. Large caliber revolvers are good for hunting in many states, although about an equal number of state game commissions frown on their use in the deer woods. Before investing in a big-bore pistol or revolver for hunting purposes, be sure to first check with state or local authorities to make sure you'll be able to use it without risking possibe penalties. Again, the basics should first be mastered with a 22.

Shotguns are regulated far less stringently than handguns, and are legal in most places. Still, be very careful not to run afoul of local, county or state laws. Laws governing the manner in which the gun may be transported are particularly easy to violate — even with no criminal intent whatsover. For example, in one western state, it is perfectly legal to carry a loaded gun in your car, provided that the gun is in "plain sight." In that same state, an unloaded, disassembled gun cased and locked in your trunk could be a ticket to jail. While that

sort of thing makes no sense, it is better to be informed rather than telling the judge why you violated the law.

Rifle, handgun or shotgun? You can choose to specialize in using just one type of firearm. A few shooters do. Better to try all three — it's three times the fun.

Buying a Gun — New or Used

For the beginner, the incentive to buy a used gun is price. It is possible to buy guns with very little use and only superficial wear for substantially less than the price of a comparable new gun. It is also possible to buy a hunk of dung for a very hefty tariff. Does that mean that the beginning shooter should only buy used guns or avoid them totally?

Above left — The shotgun is normally used to hit a moving target, like the clay pigeon this gunner just shattered (circle).

Left — Years ago, Skeet shooting was originally designed to provide shotgunners with some offseason wingshooting that would keep their eyes sharp. Today you'll find Skeet ranges throughout the country.

Below—Trap shooting is another organized shotgunning sport that can prove to be both exciting and challenging. Some trap shooting (Skeet too) is done at night under the lights. The white "specks" in the right-center area of the photo are actually pieces of a broken clay bird.

No, but there are a few categories of used guns that the beginner should avoid.

It is difficult to differentiate between a "collector gun" and a "used gun" without starting an argument with someone. Someone, someplace, is sure to collect virtually every type of gun ever made — new, obsolete or antique. For purposes of this book, assume that a "used" gun means a commercial gun identical to a gun still in production or any gun made by a manufacturer still in business and built within the last 20 or so years. Collector guns of advanced age, or made by defunct manufacturers, are not a good beginner's bet. Gun collecting is a wondrous hobby, but collector's guns are not always the best bet for the beginner whose goal is to learn to shoot — not necessarily to collect. Parts may be either impossible or very expensive to obtain. Time required for repairs will certainly be much greater than for modern guns made by current manufacturers.

There are also pitfalls involved in buying modern used guns. A few gun types are similar to the travelling salesman's 1-year-old used car with 150,000 miles on the odometer. Used trap and Skeet shotguns come to mind. These guns are generally superbly cared for, but fired a great deal. Wear is not necessarily obvious. If the gun happens to be current — like a Remington 870 pump shotgun — and the price is right, even a used 870 trap gun might be a good buy. If the gun is an expensive and discontinued number — like a Winchester Model 12 — the gun should be considered a collector's item.

Police departments once kept the same guns at least as long as it took cops to be hired and retired — 20 or 30 years. The progress of gun development is now such that many departments have replaced their guns several times in the past 10 years. The most recent trend has been to replace 38 and 357 revolvers with 9mm autoloaders. To the beginning shooter, the PD cast-offs represent good buys with some holster wear but little mechanical wear. A PD trade-in could be a "best buy" for the beginner.

In 1968, the Federal Government passed laws which banned the importation of surplus military guns, even though the guns were otherwise totally civilian legal. Those of us who are among the "baby boom" generation, and bought surplus guns prior to 1968, cut our shooting teeth on the $15 British Enfields and $30 German Mausers and American Springfields which were so readily available prior to '68. Recently, Federal gun laws were changed to again permit old military guns to be imported. We are now seeing a new crop of interesting, inexpensive G.I. guns. The only difference between the old and the new crops is the fact that many of the current surplus guns are semi-autos. While most are very good buys, note that semi-autos are not ideal for the beginnner. On the plus side, ammo is also coming in by the boatload. While not suited for hunting, military ammo is great for plinking and learning to shoot. It is possible to learn proficiency with a high-power rifle using military ammo — for pennies on the dollar — as compared to acquiring the same skill with new commercial ammo.

As expressed in the introduction to this book, we get back to where to buy guns. Especially in the area of used guns, buy from a reputable gun shop. The people in the gun shop will be able to advise you. They will expect you to be back to buy the megabuck guns after you graduate from the cheaper, used plinkers, so they are less likely to stick you with a clunker. The kid who sells nail polish at the discount emporium and fills in at the gun department will be far less concerned with your progress as a shooter than the owner of a gun shop.

While the used gun can save you some money, a new gun is hardly likely to shoot you in the wallet. New guns are not only new, they come with warranties. I must again get back to the concept of buying from a gun shop. Please note that I buy garden hose and T-shirts from discount stores — my mother was not frightened by an "Attention K-Mart Shoppers" announcement. Guns bought from discount stores come "out of the box" and into your hands. Minor problems encountered with a gun bought from a discounter will result in the advice, "Send it to the factory, we aren't liable for..." Minor glitches with new guns may be resolved by the gun shop. It may be, in certain instances, that the gun shop is an "authorized repair station" for a gun maker.

As the new shooter gains skill, the first thing to do is to "trade up" to "better" guns. The discount store seldom deals in trades and used guns — the gun store does. Please note that I do not own or hold an interest in a gun store. But after being involved in shooting sports for 30-plus years, I know where I feel most comfortable and most assured of buying the most appealing guns and getting the best deal. It isn't at the discount store.

chapter 4

CHOOSING YOUR FIRST RIFLE

RIFLES COME IN many sizes, shapes, calibers and prices. When the domestic guns and the raft of imports used in plinking, target work, varminting and hunting large and small game are considered, the selection is staggering. Even choosing a particular caliber can be a big job, as there are nearly 100 separate rifle calibers in use. The total is arguable, as the term "obsolete" is debatable in the case of some of the really old timers. For example, the old 50-70 Government was replaced by the 45-70 in 1873. It *must* be obsolete. But, to a shooter who just bought a brand-new replica Sharps rifle, the 50-70 is high-tech.

The task is simple for the beginning shooter. Whether the beginner is 6 or 65, the only logical caliber choice is the 22 Long Rifle rimfire. Rimfire 22 ammunition is inexpensive and available most everywhere sporting goods are sold. (Try buying a box of 460 Weatherby Magnum ammo at the local country bait store . . .) Rimfire ammo fired from a 22 produces nearly no recoil, and little objectionable noise. This means that the beginner can concentrate on proper holding and sighting techniques and proper trigger squeeze without living in fear of teeth-rattling recoil and blast heard in the next county. Learning with a big bore is likely

to produce some very bad habits born of fear. Even most military organizations have typically issued 22 rimfire rifles for basic training. Our own 5.56mm NATO (military version of 223 Remington) does not kick badly, but 22 rimfire conversion units for the M16 G.I. rifle have long been issued for training.

The fact that 22 rifles are soft-spoken and not likely to frighten the children and horses does *not* mean that ear protection isn't necessary. This topic will be mentioned several times, simply because it is important. Hearing loss can come from prolonged exposure to moderate noise levels. Particularly if the new shooter is using one of the indoor ranges typically available in metropolitan areas, hearing protection is still more important.

I've already mentioned that 22 rimfire ammunition is accurate, but I'd like to go a step further and point out that some specially manufactured 22 match (target-type) cartridges just may be the most accurate rifle ammo you can buy. Even standard off-the-shelf 22 ammo is capable of more accuracy than the average rifleman can muster.

For plinking (informal shooting) and target shooting out to a maximum of about 75 yards, the little 22 is hard to beat. What's more it makes a

Rimfire rifles come in four basic action types. From left: bolt action, lever action, pump and autoloader.

fine cartridge for hunting small game like rabbits and squirrels. You'll never outgrow a 22.

Because of these and other reasons, the ubiquitous 22 is by far the most popular rifle caliber in use, and more 22-caliber ammunition is sold from year to year than any other kind. Because it is so popular, there is a huge assortment of rifles chambered for the 22 Long Rifle cartridge. So even though you've narrowed down your choice of calibers for that first rifle, there are still some decisions to be made.

Rifles chambered for the 22 Long Rifle round can be found in almost every action type imaginable. You can buy lever-action carbines that look very much like the "cowboy" rifles you see on television, slide-action pumps which are cocked for firing with a trombone movement, or a semi-automatic selfloader that requires only a pull on the trigger for each shot to keep the rifle loading and firing. You can get a whole raft of 22s, ranging from Olympic-grade match rifles and top-quality sporters on down to low-priced single shots. Bolt-action repeaters are available to fit almost any budget. There have even been 22 rimfire machine guns built for police use. Whatever sort of rifle action you can name, you can bet that a 22 either has been or is being built somewhere.

So what kind of 22 rifle should you, as a begin-

Bolt Action

Lever Action

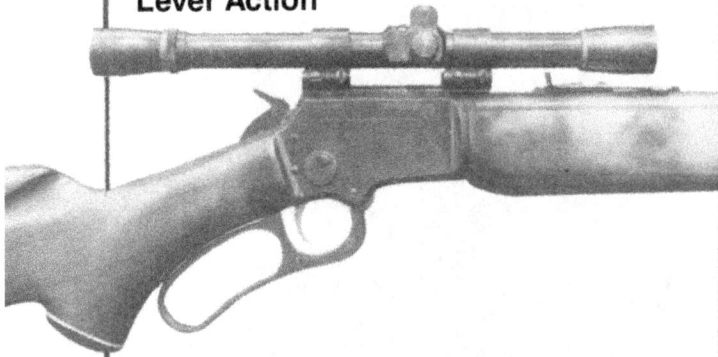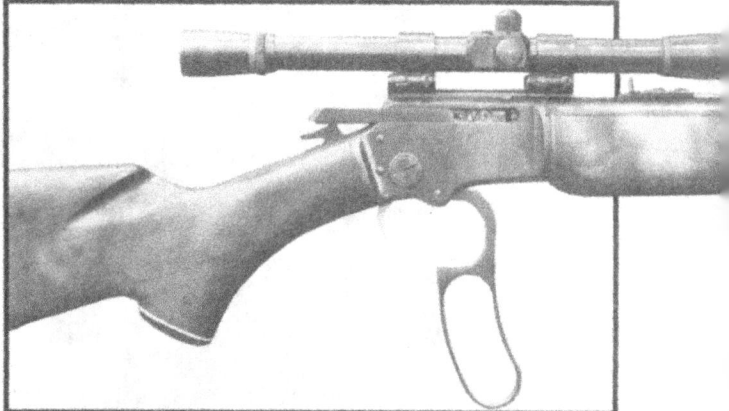

Pump Action

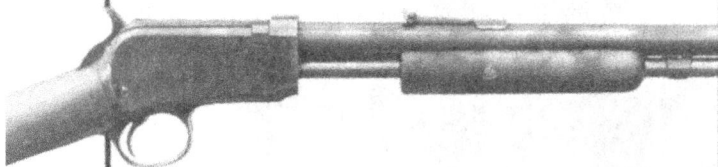

Autoloader

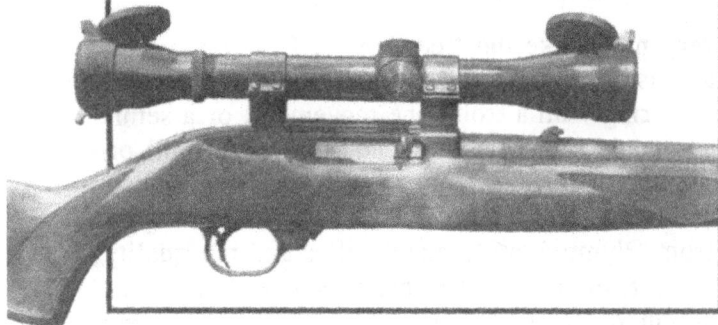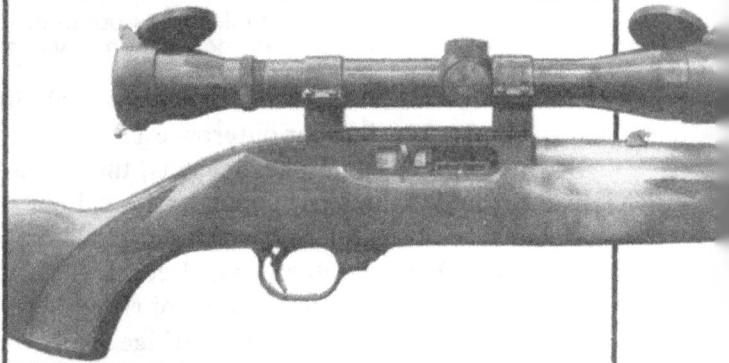

Here's a closer look at the four basic action types used in 22 rifles. They are shown with their actions closed, on the left, and their actions opened, on the right. Top to bottom: bolt action, lever action, pump and autoloader.

ner buy? The selection will be guided partly by the amount of money you're willing to pay. The rule-of-thumb to follow here is to get the best you can afford — you can learn to shoot with the cheapest rifle available, but as your skills improve you're certain to want to move up to a better firearm. When this happens — and it will if you've skimped too much on your first purchase — you'll find that you would have been far better off in buying a higher quality arm to begin with. Many sporting goods dealers are reluctant to take a low-priced 22 rifle in on trade, and even if they agree to do so, you'll only get back a fraction of what you originally paid for the gun. So it pays to buy a rimfire rifle that you can be happy with for a long time — a good 22 will last several lifetimes if properly cared for, and I can promise it will get lots of use over the years.

That doesn't mean you have to spend a fortune on your first rifle. You can get a good, usable new bolt-action repeater for prices that begin under about $150. Used ones, naturally, are less — depending on your ability to make a deal. Some of the less expensive autoloaders are in the same price league. If you have the bank account to support you tastes, it is possible to spend a thousand bucks or more, but this sort of expense is hardly necessary for the new shooter. Many of the solid, high-quality guns which can be expected to hold their value long after price has been forgotten are in the $300-$400 range.

Don't let a limited budget keep you away from shooting. There are many used bolt-action and single-shot 22s available for $100, or maybe a bit less. The key is to buy the best rifle you can afford — and if the budget won't stretch to pay for that fancy model that really caught your eye when you started shooting, settle for something less and trade up later when your wallet is as big as your desires.

What Rifle is Best for You?
AUTOLOADING 22 RIMFIRE RIFLES

Let's set budgetary considerations aside for a minute and see what kind of action would be best for you. Autoloading 22s are highly popular among experienced plinkers, but they're not the best choice for a beginner to start off with. In the first place, it's easy to forget that the rifle reloads itself every time you pull the trigger. That means there is always a bullet in the chamber, ready to fire. So until you've learned the safety rules — and I mean *really learned* them — so that you automatically keep the gun barrel pointed in a safe direction *at all times*, it is best not to use a self-loading rifle. Remember, an accidental discharge is not only embarrassing, it could be fatal.

Another problem with autoloading rifles is that

(Above) Autoloading 22's like this Ruger 10/22 are fun to shoot and highly popular among rimfire fans, but they're not the best choice for beginning riflemen.

The Marlin Model 39 is one of the most popular lever-action 22 repeaters on the market. This is a well made rimfire that can be expected to last for many years if given proper care.

69

many of them feature a tubular magazine. It's easy to overlook a stray round in one of these hard-to-inspect magazines when the rifle is put away, and the next person who picks up the rifle may be in for an unpleasant surprise!

I have still another objection to autoloading rifles for beginning shooters. In order to become any kind of a marksman with a rifle, you need to develop a high degree of shooting discipline. That means the shooter taking a great deal of care to see that each and every shot goes where it was intended to and concentrating on only one shot at a time. With an autoloader, it's too easy to plunk another quick round if the first shot misses.

LEVER-ACTION AND SLIDE-ACTION RIFLES

Lever- and slide-action rifles make better choices for the beginner because they are operated by hand, and a fresh round isn't chambered until you're ready for it. Too, lever and pump repeaters have a lot of shooter appeal — they can be operated almost as fast as an autoloader, and lever 22s in particular have a "big gun" look.

However, both slide- and lever-action rimfires have at least one point in common with certain autoloaders — they all feature tubular magazines, and this kind of magazine is harder to check to make sure it's unloaded. Most modern 22s fitted with tubular magazines now have a bright orange-colored piece of plastic that pushes the cartridges through the tube and into the action, and if this is visible when the action is open, you know that there are no stray rounds hung up in the tube. However, powder residue tends to discolor the bright plastic after prolonged firing, and it eventually takes on the same metallic hue as a 22 cartridge, making quick identification difficult.

SINGLE SHOT OR CLIP-FED BOLT-ACTION 22 RIFLES

For these reasons, my own first choice for the beginning rifleman is any kind of single shot or clip-fed bolt-action repeater. There are a number of tubular magazine bolt guns on the market, but I've already explained why I dislike this kind of feeding arrangement for first-time shooters.

The very safest rifles for a beginner to start off with are the single shots. These models must be loaded one round at a time, as they have no magazine. There are several different kinds of single-shot 22s on the market, and while you can learn to shoot with any of them, some are better choices than others.

The most common type of single-shot rimfire available is the bolt action, but you are not totally limited to the bolt gun. In recent years — sadly no longer in production — Ithaca and Savage offered falling-block single shots with the action activated by a lever underneath the receiver. The Ithaca and Savage 22s were true beginner's guns in that while they were single shots, they were designed to look much like lever-action repeaters. These look-alikes featured exposed hammers which indicate whether the gun is cocked or not. The hammers must be manually cocked before shooting — the levers only open and close the action. The shooter must make two separate and conscious steps (loading and cocking) before the gun will fire, as opposed to an autoloader which fires with only a tug on the trigger. While these guns have been discontinued, they are mentioned because they have been in and out of production several times over the past 20 or so years.

Another single-shot — and one of this author's personal favorites — is the Stevens Model 72 "Crackshot" from Savage. Also presently discontinued, this rifle is being mentioned because a few are still available new in dealer stocks. The Crackshot is a factory reproduction of a rifle Stevens built 80 years ago. At that time, it was one of the most popular "first rifles" for the very same reasons it could be today. One disadvantage was that it

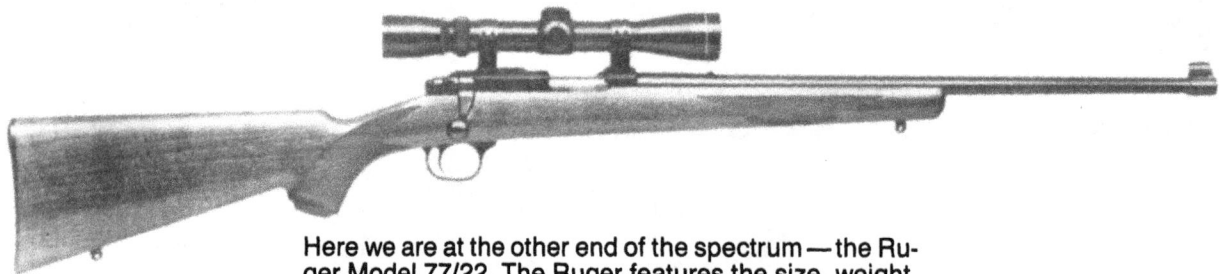

Here we are at the other end of the spectrum — the Ruger Model 77/22. The Ruger features the size, weight, fine walnut and quality associated with the invariably more expensive centerfires. A rifle of this sort is often an excellent understudy to the centerfire, in that handling qualities and safety features are similar.

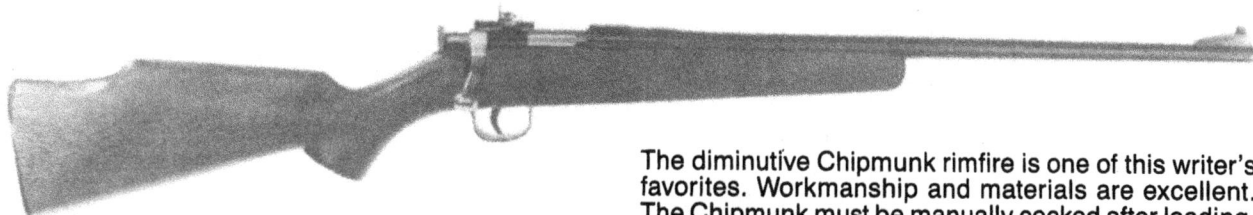

The diminutive Chipmunk rimfire is one of this writer's favorites. Workmanship and materials are excellent. The Chipmunk must be manually cocked after loading, which is the safest of all possible shooting worlds. It is hard to go wrong buying this rimfire.

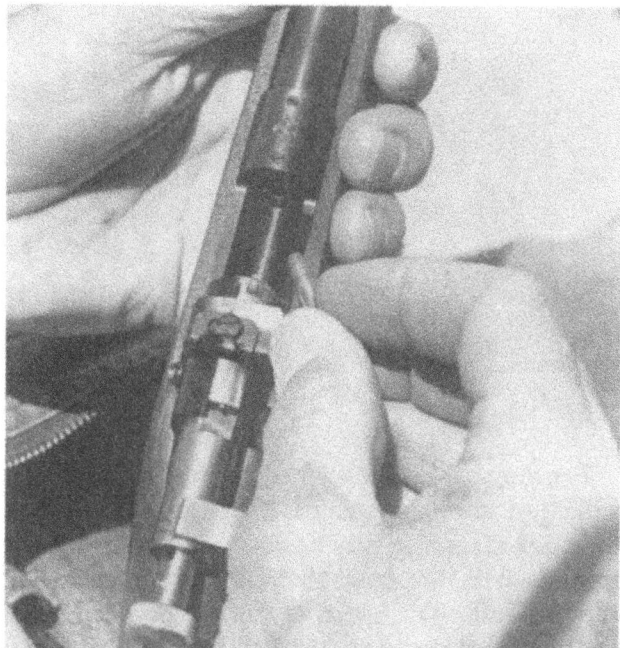

The Chipmunk is tiny—ideally sized for young children. Note the size of the parts compared to this adult's hands. Loading and unloading is simple. The knurled knob must be pulled back after loading and closing the bolt. The Chipmunk's standard receiver sight is simple and readily adjusted.

was fairly expensive in its most recent incarnation —but it was also very well made.

The bolt-action single shot is by far the most popular type among rimfire one-shooters, and this is the one the majority of new riflemen start out with. As a matter of fact, some of the most costly target rifles on the market are single shots, because cutting into the action assembly to make room for magazine feeding reduces the "stiffness" of the action and can impair accuracy. Serious match competitors generally *prefer* single shot firearms over magazine-fed models.

If you are seriously taking up target shooting as a sport, there are a number of moderately priced target models available as used guns from Winchester, Stevens, Remington, Mossberg and others. Most of these guns have been discontinued, simply because organized paper-target shooting competition has suffered a loss of popularity in recent years as range sites, like small airports, turn to

housing developments. On the plus side, target rifles generally receive better care than most any other gun type, and are hardly worn-out in use. In looking through the latest *Gun Digest* catalog section, some of the really super new 22 rimfire target rifles range up to $2,300(!). For the beginning target shooter, who does not have immediate Olympic aspirations, a good used target rifle at perhaps $250 to $500 is a good bet. That sort of money will buy some superb equipment.

Most of the less expensive guns use a detachable box magazine. Interestingly, the really pricey numbers are often single-loaders or are equipped with a magazine block to enable them to be loaded singly. That feature relates to the fact that many target shooters believe that the soft lead match bullets may be shaved or slightly deformed when being fed from a magazine, thereby reducing accuracy. For the beginner, that sort of distinction is not worthy of discussion.

These target-style rifles are heavier than most rimfire sporting rifles, and you'll use them with micrometer-adjustable peep sights and slings. Their extra weight and overall design make these firearms less versatile for hunting and some other uses, but for target work they're great. The trigger pull will be much lighter and crisper than you'll find on other single shot 22s, and they'll be capable of far greater accuracy.

While a single-shot target rifle sounds like a great beginner's tool, many beginners—especially young shooters — find them cumbersome, heavy and clumsy. Beginners and smaller folks simply do not have the experience to handle, appreciate or need the weight and refinements for plinking or small-game hunting. In fact, the heavy barrels and stock configurations are most suited — in most of the older guns — for prone shooting. Not many of us plink from that position. Until recently, most of the major manufacturers offered single shots basically similar to their upscale repeaters. Based on current catalog information, only Marlin is in that

Right — Rimfire rifles come in a surprising variety of shapes and styles, but selecting one is an easy task if you know what to look for, and what to avoid.

Below — Clip-fed 22 rifles are safer than tubular-fed repeaters because the magazine can be removed entirely from the rifle.

market today. Again, that may well change, and probably will.

The only problem with these rifles — or with any single-shot plinker — is that they're slower to operate than a magazine-fed repeater. While that's a *good* feature to have on a first rifle because it trains the shooter to be sure and deliberate, the beginning rifleman as he or she gains in skill and confidence may grow impatient with the need to load one round at a time. This is the only drawback the one-shooters have. Most shooters soon feel they've "grown out of" their single-shot 22s and begin looking around for a repeater to trade up to.

For this reason, a bolt-action repeater might be a better choice to start out with — particularly for a more mature beginner. For young shooters, the single shot makes the best possible rifle to learn with, and "trading up" can be put off for a few years until the safety rules are really mastered. Too, many single-shot models are offered with smaller-than-standard stocks to fit youthful frames. A diminutive teenager or pre-teen shooter can have trouble using a full-sized rifle stock, and that's what all repeaters come equipped with. For a shooter who is still several years away from reaching full growth, a "youth's model" single shot with a scaled-down stock is the way to go.

But for the older shooter, stock size is less important (most off-the-shelf 22 rifles come equipped with stocks that are pretty standard in dimension). In addition, a mature beginner is capable of learning the necessary safety rules faster and is generally more responsible than an adolescent gunner. For these reasons, you might want to consider a repeating bolt rifle as your first choice.

For the reasons mentioned earlier, I would limit selection to box-magazine-fed rifles. The removable box magazine may be easily inspected for live ammo. With the magazine entirely removed and the bolt left open, the rifle can not be fired. With the magazine removed, the rifle may be considered a single shot. The young shooter may master the basics, then be permitted to gain the advantages of a repeater simply by taking the magazine out of the drawer and putting it back in the rifle. The same may be said of the tube magazine, in that the tubular magazine insert may be removed with essentially the same result. The only disadvantage is that the long tubular magazine insert, unlike the box magazine, is easily damaged when not protected by the magazine housing.

What to Look for in Buying a Rifle

There are a number of clip-fed bolt-action 22s around — price has already been covered. How do you select the right rifle for you from the large assortment available? After first checking the actions to make sure they're empty (one of the first safety rules you'll learn in the chapter on rifle shooting), get permission from the store clerk to test the trigger pull. Look for a trigger that "breaks" (releases) cleanly and crisply — a long, draggy pull doesn't do much for a rifle's accuracy. A rifle with a trigger that's too "heavy" should also be discarded. (By "heavy" I mean in excess of, say, 6½ pounds of pressure [by the finger] on the trigger. If it feels too heavy ask the clerk to check the gun with a trigger pull gauge.)

Work the action (the bolt in this case) to see how smoothly it operates. Most high-quality rifles will function with no tendency to bind or stick. Then open the bolt (another safety rule) and bring the rifle to your shoulder. When the gun is properly shouldered, and feels comfortable, ask the clerk to help you check the fit. Look at the sights—can you see both the front and rear sight without forcing your cheek down upon the stock? Do the sights appear clear and sharp? Are they adjustable for elevation? Hopefully, the answer to all of these questions is, "yes."

Finally, check the rifle's overall appearance. Nearly all new 22s feature barrels and actions that are polished and blued (to protect against rust), but some will have a better finish than others. How about the stock? Is the wood genuine walnut (more costly and desirable than "walnut-finished" hardwoods)? Most important of all at this point is which of the rifles look best to *you*? Choosing a firearm is a personal thing, so you shouldn't let someone else's opinion sway you too much in the final selection. Once you've narrowed your choice to two or three models of the same general style and price range, *you* make the final decision.

While I favor clip-fed rifles over the tubular fed variety as a general rule — particularly for beginning shooters — to be fair I must note that tubular magazines do have certain advantages. In the first place, removable clips can be mislaid or even lost in the field, while tubular magazines are fastened permanently to the rifle. It's *possible* to lose the entire magazine-tube insert (which must be drawn at least part way out of the magazine before the magazine can be loaded) but this really isn't likely to happen.

Too, tubular magazines have larger capacities than clip or box magazines. A typical clip for rimfire ammunition will hold from five to seven rounds, while a tubular magazine will hold more than twice that amount. That means you can shoot longer before reloading — and when you do need to reload, tubular magazines are somewhat more convenient to feed fresh cartridges into. On the other hand, you can buy extra magazines for most clip-fed rifles, and carry some pre-loaded spares in your pocket for super fast reloading.

If you really want a pump, lever or autoloading rifle, it is possible to learn to shoot these firearms well and safely. Extra care must be taken with these repeaters to make sure the action isn't unknowingly closed on a live cartridge — they must be examined a little more closely when it's time to put them away. If you're willing to take this extra care right from the beginning, go ahead and buy that fancy pump or lever gun.

And yes, you *can* learn to shoot with a semi-automatic, or self-loading rifle, but this would be my last choice for a beginner to use for reasons stated earlier. It's just too easy to forget that an autoloader is always ready to fire (by just pulling the trigger). Having that kind of quick-repeat firepower on tap isn't a real aid to shooting discipline, either. Young shooters in particular are too prone to see just how fast they can empty the magazine.

As I've said, keep these guidelines in mind and use them to pick the 22 rifle that suits *your* tastes. You'll be the one paying for it, and you're the guy or gal who's going to be using it.

One other thing to watch out for is that you buy a rifle chambered for 22 Long Rifle ammunition. There are other rimfire calibers available, including the 22 WMR (Winchester Magnum Rimfire). This is a more powerful, noisier round than the 22 Long Rifle, and ammunition costs nearly three times as much. Shooting a 22 WMR can get expensive in a hurry, so make sure you don't get talked into a 22 Magnum rifle! That's a firearm you can graduate up to later if you feel the need for something a bit more powerful — but buying a 22 WMR to learn with would be a mistake. There are also certain rifles chambered for the 22 Short only. These are intended for gallery target use only, and the Short rimfire isn't powerful enough for small game hunting and other applications. Rifles chambered for the 22 Long Rifle will also shoot 22 Shorts, but the reverse isn't true. The 22 Long Rifle rimfire cartridge is versatile, inexpensive and available wherever ammunition is sold. It's the logical choice for anyone who wants to get started in rifle shooting.

chapter 5

RIFLE SIGHTS AND HOW TO USE THEM

BEFORE YOU CAN start shooting your new rifle, you need to know something about rifle sights and how to use them. Most new rifles — and nearly all rimfires — come equipped with some kind of sighting equipment as they leave the factory. These factory sights are usually simple, open sights that can be roughly adjusted vertically for elevation and may or may not be adjustable for windage (lateral or sideways movement). In the case of most rimfires, the sights come factory-adjusted for about 25 yards. However if you are going to be shooting at longer ranges (50, 75, or even 100 yards), you must adjust your sights for the desired range; otherwise you will probably miss your target completely.

While this type of sight will do the job, a variety of other, more sophisticated, sighting equipment is also available to the beginning and more experienced shooter alike.

The kind of sight you choose will depend on your particular shooting needs, the type of rifle you own, and the size of your shooting budget. At the moment, there are four different and distinct types of rifle sights in current use: 1. The simple, open iron sights already mentioned; 2. The peep or aperture receiver sight; 3. The telescopic or "scope" sight (rapidly becoming the most popular type of rifle sight in ex-

istence); and 4. A relatively new category of non-magnifying optical sights — these project a bright orange dot that is a superimposed over the target by means of an optical illusion to serve as an aiming point. We'd call these "dot-point aiming" sights.

Open Sights

Open sights consists of a flat blade or beaded post mounted at the front or muzzle end of the barrel, paired with a V- or U-notched blade mounted farther back along the barrel, usually just in front of the receiver. The shooter looks through the notch of the rear sight blade and fits the top portion of the front post into the center of the V- or U-shaped notch. While maintaining this front-to-rear sight alignment, the shooter then positions the front sight so that it either covers the target or sits just under it. Most experienced target shooters prefer to have the bullseye, or center part of the printed target, appear to sit just on the top of the front sight post and adjust their sights accordingly — this is what is known as a "6 o'clock hold."

While open sights are relatively crude aiming aids when compared to most other sighting systems available, they will do the job and are satisfactory for short-range hunting, plinking and casual target shooting. Their usefulness is limited by the fact that they can't be adjusted as easily as the more sophisticated sighting devices, and the sight picture they present is really too coarse for long-range precision work. Too, the fact that the shooter must concentrate on three different sighting elements — the rear sight, the front sight, and the target itself — means the focus of the eye must rapidly change to keep all three elements in some kind of register. This is a trick young, healthy eyes seem to master easily, but it gets harder to keep three different things in seemingly simultaneous focus as the eyes get older.

Receiver Sights

For only a few dollars extra you can add a set of receiver, or aperture sights to your new rifle. The

Above — Here's a European rear sight with twin sighting leaves. The front leaf is higher than the rear one and is used for long-range shooting. It folds forward out of the way to allow the lower rear leaf to be used at shorter ranges. Seldom found on U.S. rimfires.

Here's what a proper sight picture looks like through open sights. Front sighting blade or post is centered in the V-notch or rear sight, with the top of the post just even with the flat top of the rear sight. Then the target is aligned just over the front sight so that it looks like the bullseye is barely sitting on it. This is what is known as the "6 o'clock hold."

Below — The most common type of sight found on rimfire rifles is the open iron variety supplied at the factory. The rear blade is usually step-adjustable for elevation, but sideways adjustment (windage) often must be supplied by drifting the sight sideways in its dovetail notch.

Below — This "bead on post" front sight is used with either open rear sights or adjustable "peep" receiver sights.

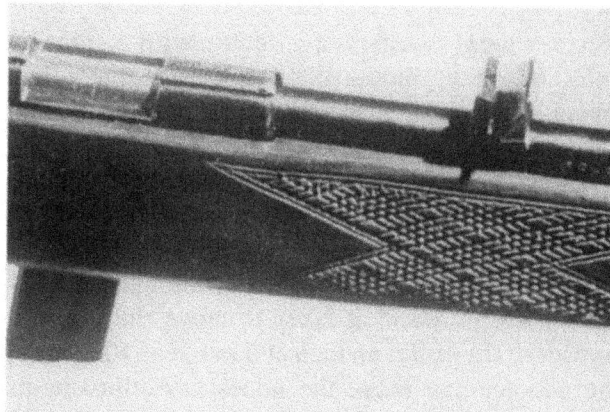

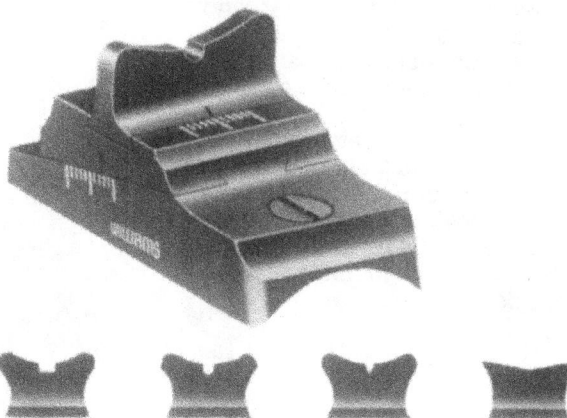

Above — The open rear sight has interchangeable sight blades with varying shaped sighting notches and is fully adjustable for both windage and elevation.

receiver sight consists of a device with a sighting hole, or peep, mounted on a steel frame. The frame allows the aperture to be moved up and down and sideways by turning some adjusting screws. These screws are usually calibrated to allow precise adjustment of these movements and may be designed to produce a clicking noise for each increment of adjustment. By knowing exactly how many "clicks" it takes to move the point of impact of the bullet an inch at a range of 100 yards, the shooter can make the necessary adjustments without a lot of trial and error.

The receiver sight is also used in conjunction with a front blade or post, and it's sometimes possible to use the front sight that the rifle originally came equipped with when you install a rear-mounted peep. Often, though, it's necessary to install a taller front sight because the aperture of most receiver sights sits higher above the barrel than the open sight it replaces.

While the receiver sight can be adjusted much more accurately and conveniently than the open iron sights supplied by the factory, its chief advantage stems from the fact that it is also much easier to sight through. Rather than concentrating on the rear sight, the shooter merely looks *through* the aperture or peep at the front sight and target. The aperture ring appears as an indistinct blur, and experienced marksmen pay no conscious attention to it. The eye naturally centers the front sight (or target) at the strongest point of light — and the light is strongest at the exact center of the aperture. Centering the front sight in the rear peep is an automatic response, and once you get used to this idea

Above and below right — Receiver, or "peep" sights are faster and easier to use than open sights, and they can be easily adjusted for changes in windage and elevation. Such sights cost only a few dollars and make an excellent investment. Both of the sights pictured are William's Foolproof Receiver Sights — the one above is for mounting on the left side of the receiver; the one on the right is for mounting on the right side of the receiver.

(Above) Here's an example of an extremely sophisticated aperture sight sold for serious target use. Note how the aperture is shielded from light by a rubber eyepiece. Knurled adjusting knobs make sight changes easy.

Above — Open iron sights are standard on most 22 rimfire rifles, but some models, like this bolt-action target-style Anschutz Mark 2000, are available with fully-adjustable receiver sights.

Right — When looking through a receiver (peep) sight, this is the image you should see. The rear aperture should appear indistinct while the front sight and target are much sharper.

you'll find the receiver sight much faster and easier to use than open sights. (Some shooters mistakenly insist on looking *at* the aperture ring rather than simply *through* it).

Most receiver sights feature screw-in discs with varying sizes of apertures for precision target work, but the "strongest-point-of-light" principle works so well that many hunters (and other shooters) remove this disc and simply sight through the threaded tube used to hold the discs in place. I've done this on several different occasions and found that using the oversized "aperture" didn't affect accuracy all that much. With the disc removed, the receiver sight becomes one of the fastest, most practical hunting rifle sights in existence.

Again, while looking *through* the rear peep, the shooter places the front sight blade (or post) either directly under the target (6 o'clock hold) or superimposes it over the aiming point. Most serious target shooters adopt the 6 o'clock hold, while some hunters and plinkers prefer to put the top of the front sight directly on the point they want the bullet to strike. This latter tactic blots out part of the target from view, but works okay for relatively short-range hunting and tin can plinking sessions.

When adjusting either open or receiver sights to get them properly "zeroed" on target, the rear sight is moved in the *same* direction you want the bullet's point of impact to move. In other words, if the rifle is shooting 2 inches high and an inch left of center bullseye, the rear sight should be moved downward and right the appropriate number of "clicks" or increments to bring the next bullet on target. Very rarely is it desirable to move the *front* sight (usually when there's not enough adjustment left in the rear sight to move the bullet impact all the way on target) — and in this case, the front sight is moved in the *opposite* direction you want the bullet strike to move.

Telescopic Sights

Sooner or later, you'll probably want to mount a magnifying "scope" sight on your rifle. For the kind of plinking, hunting and target shooting you'll do with your 22 rimfire a scope isn't really needed (and such sights aren't allowed in most target competition). But if you decide to graduate to a centerfire rifle later on, you're almost sure to want to add a telescopic sight to your rig.

Riflescopes offer several important advantages to those who take up hunting as a sport. The optics of a good scope not only magnify the target image,

These Redfield Widefield scopes provide the shooter with — as the name would imply — a larger field of view with which to scan the target. These scopes, with their one-inch diameter bodies are stronger than typical scopes designed only for rimfires, but are worth the extra cost. These fixed-power scopes are featured in the degrees of magnification most popular with experienced shooters.

A

CH ,FCH, MCH

4P CCH

4 PLEX

B

C

PCH

DOT

D

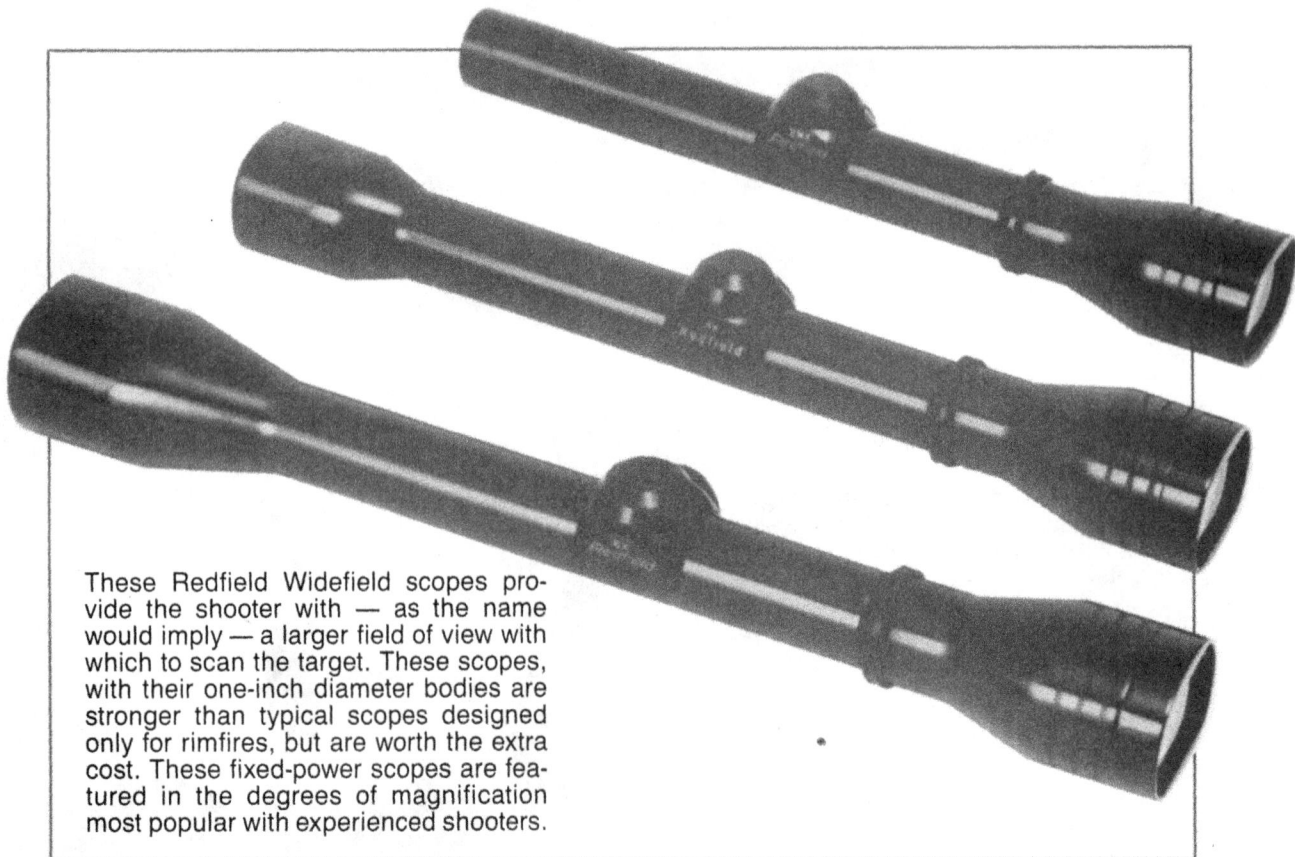

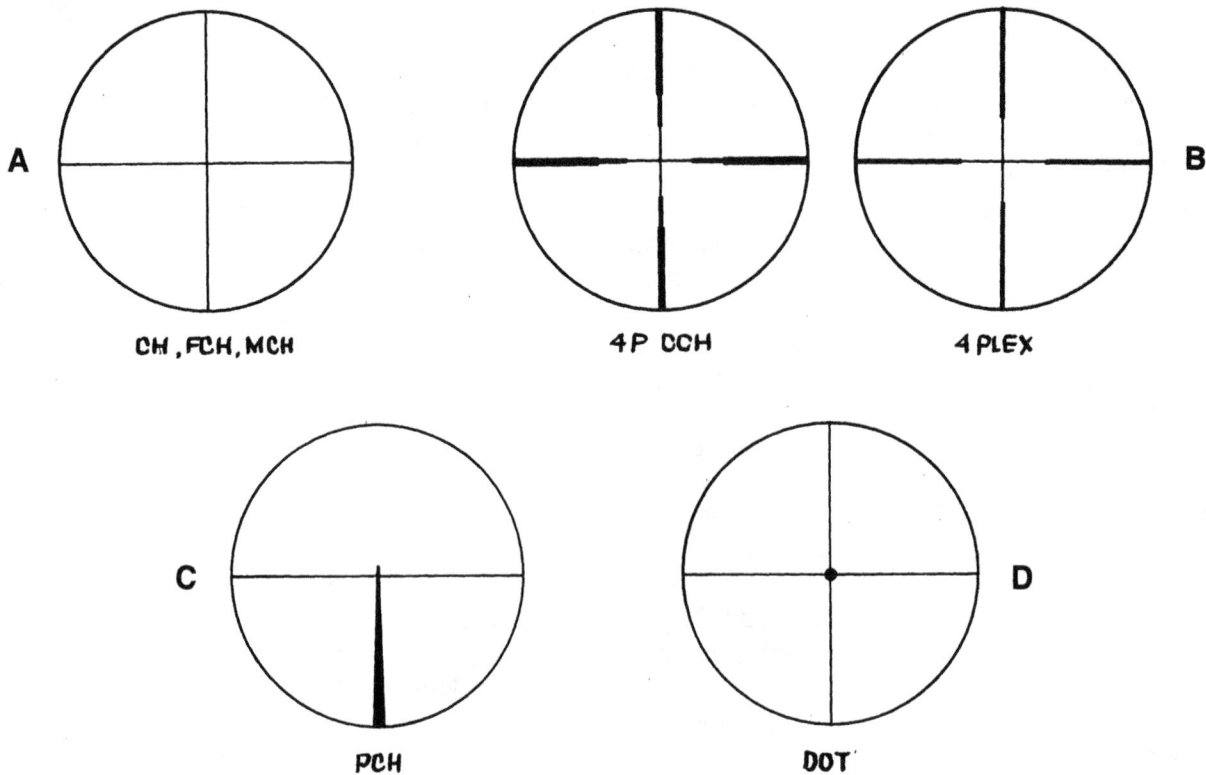

A is the standard "Crosshair" reticle most commonly seen in the hunting field and target range. B features finer inner crosswires and heavier outer wires — popular for hunting use in fast shooting conditions. C is the "Post and Crosshair," which is ideal for dim light and close range, but the coarseness of the post can obscure the target considerably at longer ranges. D the "Dot", consists of very thin crosshairs centering on a dot. This type is often used by target and varmint shooters.

There are high-riding scope bases available that allow you to see *under* the scope and use auxiliary or standard iron sights if necessary.

Scopes used on high-powered centerfire rifles should have adequate eye relief and be mounted far enough forward that recoil won't slam the eyepiece into the shooter's eyebrow. This shooter has his eye too close.

all shooting conditions. Most high-quality optical sights are effectively sealed against weather and are waterproof, fogproof and reasonably impact resistant. But that doesn't stop water from collecting on the outer surfaces of the lenses during a heavy rain, and any optical device with its carefully spaced internal lens system has to be more fragile than an iron open sight. For this reason, many hunters attach scopes to their rifles in quick-detachable mounts to allow the use of iron sights as a backup system in case the scope goes haywire. Or there are high-riding see-through mounts you can buy that let you see the iron sights *beneath* the scope, and removal of the glass sight isn't necessary before the open sights can be used.

Another common problem with scopes comes when the sight is mounted on a heavy-recoiling

but they improve its clarity and definition as seen by the shooter. The scope gives the hunter a final chance to evaluate a trophy head before he pulls the trigger, and the crosshair reticle most commonly used makes very precise bullet placement possible. Too, the lenses of a quality riflescope can serve to "gather" light and clarify the target enough that the hunter can see to shoot several minutes earlier and later in the day.

Another big plus for scope sights is the fact that a shooter doesn't have to concentrate on several separate images at the same time. Looking through a scope's reticle, he simply sees a magnified image of the target with a set of crosshairs (or other aiming device) superimposed over it.

While scope sights are in wide general use, they're not the perfect answer for all shooters or

centerfire rifle with the eyepiece too close to the shooter's eye. This is compounded if the shooter doesn't hold his rifle in the proper position, and "crawls the stock" (moves his head too far forward) to look through the sight. This combination can be a painful one when the rifle (and scope) recoils rearward and results in what is known as "shooter's eyebrow." This condition is identified by a crescent-shaped scar located immediately over the shooting eye, often bisecting the eyebrow.

A little intelligent prevention is the best "cure" for shooter's eyebrow. In the first place, when you select a telescopic sight for a centerfire rifle that has moderate to heavy recoil, be sure to choose a model that has sufficient "eye relief" — this is the distance the eye must be positioned behind the eyepiece in order to see through it properly. All rifle-

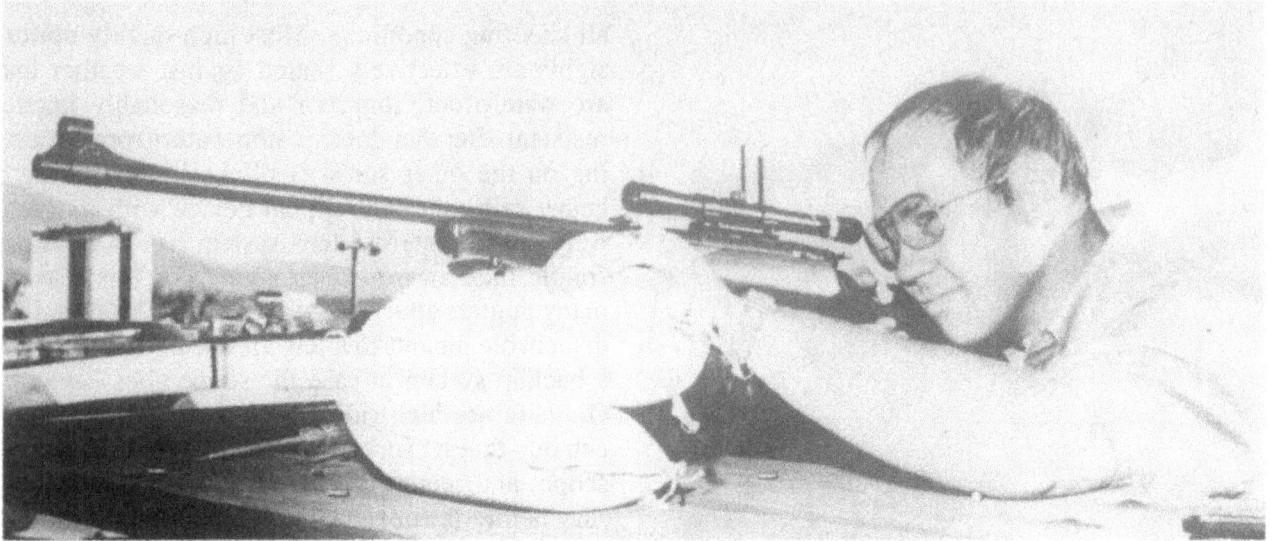

Some inexpensive scopes are scaled down for use on rimfire rifles only. These miniature scopes have tubes measuring 3/4- or 7/8-inch in diameter and come equipped with rail bases for mounting on ready-grooved receivers. These scopes aren't built to hold up under the recoil of a centerfire rifle, but work okay on 22 rimfires.

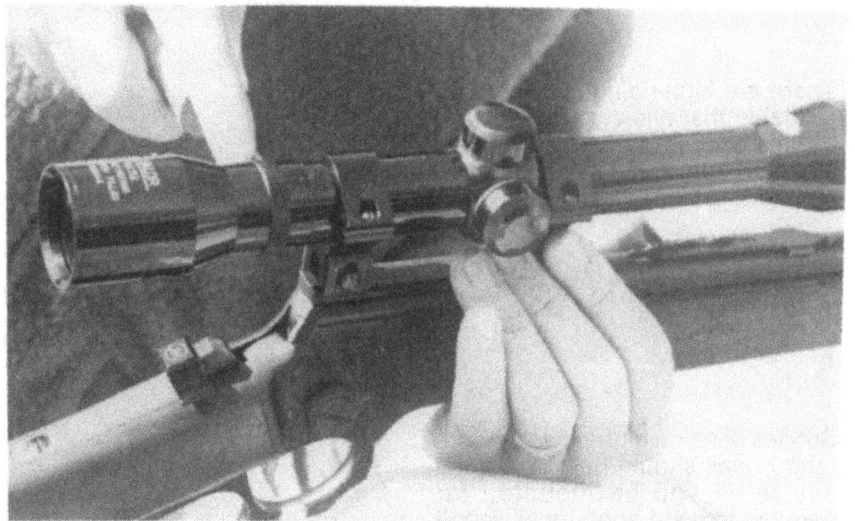

Mounts are available to allow full-sized scopes with 1-inch tubes to be mounted on the grooved receiver of rimfire rifles.

scope catalogs should list the eye relief for each model that a maker offers, and a simple comparison of figures will soon show you which scopes are and aren't suitable for hard-kicking rifles. Generally speaking, you'll want a sight with an eye relief measuring at least 3 inches (and preferably 3½ inches or better) for a big-bore centerfire. And then make sure the gunsmith mounts the sight far enough forward on the rifle to prevent "shooter's eyebrow."

For rimfire rifles, eye relief isn't critical, and you can get by with a much less sophisticated (and expensive) scope than you'd need for centerfire firearms. Most rimfire rifles come equipped with a pair of parallel grooves cut into the top of the action to accommodate certain inexpensive scopes designed for use on 22 firearms only. These scopes

come complete with the mounts needed to attach them to grooved rimfire receivers, and you only need a screwdriver (or in some cases, a coin) to fasten the scope in place.

While these inexpensive scopes are fine for plinking, they aren't as well made as the more costly (and larger) sights designed for centerfire use. Too, they lack the brightness and definition of the bigger scopes. For this reason, many shooters prefer to install a sight designed for centerfire use on their 22 rifles. There are special 1-inch mounting rings with grooved bases available to simplify this chore.

If you're mounting a scope sight on a 22 rifle that has a grooved receiver, you can usually handle the job yourself with no trouble. But for the more sophisticated scopes usually mounted on center-

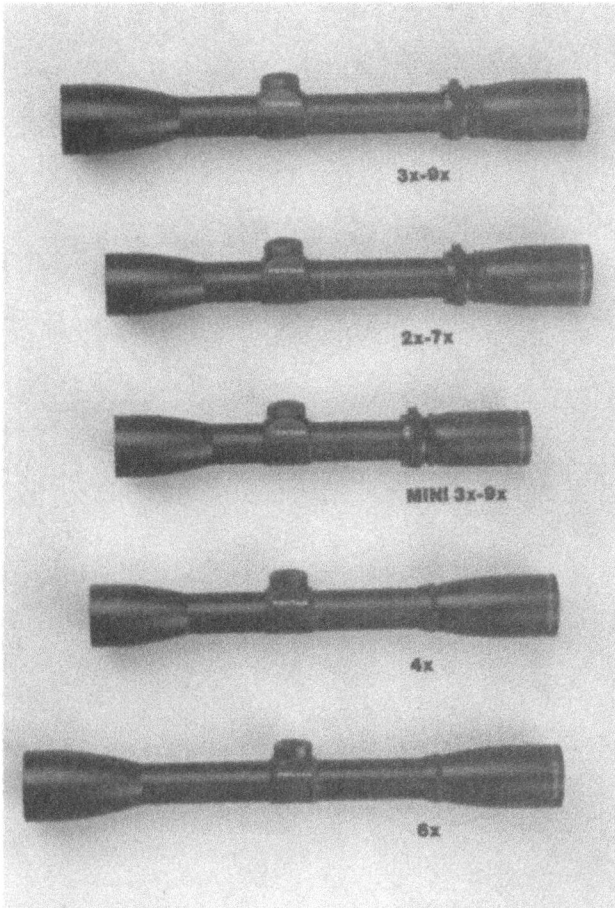

This group of Pentax Pro Finish scopes represent a typical line of modern scopes. The variable power scopes sound like an irresistable idea, until the new shooter realizes that the high-power settings are not all that practical most of the time. For example, try using a set of 10× binoculars — which have a much larger field of view — and see how every twitch throws you off target. These modern scopes are nitrogen filled and waterproof to prevent fogging in cold weather.

Collimators are optical bore sighters. This particular collimator is the Bushnell Model 74-4001 TruScope bore sighter. You carefully mount this unit in the muzzle of your rifle, look through the scope and adjust the scope's crosshairs to coincide with the crosshair graph contained inside the collimator. These units cost around $25; however, your local gunsmith or gun shop will probably be happy to "collimate" your rifle for only a couple of dollars. What's the benefit? Using a collimator to bore sight a rifle usually takes 1 or 2 minutes — the conventional methods may take 10 minutes and require a target be set up at some distance. Once your gun is properly collimated you must, however, give it a final, more precise sighting-in at the range.

fire rifles, let your local gunsmith or sporting goods store take care of the mounting at the time of purchase. Even this job isn't difficult, but you need some kind of padded vise and maybe some specially ground screwdrivers to do it right. Too many home-mounting jobs end up with gouged stocks and marred retaining screws, and it's even possible to damage the scope itself.

Another problem with mounting scopes at home is that your gunsmith has a gadget called a collimator that he'll use to adjust your scope so that it's approximately in alignment with your rifle's bore. That doesn't mean your rifle is now sighted-in and ready to go — you must still take it to the nearest range and make the final necessary adjustments while actually shooting the rifle. But it does save a lot of time and ammunition, since you're usually assured that your first shots will at least be on target somewhere. Without a collimator, you must bore-sight the scope by looking through the rear of the receiver down the bore, and then adjusting the scope crosshairs so that they intersect a distant object you can also see through the bore.

Riflescopes come in two basic varieties: fixed power, with the image magnified a set number of times — say 3 times or 4 times what the eye would otherwise see; and variable power — that is, the magnification can be varied by moving a power selector ring. In these latter scopes, the highest power available will be roughly 3 times that of the lowest magnification offered. While variable scopes are more expensive and slightly more fragile than the fixed power variety, shooters seem to like them, and the most popular riflescope now being sold is the 3-9x variable. Variable-power

One way to conserve ammo, and to make your sighting-in chore with a scope less frustrating, is to use a scope collimator. Making sure your rifle is unloaded, the collimator is placed in the muzzle as shown here. All you have to do is adjust your scope's crosshairs to converge with the center of the optical display you'll see through the scope. This will virtually guarantee your first shots will be on the paper if not in the black.

scopes are versatile and nice to have, but they're also considerably more costly and usually larger and heavier than the fixed power models.

Most beginning shooters seem to like the idea of getting the most powerful scope possible to put on their rifle, reasoning that if a little magnification is a good thing, a lot must be even better. After all, who could miss when the target image is magnified to 10 or 12 times its normal size at a given range? That sounds good in theory, but it doesn't work out too well in actual practice. For one thing, no rifleman holds his firearm perfectly still when he sights-in on a target — muscle tremor creeps in, and even the pulse of blood pounding through your veins affects your "steadiness." Much of this involuntary movement isn't apparent when you're using iron sights or a scope with low magnification

— but switching to a high-powered optical sight magnifies not only the target, but any movement on your part. Looking through a 12x telescopic sight (12x means the image is magnified 12 times) at a distant mark is a little like riding a bicycle over a bumpy road and trying to read a newspaper at the same time. Every bump and jump of your nervous system shows up as the target appears to bounce all over the place. Really high-powered sights (like 10x and 12x scopes) are intended to be used on a rifle that's solidly rested against sandbags or some other solid, unmoving support.

Another problem with a high-magnification scope sight is that the field-of-view (the size of the area you can see through the eyepiece) narrows as the magnification goes up. The field-of-view is so narrow on an 8x or 10x scope that you can have

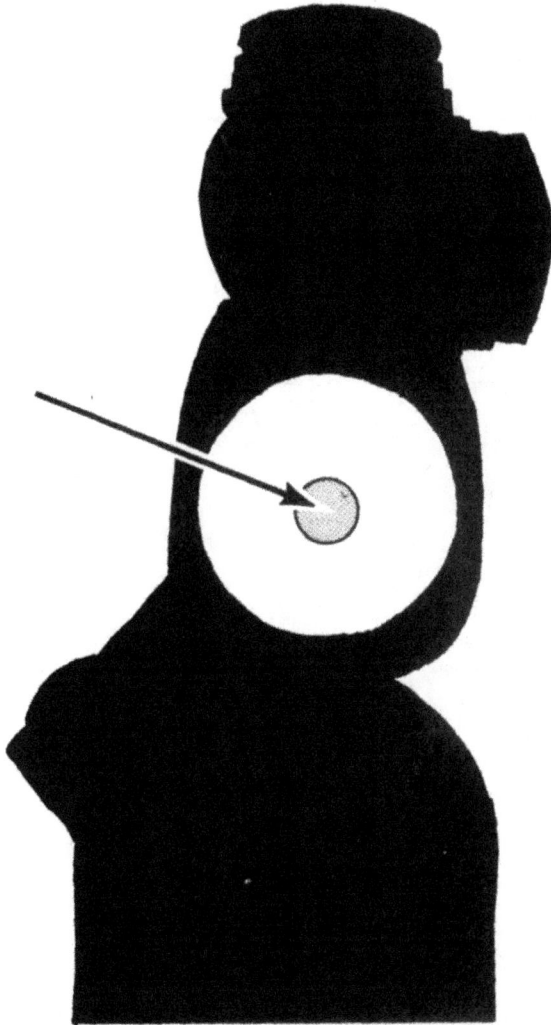

When using a "dot-point" aiming device, this is the sight picture you will see — merely place the colored dot (arrow) on the intended target.

which direction the crosshairs move when you turn the screw in the direction the arrow points. If your bullets are striking the target low and to the left, simply turn the screws to move the crosshairs higher and to the right. Don't forget to replace the metal caps when you're through adjusting.

Dot-Point Aiming Sights

The other sighting system I've mentioned — dot-point aiming — doesn't usually magnify the target, but serves to "project" a bright orange dot that is used as the aiming point. This dot is projected as an optical illusion, and it appears in the same sighting plane as the target. You merely place the dot on the target and pull the trigger.

The advantage of dot-point sights is that they are very fast to use, and they work well on moving targets shot at close range. The dots these sights project are too large to make long-range precision marksmanship possible, but they can be helpful when used on a 22 (or shotgun) in hunting small game.

On the other hand, since these sights don't enhance or magnify the target image, they're not as versatile as a good scope sight. Too, some of these sights are not transparent and thus require the shooter to keep both eyes open. With this particular variety, the master — or shooting — eye sees only the bright, glowing dot. It's the other eye that sees the target, and when the two images are "blended" on the optic nerve, the brain "sees" the dot superimposed over the target. These sights have their uses, but they're not the best choice for the beginner.

Which Sight to Choose?

While there are several different kinds of rifle sights available, it's my opinion that the beginning rifleman should first learn to shoot with simple iron sights. These can be the open variety furnished on each 22 by the manufacturer, or a more sophisticated aperture sight purchased as a separate accessory and installed on the rifle's receiver. These are very basic rifle sights, and the new shooter should begin by learning the basics. If you should add a scope to your rifle later on, there's always a chance of having it fail — and if you don't know how to use the iron sights, you're all through shooting. Start with iron sights first. Making the transition to a scope comes easy later on. What's more, you may never feel the need for an expensive telescope sight — millions of shooters get along fine without them, and you may be one of

real problems even finding the target in the scope — and that's particularly true if the target is a running deer intent on escape.

For these reasons, a 3x or 4x glass sight is a better choice for most hunting and plinking uses. The field-of-view is usually large enough to let you *find* the target in a hurry, and when you're ready to shoot, the magnified image won't bounce around quite so much like a hyperactive jackrabbit.

Adjusting a scope sight to bring the crosshairs to rest at the same spot the bullets are hitting downrange is a simple matter. A pair of adjusting screws are located in the turret, usually at right angles to each other. Unscrewing the metal caps covering these twin screws gives you access, and somewhere on the face of each screw you'll find a small arrow and an identifying letter that shows you

Right — When a rifle is fired the bullet's path is not "straight" — it follows an arc. In this illustration a rifle-scope sighted for a 200-yard "zero" has just fired a 150-grain, 30-06 Silvertip. The bullet actually crosses the line of sight at two points—50 yards (on its way up) and at 200 yards (on its way down).

those who are content with the sights your rifle came equipped with when you bought it.

Regardless of which type of sight you decide to use, there are some important rules to apply. I've already mentioned a couple of different sight pictures you can use with iron sights and the fact that scope crosshairs are used to intersect the target at the exact point you want the bullet to strike. But there's more to it than that. In the first place, you must take care not to tilt or "cant" the rifle so the sights slant to one side. If you cant the rifle when the trigger is pulled, the bullet will strike high and to one side of the intended target. The sights must be seen as straight and as level as possible if you want to hit what you are shooting at.

You should also be aware that the bullet from a rifle doesn't fly in a perfectly straight line, but in fact travels in a shallow curve or parabola that rises to cross the line of sight (as seen through the top-mounted sights) at a point fairly close to the muzzle, and then crosses it again farther downrange as it falls toward the earth. (This isn't strictly true, as the bullet begins dropping in response to gravity from the moment it leaves the muzzle. But the sights are not aligned exactly parallel with the bore, so that the path of the bullet describes an upward and downward curve *in relation to the line of the sight.*) This means that if your rifle is sighted-in to put the bullet's impact exactly where the sights meet the target at 100 yards, it will strike slightly higher at closer range, and lower as the range increases.

The parabola the bullet travels after it leaves the muzzle is called its *trajectory*. Different loads and cartridges have different trajectories, and sights must be adjusted differently to accommodate these variations.

Without some kind of sight, it would be difficult to achieve any high degree of accuracy with a rifle, particularly at long range. It's *possible* to learn to hit even moving targets thrown close to the shooter without the use of sights — this kind of stunt is called "instinctive shooting," and can provide the beginner with valuable practice if properly done. But for normal marksmanship with a rifle, an accurate set of sights is a real necessity.

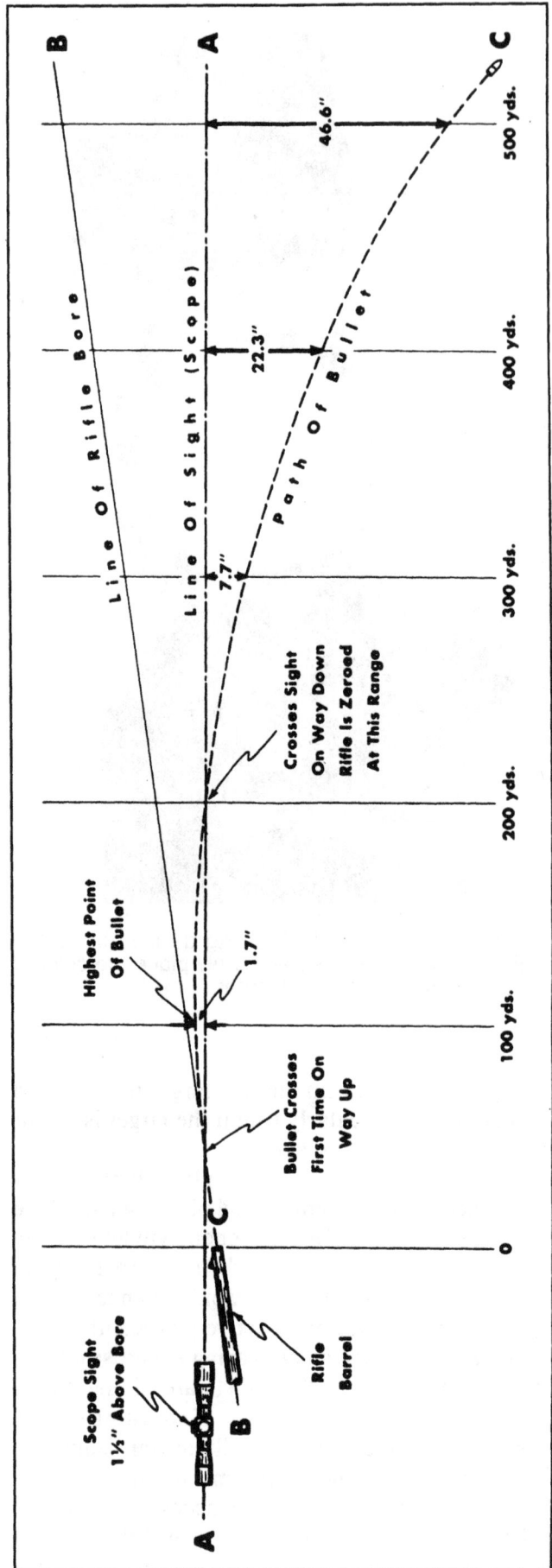

Line Of Rifle Bore

Line Of Sight (Scope)

Path Of Bullet

46.6"

22.3"

7.7"

1.7"

Crosses Sight On Way Down Rifle Is Zeroed At This Range

Highest Point Of Bullet

Bullet Crosses First Time On Way Up

Scope Sight 1½" Above Bore

Rifle Barrel

500 yds.

400 yds.

300 yds.

200 yds.

100 yds.

0

A B C

chapter 6

BASIC RIFLE SHOOTING

BEFORE YOU'RE ready to head for the range—whether it's a formal rifle range with full facilities or simply a nearby hillside that you can shoot into safely—you should become familiar with your rifle and know how it operates. Most new rifles come equipped with an instruction manual, and this should be carefully read before you attempt to operate the firearm. Make sure you know where the safety is located and how to work it. Also make sure you know how to safely load and unload the rifle. Before you go to the shooting range, be sure to check the bore for obstructions or excess oil or grease—this can be removed by running a dry patch on a cleaning rod through the bore.

If this is the first time you've used your rifle, it might be a good idea to have a more experienced shooter help show you how to operate it. Have the store clerk or salesman show you how the gun works at the time of purchase. Of course the gun should never be loaded during this learning period, and even though you know the gun is unloaded, you should still obey all the safety rules. Remember the First Shooting Commandment: *Treat each firearm as if it were loaded.* That rule applies whenever you're handling a gun.

During this familiarization period, you can practice "dry firing" your rifle at some mark or target fastened to a basement wall or some other safe location. Dry firing means pulling the trigger of your rifle on an *empty* chamber.* Check and double check to make sure there are no cartridges *anywhere* in the gun, and then stand or sit across the room. Hold the rifle in one of the recommended firing positions (which we'll get to later on in this chapter), align your sights with the target, and then slowly squeeze the trigger until the firing pin falls. But be sure to observe all the safety rules, just as if the rifle *were* loaded (remember?).

Basic Shooting Instruction

Once you know how to operate your rifle and are reasonably familiar with it, it's time to buy a few

---

*On some 22s (when dry fired) the firing pin will hit against, and tend to peen, the edge of the chamber. A great deal of damage could be done to the chamber itself. If it appears this "peening" is taking place, have a good gunsmith correctly alter the firing pin. A simpler solution is to chamber an *empty, once-fired* cartridge case when you do your dry firing. The case rim will act as a buffer between the firing pin and the outer areas of the chamber surface.

boxes of ammunition and go to the range. If you don't have formal range facilities nearby, look around the countryside for a safe shooting area with a high hill or bank to shoot into. If you're on private ground, be sure to look up the owner and get his permission to shoot in advance.

Before you head for the range, make sure you've got some kind of support to take along to use as a rest for your rifle—this can be a blanket, a rolled up sleeping bag, sandbags or anything else flexible enough to mold itself to the contours of the rifle stock. If you're going to a regular rifle range, chances are there will be sandbags available for you there. But if not, be prepared to improvise.

When you arrive at the shooting area, the first thing you need to do is find someplace you can set up shop. By that, I mean you need some kind of support to help hold your rifle while you're shooting it. While you'll eventually progress to the point that you're supporting all the weight of the rifle yourself, you'll need something to rest it on this first time out. Again, if you're at an established range, there should be shooting benches at hand that you can use. In the absence of regular shooting benches, you can bring along a card table (shaky, but better than no rest at all). Using a car hood or other part of your automobile is not really a good idea. Muzzle blast from larger calibers can cause damage to the paint or even blow out windows. There have even been mirrors and windows shot out!

Once you've located a flat, reasonably stable surface that will elevate the rifle at least a yard or so off the ground, bring out your sandbags or blanket roll. Use these to support the forend of your rifle. Place the rolled up blanket or sandbags on top of your improvised "bench," and then put the rifle forend on top of these semi-soft supports. Again, make sure the rifle is unloaded at this point and that the action is left open.

That done, you can place your targets downrange. You can manufacture your own targets by simply drawing a 3- or 4-inch solid circle in the center of a piece of typewriter paper, or you can buy targets at your nearest sporting goods store. The targets can be attached to the side of a cardboard box with masking tape, or held on with thumbtacks. If you're at a regular rifle range, there will be target supports available for you to attach your targets to.

Place the targets approximately 25 yards from your firing line or shooting spot. Stepping off 25 long paces will give you about the right distance, and it won't hurt if the range is a bit shorter. Don't stretch the range at first though—long-range shooting can come later on in the program.

Opposite page — Make sure you have some kind of padded support to hold the forend of your rifle steady. A regular shooting bench with a sandbag, as shown here, is ideal, but you can get by with a rolled up blanket or jacket. Remember that rifle barrels can get hot. Be careful of using nylon or other materials that may melt.

Right — Always shoot into a substantial, safe backstop like this shooter is using. Even a stray 22 bullet may travel more than a mile. Heavier calibers can go much farther and still carry enough energy to cause injury or property damage.

Make a final check of the backstop to make sure it's high enough (and soft enough) to stop stray bullets and examine the area on each side of the shooting site to see if ricochets could cause any damage. Now, you're ready to start shooting.

When you return to your rifle, take up a comfortable position behind it—sitting on a chair or stool, if possible, or leaning against your car hood if that's what you're using as a "bench." The main thing is to find a position that is comfortable for you, and one that you can maintain for several minutes at a time.

Finally, make sure no one is downrange or in front of your shooting position. Then open your box of 22 ammunition, take out a single cartridge and load it directly into the chamber. This can be difficult to do if you're using anything but a bolt-action or single shot rifle. If you have a pump, lever or autoloading rifle, you may need to fill the magazine (open the action and leave it open while you're filling it) and feed each round by operating the action.

Put the rifle on safety as soon as you close the action. Then, while continuing to rest the forend of the rifle on your rolled-up blanket, sleeping bag or sandbags, grasp the pistol grip in your shooting hand (your right or left hand, depending upon your preference) and pull the buttplate into your shoul-

der. Your cheek should be resting against the comb (top) of the stock, with your eye behind the rear sight.

For most right-handed shooters, the right eye is the dominant one, although this isn't always the case. If you're right-handed and you know that your *left* eye is the dominant one, you might want to squint the left eye partly (or even entirely) closed. Most rifle shooters do this anyway with iron sights to help them sharpen their focus.

Line up the sights with the target as explained in Chapter 5. If you're using open iron sights, place the front post or bead so that it nestles down into the center of the U- or V-shaped rear sight notch, with the top of the front post even with the flat top of the rear blade. Then move the rifle until the bullseye looks like it's sitting just on top of the front sight.

Then move the safety to its "off safe" position while doing your best to maintain that sight picture. Once the safety is off, steady the sights again on the target, put your finger on the trigger and prepare to shoot. Remember to let the rifle forend rest on the blanket roll or sandbags and support the butt end of the stock with your right hand and shoulder (southpaw shooters should reverse this procedure). Your free hand can be used to hold the forend down against the sandbags or to help steady

87

1

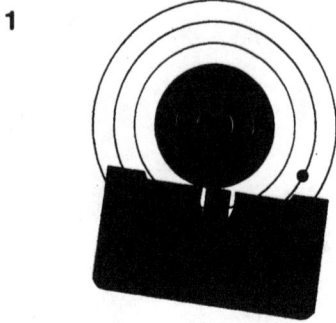

If your sights are "canted" (tipped) to the right, your shot will hit low and to the right.

3

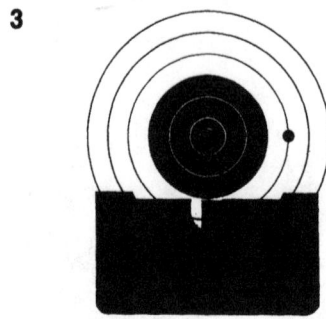

If the front sight blade is too far right, your shot will go right.

5

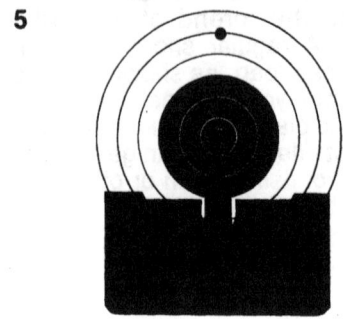

By raising the front sight blade too high, your shot will go high.

2

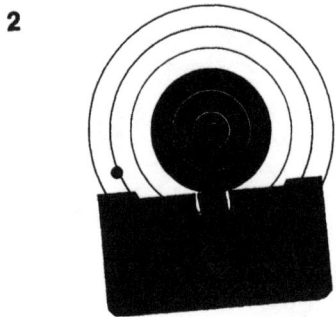

Canting your sights to the left will make your shot go low and left.

4

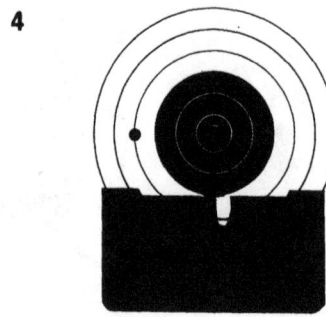

When the front sight blade is too far left, your shot will go left.

6

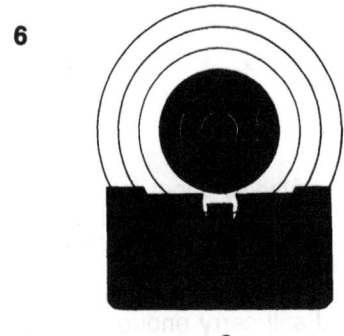

If you hold the front sight too low, the shot will go low.

the buttstock. This should give you a very steady shooting position, and you shouldn't have too much trouble keeping the sights on the target.

While holding the rifle as steady as possible, take a deep breath, exhale to let it part way out, and then apply gentle, continuous pressure to the trigger. Don't *pull* the trigger—this will make the rifle move off target just as it fires. Instead, squeeze the trigger gradually, using the tip or second joint of your index finger to apply pressure. Keep on squeezing the trigger, adding slightly more pressure until the trigger "breaks," and the rifle fires. (If you run out of breath before the trigger snaps, take another breath and start over.)

Most factory-installed sights are bore-sighted to be reasonably accurate at a range of 25 yards, so the bullet should strike somewhere close to the bullseye. To check on this, open the bolt of the rifle, lay the gun down against the sandbags (or blanket roll) with the muzzle pointing downrange, and walk up to inspect the target. Use a pencil or some other marking device to mark this bullet hole so that it can be identified later and return to your rifle. Repeat the same procedure at least twice more and compare the results. If you've done

things right, you should have three holes reasonably close together on the target—ideally all three holes should be touching each other.

If the holes are widely dispersed, you've done something wrong. You've either jerked the trigger at the last second, canted (tilted) the rifle, or varied your sight picture. Keep on practicing until your three-shot groups make a pattern no larger than 1 1/2 or 2 inches across.

When your patterns have shrunk to that size, note where they've landed with relation to the bullseye. If the group is hitting the center of the bullseye—all well and good. Your sights are properly adjusted for that range, and now you know that if you miss a shot it isn't the rifle's fault.

But if your groups are consistently "printing" (landing) in the white part of the target away from the bullseye, you'll have to make some adjustments to your sights before you can continue. Most open iron sights have elevation adjustments built into the rear sighting blade, and some can be adjusted sideways as well. Remember to move the rear sight in the *same* direction you want the bullet to move on the target. If your groups are striking low, move the rear sight up a notch and try shooting the rifle again. Fire three shots after making each adjustment (to make sure you didn't flinch or cause the rifle to move to produce a "wild" shot) and continue adjusting the sights until the bullets are making holes in the bullseye.

Once your rifle is "on target," and you've mastered the art of holding it steady on the rest and keeping the sights in proper alignment while squeezing the trigger, you're ready to try supporting the rifle by yourself.

There are many different positions you can hold a rifle in while shooting, and some of these positions are too unorthodox to be used in match competition. Hunters and plinkers are quick to take advantage of "field rests" (using a handy branch or large rock to steady the rifle against), and the smart shooter will do everything possible to stabilize his rifle. There are no formal rules for plinking, and the steadier you can hold that rifle the easier time you'll have hitting the target. That's the idea behind using some kind of bench or table and a rolled-up coat or sleeping bag to shoot from when you're first starting to practice. The first thing you must do is learn to *hit* the *target*.

After firing three shots at the same mark, check your target to see if the holes are close together. Then see what adjustments you need to make to the sight in order to bring the group into the center of the bull's-eye.

(Below) The rear sight found on most 22 rimfire rifles is drift-adjustable i.e., if the point of bullet impact is to the left, you simply drift the rear sight to the right until the point of impact is the same as the point of aim. This shooter is wisely using a Brownells nylon-tipped drift that won't mar or damage the rear sight assembly.

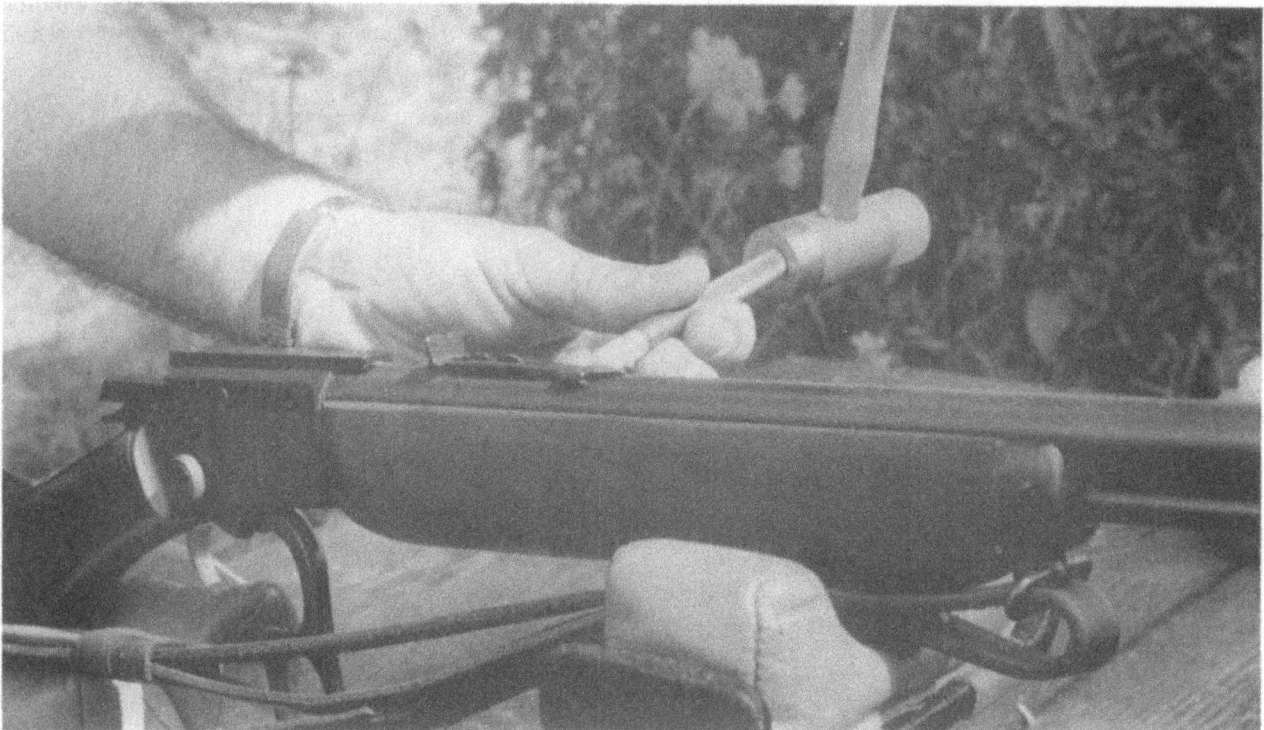

89

Before you can call yourself any kind of a marksman, though, you need to be able to hit the target *without* using any external rests or supports. In fact for any kind of target competition other than benchrest shooting, you can't use anything but your body for support.

There are four basic shooting positions in general use, and every rifleman should be aware of these and know how to use them. They are: *prone* (lying down), *kneeling, sitting* and *standing*. The prone position is the steadiest of the four, while the standing (or offhand) position is the shakiest. Most serious target shooters regularly use all four. Even if you don't intend to enter formal competition, you should know how to assume each of these basic positions.

Because the *prone position* is the steadiest, it's the one you should start with. To assume the prone position, lie on your stomach with your body making a line that points to the 2 o'clock position on a clock face with the target at 12 o'clock (left-handed shooters should face toward 10 o'clock).*

---

*This "angling" of the body is, in and of itself, a very natural, supporting position—it not only allows the shooter to properly hold his firearm comfortably, but also helps eliminate the bone-jarring recoil of most of the larger centerfire rifles.

Place your left, or non-shooting hand on the forend of the rifle and move that elbow directly underneath the rifle for support. Grasp the pistol grip of the stock with the other hand and rest that elbow on the ground to one side. Pull the buttstock into your right shoulder (left for southpaw shooters) and place your cheek against the comb of the stock. You should then be able to see through the sights comfortably.

To make your position more comfortable and even steadier, spread your legs a comfortable distance apart and press the instep of each foot toward the ground. When the rifle is pointed toward the target, it should make an angle of between 30 and 45 degrees with the line formed by your body.

This position can be hard on the elbows, and it's always a good idea to bring a blanket or some kind of pad along when you're planning to shoot from the prone position. This keeps the elbows from getting too sore and also protects your clothing from dirt.

While the prone position is the steadiest of the recognized shooting stances, it does offer some practical problems. In the first place, it takes a little time to get into this position, and a hunter, for instance, may not have the time to spare. Too, when you're lying on your stomach, your head isn't too far above the ground—weeds or other undergrowth can hide the target from view. Finally,

These shooters are using the prone position, which is the steadiest of all positions where some external support is not used.

it *is* awfully hard on the elbows.

The *sitting position* ranks next in steadiness, and it has the advantage of being very comfortable. To get into this stance, again face the 2 o'clock position (with the target at 12 o'clock) and simply sit down. Spread your legs at a comfortable angle and bend your knees. Then hold your rifle in shooting position, making sure that the left elbow is directly under the forend, and lean slightly forward until both elbows are braced against the insides of your knees.

The sitting position gives you a much higher stance than when you're prone, and with a little practice you can almost fall into the position in a matter of seconds. It's very useful for hunting, as well as target shooting.

The *kneeling position* is both less steady and less comfortable than sitting down to shoot, but it gives more support than standing—and you can get into it in a hurry. In practice, kneeling isn't much faster than sitting and most sportsmen prefer the steadier sitting position.

To assume the kneeling stance, a right-handed shooter should face 2 o'clock (again), and then kneel on the right knee. The left leg is extended slightly, and the right foot is turned inward so the shooter can sit on it. With the rifle held in firing position and the left elbow held directly under the forearm, the shooter leans forward until his left elbow is just forward of the left knee. The knee contacts the arm just behind the elbow to support both the arm and rifle. If you find it uncomfortable to

Above — The sitting position can also be used in this variation — with the shooter's legs crossed at the ankles. Some find this variation more comfortable. (Courtesy Michaels of Oregon)

The sitting position is both comfortable and steady to shoot from. This girl is relaxed, yet her rifle is properly supported by this shooting stance.

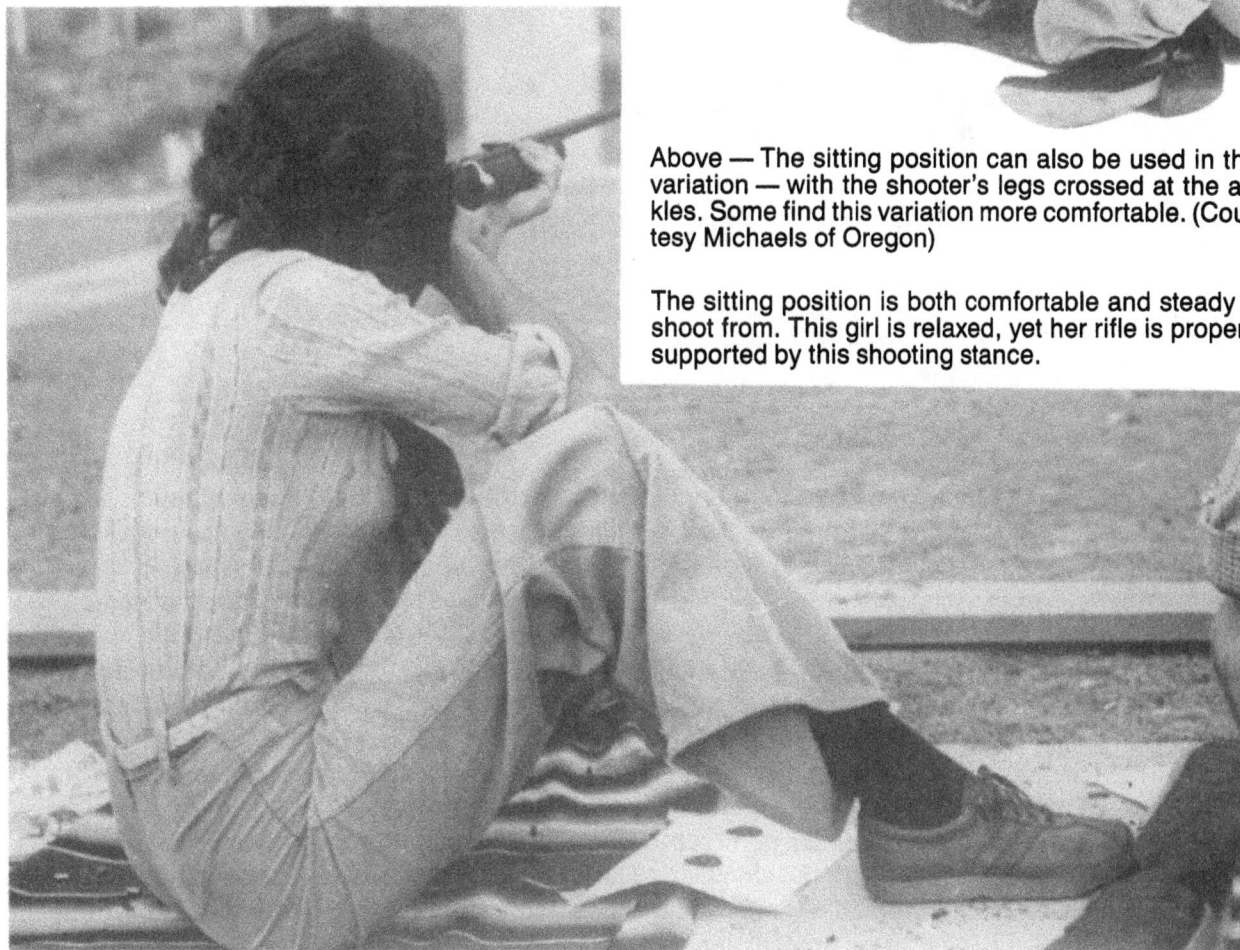

91

Left — This young shooter checks his position before closing his rifle bolt and preparing to fire. Note spread legs.

Below — The kneeling position using a rifle sling for extra support. Note shooter's elbow is placed just ahead of knee.

sit on the turned-in instep of your right foot, competition rules also allow you to hold the right foot vertically with the toe on the ground and sit on the heel.

The *offhand,* or *standing position,* is much less steady than any of the lower stances, but it's the fastest of all shooting positions to get into. Simply face away from the target at right angles (approximately 3 o'clock for right-handed shooters), turn the head to face the target and raise the rifle to the shooting position. The left arm (or the arm supporting the rifle) should be held with the elbow directly under the rifle and the forend either resting in the open palm of the hand or on spread finger-

tips. Position the hand at the point along the stock that feels most comfortable to you. In most formal competition, the left (or supporting) arm may not come in contact with the body. But during informal shooting, many riflemen find it more comfortable to tilt the left hip forward and rest the left elbow against it, with the upper arm held against the body.

The right (or shooting) hand should be firmly grasping the pistol grip, and the right elbow should be held straight out or even slightly above horizontal. The muscles of the shooting arm should be moderately tensed to pull the rifle butt into the shoulder.

can be easily adjusted to varying lengths, and the strap can't. There are basically two different types of true slings on the market: the two-piece military style with its brass claws and swivel; and the one-piece "civilian" model. The military version is better suited to the intricate arm strapping used by serious target shooters, but the one-piece sling is faster and more convenient to use.

Once you learn how to use a rifle well enough to enter target competition, you'll eventually have to use a carefully adjusted two-piece military-style sling—this gives very firm support when fastened in the target shooting position, but it's unnecessarily complicated for the beginner. The "hasty sling," on the other hand, is quick and easy to get into, even though it doesn't give the same, steady support.

Any kind of adjustable rifle sling can be used in the "hasty sling" position. First the sling is lengthened far enough to allow the elbow and forearm of the non-shooting (support) arm to be inserted between the sling and the rifle stock with the arm at approximately right angles to the stock. Once this adjustment is made, the shooter twists the sling a half-turn to the left (for a right-handed shooting stance), puts his left arm into the sling (between the sling and rifle) and, as he grips the forend with that hand, moves the center section of the sling high on the left upper arm. Then he wraps the front end of the sling around his left hand so that it lies over the back of his hand and covers part of his wrist when the rifle is raised to shooting position. When the sling is properly adjusted, it will be snug and tight when the rifle is held to the shoulder and the left (supporting) arm is positioned directly underneath the rifle.

As you gain confidence, try practicing these various positions, and shoot with and without the help of a sling. If your rifle doesn't have a sling, don't worry about it. You can get along just fine without this accessory for plinking and most kinds of hunting, although you may want to install one later. Most shooting sportsmen these days use their slings merely as carrying straps, and in fact, more

One thing to remember when shooting from the standing position is to avoid slouching. Stand erect with the spine as straight as possible. A sloppy shooting position gives little support, and the offhand stance isn't the steadiest one available anyhow. It needs all the help it can get.

Slings

Any of the above positions can be made steadier by the use of a rifle sling. There are several different kinds of gun "slings" you can buy, but all of them aren't true slings. Some are merely carrying straps and can't be used as a shooting support.

The main difference between a plain carrying strap and the more versatile sling is that the sling

straps are now sold than adjustable rifle slings. Just be aware that a properly used sling can be a big help in holding a rifle steady. That knowledge can come in handy if you later decide to take up hunting as a sport. Shooting at big game over long distances can be difficult under typical field conditions, and the use of a gun sling can sometimes make the difference between a miss and a clean kill.

As you're starting out as a beginning rifleman, just remember the basics. Safety first, each and every time you pick up a firearm. Then learn to get—and keep—the proper sight picture. Finally, learn to control your breathing while carefully squeezing the trigger. Ear protection isn't usually necessary when shooting a rimfire rifle, but it wouldn't hurt to use shooting ear muffs. Also, get into the habit of wearing shooting glasses *whenever* you fire a gun. Again, this is more critical if and when you move up to the centerfire calibers, but it's a good habit to get into. Wearing protective lenses can prevent gas or powder particles from blowing back into the eyes and will also deflect empty cartridge cases ejected from the rifle used

Opposite page and above — Examples of the offhand, kneeling, cross-legged sitting (above) and prone shooting position, using a rifle sling for additional support.

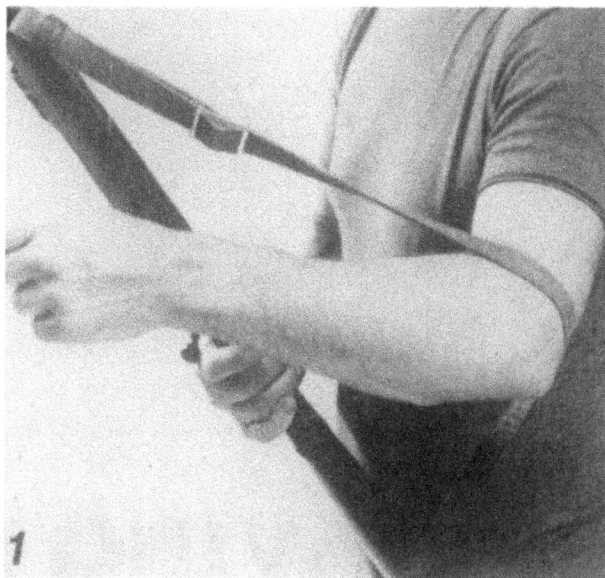

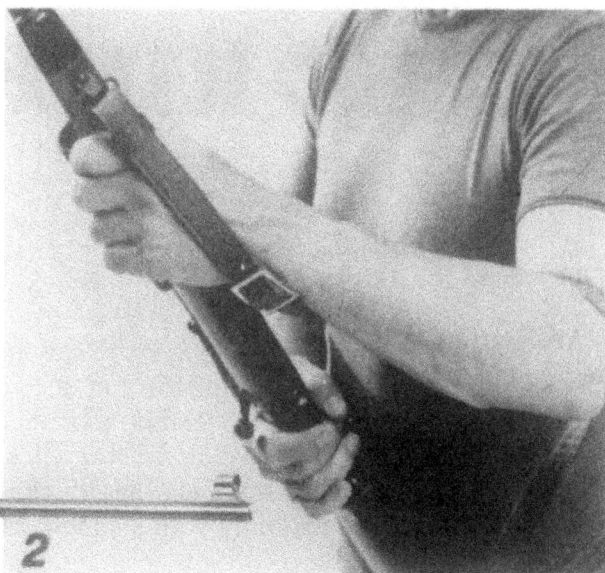

Top right, above right and above — Here's how to use the "hasty sling" to give you a steadier shooting position. (1) Adjust sling to give you enough room to allow the forearm to be inserted between sling and stock with the arm at right angles to the stock. (2) Twist the sling a half-turn to the left, insert arm, wrap upper part of the sling around back of hand. (3) Raise rifle to shooting position. (Courtesy Michaels of Oregon)

by the shooter standing next to you.

Practice shooting from all four of the basic positions described above. If you have trouble keeping your bullets on target from the standing position, move back into the sitting or prone stance to get your confidence back. If this doesn't improve your scores, set up an improvised shooting bench once again with your blanket roll or sandbags. And remember, shooting should be fun. Never continue a practice session beyond the point it *quits* being fun. It's much better to quit while you're still enjoying it—that'll help you look forward to the *next* session with anticipation.

And remember the safety rules. Treat every firearm as if it were loaded. And watch that muzzle—don't point it at anything you *don't* want to shoot.

chapter 7

HAVING FUN WITH THE RIMFIRE RIFLE

HAVING FUN is what the shooting sports are all about. There is no better way to have fun at low cost than with the 22 rifle. Few shooters, no matter how experienced they might be, can resist an afternoon of burning up ammo while chasing tin cans and other targets of opportunity with a simple 22 rimfire rifle.

Some people call the rimfire "just a 22," and neglect shooting glasses and ear protectors. While the 22 certainly doesn't compare with the thunder generated by a big-bore rifle, don't forget ear and eye protection. Prolonged exposure to even the mild crack of the 22 will not improve one's hearing. Perhaps that's why so many old-time shooters say "huh?" a lot.

Once you've become good enough with your 22 rifle to hit paper bullseyes and tin cans with consistency, you can start looking for "fun" targets to shoot at. Putting holes in printed targets *always* presents a challenge (just keep moving to smaller targets as your skill increases), but somehow it's more satisfying to hit a mark that moves or shatters when hit.

By the same token, keeping score by marking the hits within a formal target's concentric rings works fine in organized competition but isn't as much fun as seeing who can obliterate the most Necco candy wafers in a row without a miss.

Rimfire plinking (which means shooting at improvised "fun" targets) is something the entire family can enjoy, and it makes a pleasant diversion while digesting the food from a Sunday afternoon picnic. It's even a sport you can engage in all by yourself. What's more, you don't need much in the way of expensive equipment — anyone with a 22 rifle, a box or two of ammunition and a little imagination can enjoy himself for not much more than the cost of the cartridges.

Shooting at empty beverage cans is undoubtedly the most popular form of plinking — every rifleman has riddled his share of soda pop containers, and no rimfire session would be complete without having a supply of these aluminum cans along. The big advantage of these particular targets is that they're always in ready supply and cost nothing. What's more, both highly skilled and beginning shooters can have fun hitting them. At 25 yards or so, the cans present a target large enough for the most inexperienced rifleman to hole occasionally, and a really good shot can show off his skill by hitting the lower edge of the cans to make them bounce high in the air. Placing cans a few feet up a

New shooters need encouragement; and, nothing "encourages" like a pop can full of holes. Why pop cans? They're large enough to hit with the first or second round and they jump quite nicely. That combination of "easy to hit" and "impact animation" can really bring a smile to a new shooter's face.

Animated targets like water ballons, pop cans and candy, all explode, jump or shatter when hit—they're real confidence builders, and serve as a basis for competition when two or more shooters get together. Just remember to clean up the debris when the game is over!

hillside, hitting them once, then shooting at them as they roll down also gives you a challenging target, and there are other tricks you can use to add interest to the game. (Always make sure you pick up your litter afterwards, though. Leaving a messy collection of shredded aluminum containers lying around gives shooting — and shooters — a bad name and is liable to cause the landowner to post his property "off limits" for future use.)

While tin cans will probably remain the mainstay of casual sport shooting, there are a number of other inexpensive targets you can use to have even more varied fun. If you'll use your imagination and plan ahead, you can stock up for your next plinking session with a few low-priced purchases at the neighborhood supermarket. When you stop to think about it, the shelves of a grocery store are simply loaded with targets that move, burst or shatter when hit.

Frangible targets (glass objects excluded) that explode or simply disappear in a puff of dust when hit add immeasurable interest to the sport. Even a lackadaisical teenager, who is at that awkward age when it's not considered "cool" to show much enthusiasm for *anything* soon warms up to shooting at burstable marks. The fact is that shooters of *all* ages enjoy having fun, and while plinking at unorthodox targets that show some kind of action when hit is considered more frivolous recreation than building scores on paper, it *is* a lot of fun. Plain, simple fun — and that's really what recreational shooting is all about.

Before you graduate to the "fun" targets we're about to discuss, I want to caution you *not* to side-

Above — Three popular varieties of 22 rimfire ammo are seen here. On the left is a 100-round pack of high-velocity 22 Long Rifles; in the center is a 50-round box of CCI Long Rifle Stingers (Hyper-Velocity Hollowpoints); on the far right is a 50-round box of high velocity 22 Shorts. Those Shorts, by the way, will work fine in the Colt Trooper revolver seen on the right. They will *not* function in the S&W Model 41 at the left, or in the Ruger 10/22 at the top. Don't spoil your first outing! Read the barrel legend—that's where you'll find out what your gun's chambered for!

Below — The ever-present beverage can — perhaps the most available fun target you can find. Don't leave these laying around after a shooting session — be a sportsman, clean up!

step your earlier practice in shooting at paper bullseyes. Some beginners want to start knocking tin cans over right from the first rather than "bother" with the duller job of putting holes in paper. This is a bad mistake, as you can't really tell where your bullets are hitting with relation to your sight picture when you're shooting at a tin can sitting on a dirt hillside. Missed shots are registered only as vague puffs of dust, and there's no accurate way of judging how much sight adjustment is needed to get the rifle properly sighted-in.

That's a point to remember — if a rifle *isn't* properly sighted-in you'll do a lot more missing than hitting. So before you move on to "fun" targets, make sure your sights are properly adjusted by shooting at a printed bullseye at the range you'll be plinking at — usually around 25 yards or so.

Having the right ammunition on hand is also important. Regular 22 Long Rifle high-speed or standard velocity (target) ammo works well on most plinking targets, but hollowpoint bullets are also available. These work particularly well on soft-skinned fruits like overripe grapefruit, tangerines and the like. (Get on friendly terms with your local grocer's produce man, and you can pick up over-aged fruits free — these burst nicely when struck by a hollowpoint bullet.)

While 22 Long Rifle ammo is the most popular of the rimfire 22s, don't overlook the little 22 Short. This round is less noisy than its bigger brother, and if you're paying full retail price, it can be less expensive as well. (But be aware that carton lots of 22 Long Rifles cartridges are often sold at discount during off-season store promotions; when this happens, you can buy the Long Rifle fodder for less than you'd pay for 22 Short ammunition.) The 22 Short is available in both high-speed and standard loadings, with either solid or hollowpoint bullets. What's more, it can be fired in any rifle chambered for the 22 Long Rifle cartridge. The little Short may not feed through the magazine and action of some autoloading 22 rifles, and if your rifle won't handle these rounds without jamming, you probably won't want to bother chambering them one at a time by hand. However, most 22s will accept the Short cartridges with no problem, and if this is the case, you might want to consider using these lower-powered rounds for plinking. The fact that they're quieter than the longer rimfire rounds makes them particularly attractive to shooters who plink near populated areas — the soft "plop" of a rimfire Short is

Shooting at water-filled balloons (arrow) thrown into the air is fun, but don't do this without a safe, high backstop (like the hillside shown here) unless you're using 22 shot cartridges.

less likely to draw complaints than the louder "crack" of a 22 Long Rifle cartridge being fired. And don't worry about accuracy — 22 Shorts will shoot where they're pointed and are favored over 22 Long rifle ammunition in some pistol competition because of their fine accuracy and lack of recoil. (Recoil isn't a problem with *any* rimfire 22, but even the slight nudge from a Long Rifle round can momentarily upset the aim of a rapid-fire pistol marksman.)

You can also buy 22 Long ammunition, but this is a rimfire oddity that doesn't get a lot of use. It may function better than a 22 Short through some rifle actions designed for the Long Rifle round, but other than that it has no real advantage over the more plentiful Short.

If you want to minimize the chance of ricochets, get your hands on some of the 22 Short "gallery special" cartridges sold by the major ammo makers. These little rounds feature frangible bullets that disintegrate when striking a hard surface. They're a great safety investment, and they're fun to shoot.

The CB cap is one of the most useful 22 rimfire cartridges, and also a great plinking round. The CB comes in two sizes, which look exactly like 22 Long or 22 Short rimfire cartridges, but have virtually no powder charges. The CB depends primarily on the priming charge to propel the bullet. When used in a rifle, the great advantage of the CB is low noise level — about the same as a typical air rifle. Yet, the CB delivers fine plinking accuracy up to 20 yards or so — and moderate power. Do not underestimate what the little round can do. In this writer's part of the country, CBs are used to great effect in exterminating nuisance pigeons and other small varmints. While the CB is quiet and cheap, remember that it can inflict serious injury, and the bullet will travel a lot farther than the tiny "pop" would indicate.

Now let's take a look at some of the "fun" targets you can devise, along with some shooting "games" you can play with a partner or the whole family.

I have already mentioned overripe fruit. Fresh fruit works OK too, but it is expensive and wasteful to use good food as targets. An overripe orange or grapefruit explodes spectacularly when struck squarely by a hollowpoint 22 bullet. The fruit may also be rolled across the ground to provide a moving target to test the skill of even the most experienced rifleman.

While plinking is great fun, don't forget about that warning label that points out that 22 bullets have a range of at least a mile. Be doubly *sure* of your backstop.

Another trick that's just as spectacular — and easier — is to pop water-filled balloons in the air with a 22 shot cartridge. These specialized cartridges are filled with very tiny shot and turn your rimfire rifle into a miniature shotgun. The pellets don't travel far, so you don't have to be so concerned about having a huge backstop. Because the shot pattern spreads rapidly once it leaves the muzzle, close-thrown targets are much easier to hit. But you will have to throw them close to the gun, as shot fired through a rifled barrel disperses widely within a very few yards. There are special smooth-bore 22s available, designed just for shooting shot cartridges, but they're not really suitable for accurate work with solid bullets.

If you want to get fancy, you can add a few drops of food coloring to the water in each balloon. This will give you multicolored showers when the balloons burst.

You can fill other small containers with water to get other effects. Hitting a plastic container that once held 35mm camera film produces gratifying results when it's filled to the top with water and the lid sealed tightly. Hydrostatic pressure will pop the lid in the air almost every time and will often burst the container. The same principle can be used with other small containers that can be tightly sealed — large plastic bottles filled with water explode nicely when struck with a high-velocity bullet from a centerfire rifle or handgun, but the little 22 lacks the "oomph" to produce this effect with anything but small-volume containers.

Filling paper cups with water also works well. The water will geyser skyward with every hit, and you can usually repeat the spectacle three or four times before the cup empties — if you hurry.

Necco wafers and other similar types of candy make inexpensive targets that disappear satisfyingly when hit. What's more, the debris you'll leave behind will be biodegradable so you don't worry about picking up what little litter will be

Left — On the left is a capsule-tipped round of CCI 22 WMR Shot (also available in 22 LR); in the center is a round of standard 22 LR and next to it a rosette-crimped Winchester 22 Shot cartridge on the right. The crimped shot cartridges will function through any repeater (except an autoloader) chambered for 22 LR cartridges. In a rifled bore their effective range is limited; but, they can be fun when used on water balloons at close quarters!

Right — Paper cups filled with water make fun targets to shoot at. Water geysers skyward with every hit until the cups are empty.

Candy wafers suspended from a string make interesting and challenging targets. They shatter when hit, but if there's a breeze blowing, hitting them can take some doing!

left. (You should pick up your empty brass cartridge cases, though — particularly if you've done a lot of shooting from the same spot.) Necco wafers suspended from a string move in the slightest breeze and make tricky marks to shoot at. String several in a row, and you can have a game of "miss

Clay pigeons used in trap and Skeet shooting are easy-to-hit targets that become progressively smaller marks as they break apart.

and out" with a partner. The guy (or gal) who breaks the most targets wins.

If you want a larger target that will still break up when hit by a rimfire bullet, try shooting at the clay targets thrown at Skeet and trap ranges. Don't try hitting them in the air with a rifle, though — just place them against a mound of dirt. As these targets disintegrate, they present you with increasingly smaller marks to shoot at.

Another game that two can play is to draw parallel sets of intersecting lines to make a "tic-tac-toe" board. Take turns shooting at the vacant squares, and the first to hit three squares in a straight line is the winner. Make the squares small enough to be challenging targets — that way you can lose the game by missing, as well as through

poor strategy.

The same penny balloons you filled with water to provide aerial targets can also make difficult marks to hit when filled with air and anchored to the ground with a yard-long length of light string or thread. If there's any kind of a breeze blowing, the balloons will bob and weave unpredictably, and sometimes you'll swear that the wind created by the bullet's passage actually moved the target out of the way at the last second. Balloons are inexpensive, fun-to-shoot-at targets, but remember that they leave fairly durable (and highly colorful) debris. Be sure to clean up the range before you leave.

Other fun shooting stunts include blowing out a candle flame by shooting through it — without disturbing the candle. This is hard to do outdoors unless there's absolutely no wind, but you can shelter the candles by putting them inside an open cardboard box placed on its side. (Again, this stunt is easier than it looks. The bullet passing anywhere close to the flame will snuff it out — just don't shoot low and knock over the candle.)

Still another way of showing off your skill with a 22 rifle is to draw pictures on a piece of cardboard or on butcher paper. Do this by first drawing a penciled outline (keep your pictures simple) and then shooting along the outline to form holes or "dots." With a little practice, you should soon be able to "draw" the pictures with the row of bullet holes only, without the help of the penciled outline.

You can have a lot of fun plinking at various homemade targets that move in the breeze or shatter when hit. Be sure to avoid shooting at anything made of glass, however. Ricochets can result, and leaving sharp shards from broken bottles lying around the countryside is both irresponsible and dangerous. And remember your safety rules — keep that muzzle pointed in a safe direction at all times, and shoot only when you've got a safe backstop to place your targets against.

A word of caution: fence posts, road signs, tombstones, etc., are *not* proper plinking targets. Aside from the danger from ricochets, these "targets" are either private or public property and defacing them is an act of vandalism. Such practices only serve to hurt the sport of shooting.

Shooting is a clean, healthy, and safe recreation if you follow the rules. Shoot only where you have permission, and pick up your litter before you leave. As far as "fun" targets and games are concerned, their variety is limited only by your imagination.

chapter 8

GRADUATING
TO THE CENTERFIRES

WHILE YOU CAN have all the plinking fun you might ever want with a 22 rimfire, if you ever decide to do any long-rage shooting or take up hunting as a sport you're going to need a more powerful rifle. You *can* hunt rabbits, squirrels and similar small game successfully with a 22, but for anything larger the little rimfire is seriously outclassed. By the same token you can hit targets with great accuracy out to 75 yards or so with a good 22 rifle, and you can compete in the short-range rimfire matches. But if you want to shoot at targets much beyond 100 yards or engage in big bore competition, you'll simply have to move up to a centerfire rifle.

Let me once again note that some beginning shooters—particularly those who want to start hunting deer or other game right away—may be tempted to skip the rimfire 22 and start their training with, say, a 30-06 centerfire. It's *possible* to learn to shoot with a centerfire rifle, but believe me, it's the hard way to go about it. Centerfires recoil with varying force according to the caliber and the weight of the firearm—but *any* centerfire rifle cartridge generates more kick than any 22 rimfire. And learning to shoot with something like

This hunter is standing behind graphic proof that the centerfire rifle is good for a great deal more than plinking and small game hunting. Yet, the skills that enabled him to bring down this buck with his Winchester Model 88 lever-action rifle were probably learned with a rimfire, then transferred to the centerfire.

Rifles used in hunting deer and other large game should be chambered for 6mm or larger cartridges. Centerfire rifles are easy to get used to once you've mastered the basics with a rimfire 22.

Above — This 22 centerfire is just right for long-range varmint hunting, but too small for deer and similar game. Recoil is very mild with most 22 centerfires.

a 30-06, which is one of our most popular hunting cartridges, will take a much longer time because this particular round comes back with enough force to be really distracting. It's hard to concentrate on learning the basics—trigger squeeze, sight alignment, etc.—while part of your attention is taken by recoil. In fact, anticipating recoil is likely to cause you some involuntary muscle reaction—called flinching—as you finish pressing the trigger. Some beginners even close their eyes when the rifle fires, and this combination is guaranteed to result in wild shooting that won't come anywhere near the target.

What's more, the noise created by firing a big bore rifle cartridge can be equally bothersome. Once you've mastered the basics of rifle shooting with a 22, the twin distractions of centerfire blast and recoil are less of a problem. You already *know* how to shoot. Now you just need to adapt those skills to centerfire marksmanship.

Actually, there's not all that much more to learn once you know how to shoot a 22 rifle well. The same safety rules apply when you're using a centerfire rifle. You have to keep in mind that a high-velocity centerfire bullet will travel farther than a rimfire 22 unless it hits something in its path and that it will usually penetrate deeper into the material it strikes. But aside from that, you can treat your centerfire rifle much the same as you treated your 22—watch where that muzzle is pointing, and

This shooter is doing just about everything right — including ear and eye protection. He is also illustrating an important point that many beginners forget: DO NOT get your eye as close to the scope as you might with a nil-recoil 22 rimfire. (The practice of doing so is called "crawling" the stock.) Many shooters carry a small scar above one eye by forgetting about recoil and getting smacked in the forehead by the sharp-edged scope tube.

always handle any firearm as if it were loaded.

Ideally, the way to move up from shooting a rimfire rifle to proficiency with a big-game centerfire would be to make the move gradually. That would mean starting off with a relatively mild centerfire load like the 222 or 223 Remington. Both of these cartridges shoot a 22-caliber bullet of the same general size (although a bit heavier) as that used in Long Rifle rimfire cartridges. While the bullets zip along much faster when propelled by the larger powder charges contained in the centerfire loads, the rifles that chamber them recoil mildly. The noise level is definitely greater than that of the rimfire, but not to the extent that the beginner will find it uncomfortable or distracting.

The cartridges generate much more noise than their rimfire relatives, but again the muzzle blast from a 222 Remington isn't all that terrifying. Besides, one of the first things we're going to fix you up with *before* you start shooting centerfire cartridges is a good set of ear protectors. Ear protection is necessary for two reasons when you're shooting a gun with a loud report. In the first place, the sudden, sharp noise produced by the gun firing is a distraction that could cause you to develop a flinch—and flinching is one sure way to destroy any chance of accuracy. Perhaps more important is that shooting high-intensity centerfire cartridges *without* ear protection can cause serious—and irreparable—damage to your hearing. Continued exposure to very loud, sharp noises like those produced by gunfire can and will cause hearing loss. Unless you enjoy the thought of becoming

hard of hearing in a very few years (most old-time gun writers are at least two-thirds deaf), you'd better plan on wearing some effective ear protection whenever you shoot. You'll be a better marksman, and you won't suffer loss of hearing.

There are two basic types of ear protectors available. These include plugs worn in the ear channel (and there are several different kinds of shooting ear plugs available) and the big muff-type cups that are worn *over* the ears. People who do a lot of shooting generally find the muff-type protectors handier to use. They're more comfortable than ear plugs, and they're easier to put on and take off. What's more, they offer better protection—they're usually better insulated against sound, and they cover not only the ear channel itself but the surrounding bone (which also transmits noise to the inner ear). You can buy a good set of "muffs" for less than $25, and that's a pretty cheap investment when you consider it's your hearing that's at stake.

Many shooters buy both kinds of ear protectors and use the smaller plugs while hunting. Ear plugs have one advantage in that they'll fit easily in a shirt pocket and are more highly portable. On the other hand, they're less effective and less comfortable to wear (unless you pay the money to have custom-fit plugs made).

While we're talking about protection, it's a good idea to get a pair of shooting glasses (or any kind of eyeglasses with hardened, tempered lenses) if you're going to do much shooting with *any* centerfire firearm. Cartridge head failures and leaking primer pockets in commercial cartridges are rare,

Left — Ear plugs are smaller and handier than muff-type ear protectors, but they're not as efficient or as comfortable to wear. These Hoppe's ear plugs are soft and pliable and are designed to fit any ear size.

Right — Good ear protection is important when shooting centerfire rifles or handguns. Muff-type protectors like these Hoppe's Sound Mufflers work very well.

Below — A 22-caliber centerfire like this 22-250 is an easy step up from 22 rimfire shooting, as recoil is relatively mild. Note shooting glasses and ear protectors used to guard against possible (but not probable) gas blow-backs and hearing loss.

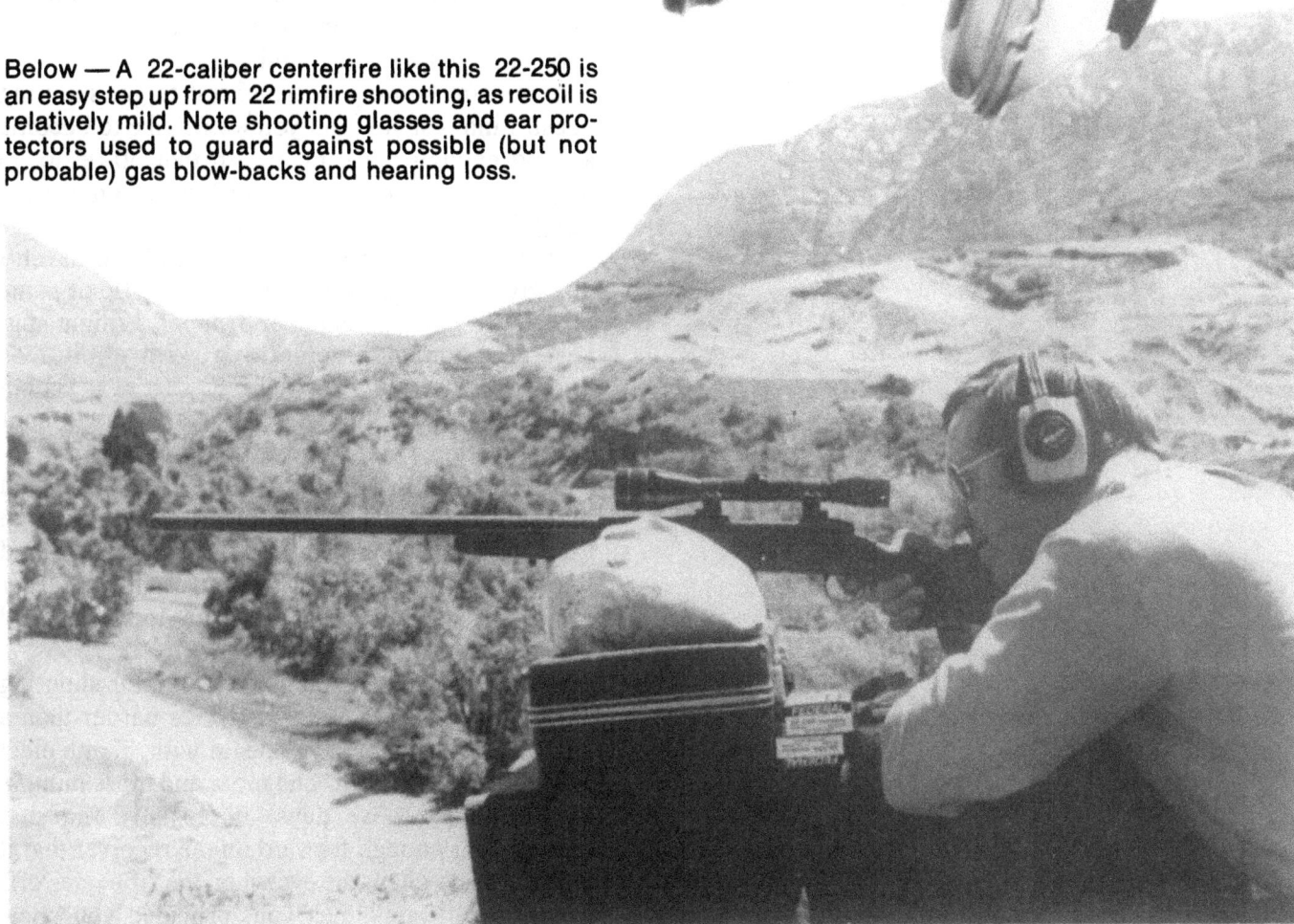

While iron sights can be used on a big game hunting rifle, most sportsmen these days prefer a scope for serious shooting.

but such accidents where powder and gases are blown back into your face *do* happen. If it's your turn to be unlucky, a pair of protective lenses could save your eyes from some painful damage.

With your eyes and ears protected, you're ready to start shooting your new centerfire rifle. As I've said, the easiest centerfire to learn to shoot would be one of the 22 "varmint" calibers—22 Hornet, 222 or 223 Remington. These cartridges don't generate much recoil, and with your ears muffled against sound, you should be able to handle one of these mini-centerfires about as well as a rimfire 22. Put a good scope sight on the rifle, and you should be able to hit targets out to 200 or 250 yards. Use a solid rest at first (remember your first lesson with the rimfires), and make sure the scope and rifle are properly zeroed in. Then you can try your hand at shooting from any comfortable position.

If you decide to try your hand at hunting varmints with a 22 centerfire (that's primarily what these calibers were designed for), practice shooting from the sitting and prone positions. The game here is to try to hit a small animal like a marmot or ground squirrel at long range, and this takes a good rifle and a steady hand.

If varmint hunting or benchrest target shooting doesn't interest you, but you *would* like to become a good enough marksman with that 30-06 to go hunting deer next season, you'll want to make the jump from your 22 rimfire right up to the larger centerfires. This can be done, but if you haven't already rushed out and purchased a 30-06 or some other even more potent deer dropper, I might suggest you consider something with a bit less "oomph" for your first hunting rifle. If you plan on hunting nothing larger than deer, I might suggest something more on the order of a 243 Winchester, 6mm Remington, 257 Roberts or 7mm Mauser. These rounds are plenty potent enough for deer if you can hit what you're shooting at, and because they generate less recoil than the larger centerfires, they're easier for an inexperienced shooter to master.

However, the principle is the same for shooting any centerfire cartridge that recoils harder than a 222 or 223 Remington. To begin with, if you elect to mount a scope sight (and more and more hunters are doing so these days) make sure that it's mounted far enough forward on the receiver that it won't whack you in the eye when the rifle goes off. Recoil itself won't hurt you, provided you know

Like rimfire rifles, centerfire longarms come in many different action types from numerous manufacturers. From top to bottom: Savage Model 99 lever-action, Remington Model 7600 pump-action, Winchester Model 94 Angle Eject lever-action, Browning BAR autoloader, Browning Stainless Stalker A-Bolt bolt-action.

When shooting from a bench, sandbags and a rest are a must. Without them your sighting-in will be less than rewarding. Cost? You can pick up an adjustable rest and sandbags from your local dealer for about $45 to $65. However, you can make sandbags from the cut-off legs of a pair of jeans — sew one end shut, fill with sand, and sew the other end tightly closed.

Above left, the shooter is "crawling" the stock and has his eye too close to the ocular lens of the scope. Above right, the shooter's head (and eye) is in the proper position. If you have to get as close to the scope as seen in the photo on the left, the eye relief of the scope *must* be adjusted before shooting. The problem? If your eye is too close to the scope, the recoil of a centerfire rifle will bury the ocular ring in your nose or eyebrow, the result being a bloody mess often requiring the services of a physician.

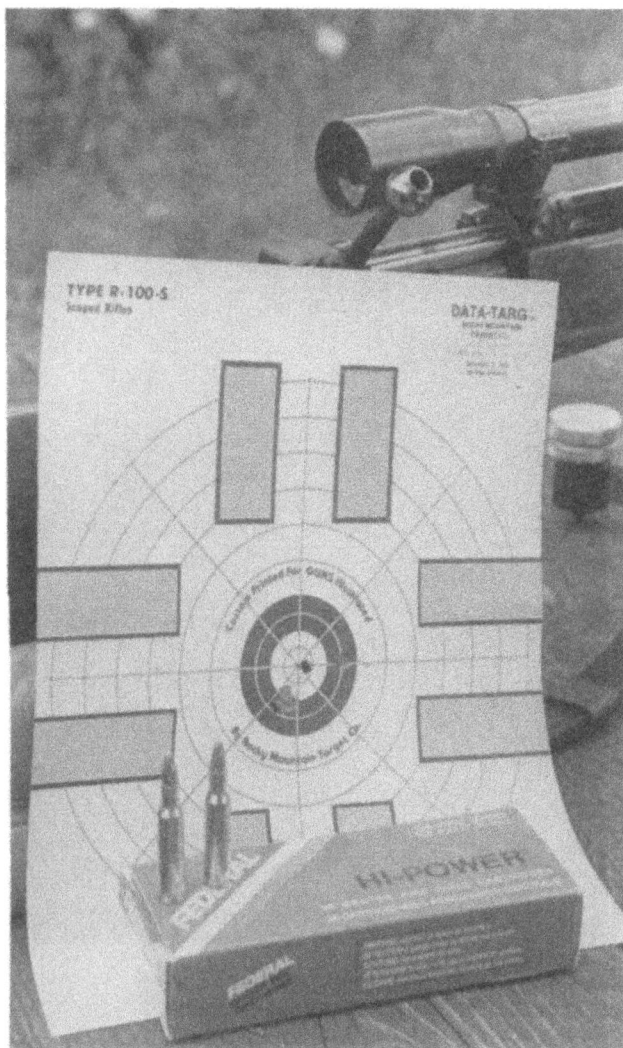

With the proper equipment and practice, centerfire groups like this (shot at 100 yards) are quite normal. When you sight-in your rifle, be sure it's adjusted for the range you intend to use it.

how to handle it. But setting yourself up for a black eye or worse isn't the best foot to start off on.

Again, follow the same safety rules as you would with a rimfire 22 and put on those ear protectors and shooting glasses. Remember to check the bore for possible obstructions, and you're ready to start shooting.

Once again I'm going to suggest that you start off with your rifle partially supported by some kind of stationary rest or sandbags piled on a benchrest or table. But *before* you shoot from this position, I want you to take a shot or two from the offhand or standing position. This is simply to give you an idea of what recoil feels like, and let you know that it's not a big deal.

There are some things to watch out for, though. Many rimfire shooters get sloppy and fail to hold the rifle tight into the shoulder. While that won't help your scores, it at least won't do you any physical harm. However, if you hold a heavy-recoiling rifle in a loose, sloppy manner, it's liable to kick you in the chops. You need to take a firm grip on your rifle, pull the buttstock into your shoulder, and lean slightly toward the rifle when you're ready to fire. Don't hold your body tensely rigid—it's not necessary, and recoil will bother you less if you're able to "give" a little and move with the punch. Make sure you haven't "crawled the stock" until the scope's eyepiece almost touches your eyebrow. That's a common mistake beginners make, but I can promise that you'll only make it once if you ever *do* get forgetful and start nuzzling the eyepiece. A bloody forehead and a throbbing headache are the penalties you can expect for your forgetfulness.

Hold the rifle right, lean into it, and you'll be pleasantly surprised the first time you fire your new centerfire slugger. With your ears protected from the sound of the shot, the recoil should be anything but terrifying. (Just for fun, try a shot with the ear muffs removed and see how much difference it makes. The rifle will "kick" a lot harder when you can hear clearly.)

Now that you've got some idea of how hard your centerfire rifle actually kicks when you shoot from the standing position, go back to the benchrest and sandbags and make sure the sights are properly aligned by firing at a target 25 or 30 yards away. If you're approximately on target (somewhere near the bullseye) at that range, set up a new target at 100 yards and try again. Make whatever sight adjustments you need to make the bullet strike correspond with your sight picture. (See Chapter 5 for instructions on adjusting your sights.) Before you shoot from a rest, let me again caution you to keep your eye away from the scope's eyepiece if you have a glass sight mounted on your rifle. Many riflemen tend to "crawl the stock" when shooting from a rest, so watch it. You'll also notice that recoil seems greater when you're seated at a bench. This is because you're shooting from a braced position, and your body has less "give" available to move with the punch.

Then try shooting from the other positions you learned when practicing with a rimfire rifle. By this time you'll be taking recoil in your stride. How recoil affects you is largely a matter of mental conditioning, and once you've developed the right attitude, you won't even be aware of a rifle's kick.

Most experienced riflemen prefer the sitting position over the standing (offhand) position, if time permits — especially for long range shots. This shooter has very wisely opted for the sitting position using a nearby tree for a back rest. Excellent support.

A rubber recoil pad installed on the end of the butt-stock can make things even easier, if you're using a rifle that digests magnum cartridges like the 300 Winchester or 7mm Remington Magnums.

That's all there is to it. The centerfire rifle's recoil is harder and it makes more noise than the 22 rimfire you started out with, and you have to do certain things to compensate for these differences. Ear protection takes care of the noise problem, while holding the rifle properly and developing the right mental attitude minimizes the effect of recoil. From this point on, it's simply a matter of getting enough practice to become proficient with your centerfire rifle. There's one other point you should be aware of before you buy a centerfire rifle and ammunition—and that's the fact that this equipment will be more expensive than what you probably paid for your rimfire shooting gear. Ammunition costs will be considerably higher when you're buying centerfire cartridges, but there's a way to economize here without sacrificing either safety or quality. It's called reloading (or hand-loading), and you'll learn more about it in a later chapter.

chapter 9

RIFLE CARE AND CLEANING

A GOOD RIFLE represents a substantial cash investment. So does a new car. The big dollar difference is that no matter how well you care for a new car, its value drops like October leaves from the day you drive it home. A good rifle will hold its value—or increase in value as the years pass. To protect your investment, learn how to care for your rifle. Though a little care goes a long way, a little neglect can go a long way in the wrong direction.

Beginners, with the best of intentions, often try to overdo the cleaning task. Not long ago, I talked with an engineering executive of a major American gun manufacturer. He pointed out that one of his company's biggest problems involved new shooters who did a fine job of taking their guns apart, down to the last screw—but could not get them together again. The guns ended up back at the factory for reassembly. Unlike the king's men's efforts with Humpty-Dumpty, the factory can put things together again. The point to be made here is that routine rifle care does not ordinarily require total disassembly. If the gun has been dropped in mud or subjected to a desert sandstorm, the beginner should find a qualified gunsmith to do the teardown and cleaning. Don't be intimidated—routine cleaning, after shooting, can be accomplished by most anyone coordinated enough to tie shoes.

The shooter's home area affects the sort of care necessary to preserve guns. Those who live in the dry New Mexico or Arizona desert have slightly different problems than shooters who call Biloxi, Buffalo or Indianapolis home. During the summer months, ambient humidity in the Midwest, the South and the East is much higher than in the West. Humidity equates to rust on external metal surfaces and bore corrosion and pitting on neglected guns.

Ammunition type—or more correctly primer type—relates to gun care. Decades ago, all ammunition used primer chemicals which promoted heavy bore corrosion. American ammunition has, in recent memory, been loaded with non-corrosive primers; that is, primer chemicals which do not, of themselves, promote rust. The problem we see today involves centerfire military surplus ammunition which is inexpensive, readily available and ideal for practice and plinking. Years ago—in this writer's youth—most shooters assumed that all military ammo was corrosive and reacted appropriately, by quickly and thoroughly cleaning their guns immediately after shooting. Today, we see a great deal of surplus ammo advertised as "non-

Above — You can't get a rifle clean unless you have the proper equipment. Fortunately, your local gun shop will have most of what you'll need. (Be sure to specify the caliber of the rifle you'll be cleaning.)

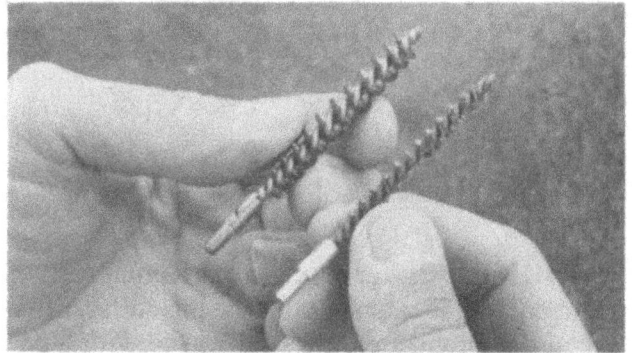

Above — Be sure you get a quality bore brush of proper caliber. The brush on the top is for 30-caliber firearms, the brush on the bottom right is for 22s. Below — The Belding & Mull one-piece cleaning rod seen here is considered to be a top-choice among experienced riflemen. Note the double-loop tip next to the bore brush. Again, make sure your rod is designed for the caliber of rifle you'll be cleaning.

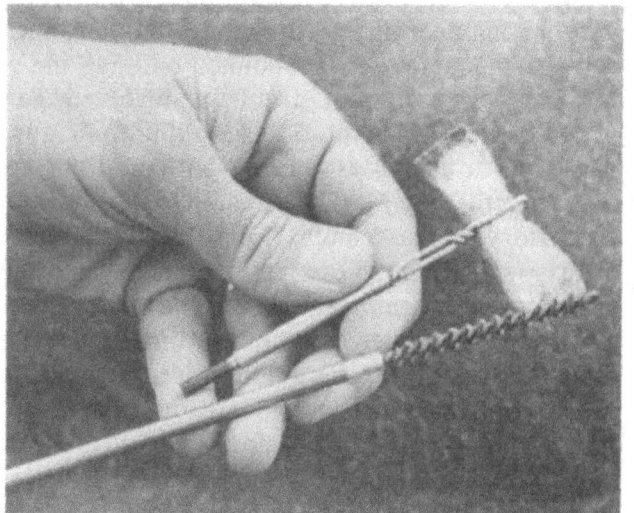

corrosive" or "mildly corrosive." Note that "corrosive"—"mildly" or otherwise—is corrosive. That means rust. It is a very good idea to assume that *all* non-American military surplus ammo, and *all* Amercian G.I. ammo made prior to 1960 is corrosive. The result of allowing a gun fired with corrosive ammo to stand without cleaning is a study in heavy, permanently damaging red rust.

Even non-corrosive ammunition can be a problem in humid areas. While the priming compounds and powder residue may not be corrosive of themselves, they do attract moisture from the air. Moisture and polished steel, as found in a rifle barrel, equal rust. This sort of damage is seldom as nasty as that caused by corrosive primers, but it can be permanent and can damage guns. I once made that mistake with a fine antique 22 and I hope the reader does not follow my bad example. By the time I "got around" to cleaning my rifle, it looked like it needed Roto-Rooter more than a cleaning rod. Most of the grunge and mung came out, but the bore was not quite as shiny as it was before my neglect. Even though the ammunition was non-corrosive, atmospheric moisture made friends with the bore to the point that I had a mess on my hands. The point to be made is that neglect is *not* benign. It is hardly necessary to detail strip your new rifle every time you shoot, but don't be bash-

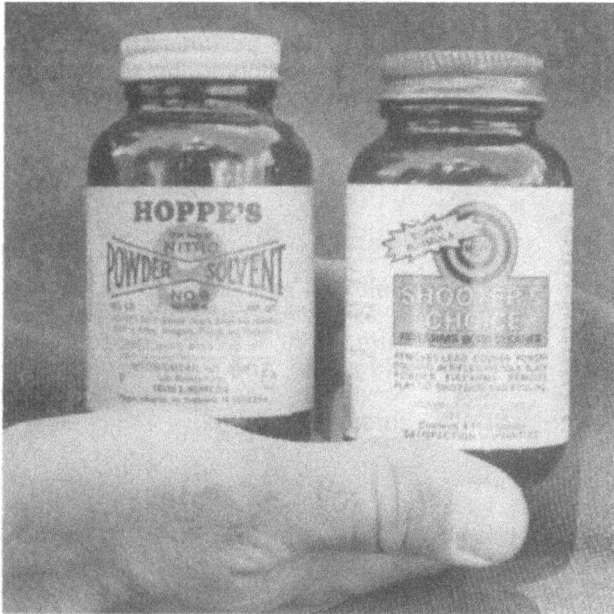

Above—On the left is a bottle of Hoppe's No. 9 bore solvent. On the right is a bottle of Shooters Choice bore cleaner. Most shops carry Hoppe's; but, you may have to look harder for the Shooters Choice. Which is best? Both do the job; however, according to those who've used it, Shooters Choice is the sort of bore cleaner by which all others should be measured.

Above and below — Powder and primer residue collect on the bolt face of any firearm. Birchwood Casey's Gun Scrubber and a stiff solvent-soaked brush will remove most of that fouling.

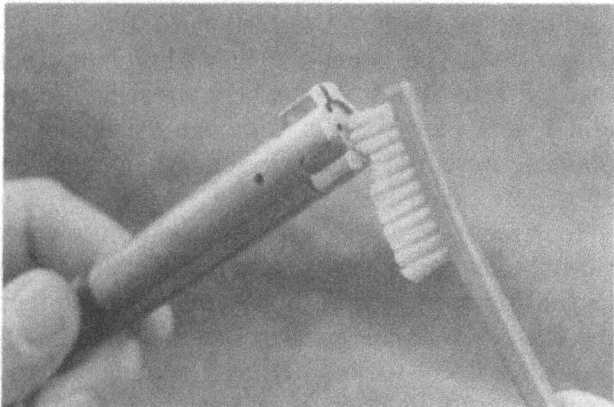

against rust with a light coat of oil.

Leading is still another problem. Bores can lead up, in that pieces of lead, bullet jacket metal and powder residue (fouling) may accumulate in the rifling grooves. That sort of fouling interferes with accuracy. In the action, fouling, dirt, and powder residue, combined with the rust that typically accompanies fouling, may interfere with reliability to the point that your gun ceases to function. This problem is quite common in 22 rimfire semi-automatic rifles, simply because they are fired far more often than centerfire rifles and, because of low ammunition costs, digest much more ammo.

Let's talk about the care your rifle really needs. Starting with the 22 rimfire, because (if you have followed this book's advice) the 22 should be the first rifle you buy and partly because the 22 rimfire is the easiest of any firearm to clean.

Until the 1930s, most rimfire ammunition used corrosive priming. Because the ratio of priming compound to powder was so high in these tiny cartridges, the shooter was forced to do a thorough cleaning job after firing only one shot. Otherwise, a few days after shooting, the rifle barrel resembled the inside of a chimney in springtime. Those days are over. The 22 ammunition produced today is non-corrosive. In addition, the plated bullets used in most 22 ammo leaves little in the way of lead deposits and the dry lubricant with which the bullets are coated does not attract dirt and actually lubricates the bore.

This means that to some extent you can neglect the bore (as long as you're using regular rimfire ammunition loaded with dry-lubricated bullets). You should run a lightly oiled patch through the bore every few months if you shoot the rifle regularly, but there's absolutely no need to clean it after every target and plinking session. As a matter of fact, the bore will hold up better if you *don't* clean it too often. The steel used in rimfire rifle barrels is softer than that used for the high-intensity centerfire tubes, and the protruding lands are more prone to wear. Shooting the relatively low-powered 22 bullets through the bore won't damage the lands, but a mishandled cleaning rod *can*.*

---

* Damage could come as a result of a bent rod and/or cleaning tips being forced down the bore thus scratching or marring the bore's rifling. In addition check you cleaning rod for any burrs (rough areas) along its length—this too will do a great deal of damage. If on examination you find rough areas, smooth out the problem spots by using 4-ought steel wool.

Left — A solvent soaked brush may also be used to clean out the interior of the action. After you're through scrubbing, be sure to wipe out the action with a clean rag.

Below — Your first step in cleaning a bolt action centerfire is to make sure the rifle is fully unloaded (the magazine and chamber *empty*). Next, run a solvent soaked patch through the bore.

I always try to at least give my rifles a bit of bore rust protection after shooting. I happen to be lazy, but I know that this task is good, cheap insurance. A lick and a promise will often suffice, but total neglect is a bad idea. After a typical shooting session, a bit of dirt is obvious in the bore. If the bore has been neglected, it is difficult to tell what might be in there. Other than a bullet, note that the *only* thing that really belongs in a rifle bore is sunshine. The bore should reflect light brightly and clearly. It's easy to say that some amount of fouling, dirt or corrosion is harmless, but don't believe it.

If I'm using a bolt-action rifle (or some other rifle that can be taken down and the bolt removed), I remove the bolt completely and do all my cleaning from the receiver end. If you can't get a cleaning rod into the bore from the receiver end, you'll have to work from the muzzle—but take extra care, as those final few inches of rifling at the muzzle represent the last portion of the bore the bullet will be touching as it exits the barrel. A dent, scratch or gouge near the muzzle will, I assure you, affect accuracy.

Get yourself a 22-caliber cleaning rod, along with slotted or jagged tips to hold cleaning patches, and a 22-caliber brass bristle brush. You'll also need some gun oil (Break-Free is one of the best.), some powder solvent (Hoppe's Number 9 is an old favorite, but there are a number of newer solvents on the market that also do a great job) and a package of pre-cut flannel patches of the right size to fit the bore. (These supplies can be purchased all together in kit form.) The cleaning rod should have a handle that turns on ball bearings to allow the brush to turn and follow the rifling twist. The screw-together aluminum rods that are sold almost everywhere firearms are retailed work okay for 22s, although the one-piece plastic-covered steel rods are better (and considerably more expensive).

To remove lead deposits left in the bore from rimfire shot cartridges, first screw the slotted (or jagged) tip to the end of the cleaning rod and then dip one of the flannel patches into the solvent. Attach the solvent-soaked patch to the cleaning rod tip and carefully insert it into the bore from *the*

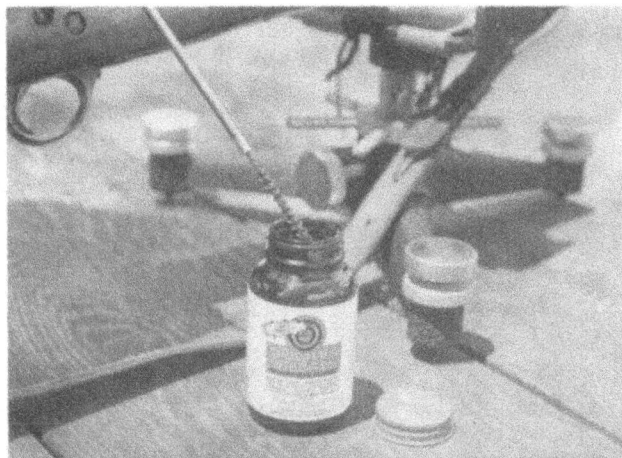

Left and below — run a solvent-soaked brush of proper diameter through the bore a number of times. This will remove the powder and bullet fouling that was softened up when you ran that first solvent soaked patch through the bore.

receiver end (if possible). Push the patch completely through the bore, remove the cleaning rod and set the rifle aside for 15 or 20 minutes to let the solvent work. Don't stand the rifle with the muzzle elevated, as solvent running back into the action can gum it up and won't help the stock if any reaches the wood.

Then remove the jagged or slotted tip, and screw the brass bristle brush to the end of your cleaning rod. Dip the brush in solvent (never use a bristle brush dry) and—again working from the receiver end—make a half-dozen passes or so through the bore. Finally, replace the brush with a patch-holding tip and run a dry patch or two through the bore. This should take care of any light leading left by rimfire shot cartridges, and all you need to do now is run a lightly oiled patch through the bore.

If you haven't been feeding shot cartridges to your pet 22, please don't go through this bristle brush routine on the rare occasions you feel the need to clean the bore. A lightly oiled patch will do—or if you insist, you might precede it with a solvent soaked patch, followed by a couple of dry ones to remove the solvent and any residue it loosens.

As you keep the bore clean and protect it from rust, you should also take the action apart far enough to allow you to give the moving innards a good wiping with a lightly oiled rag. This is a job that should be done often to keep things moving smoothly and is especially important for autoloading rifles. An old toothbrush dipped in solvent will help you remove carbon and burnt powder deposits from those nooks and crannies a rag simply will not reach. As a last step, wipe the entire rifle down with your trusty oily rag to guard against the rust raised by your fingerprints.

Centerfire rifles require more careful and regular bore cleaning. Neglect will cause accuracy loss and ruin the barrel. Centerfire rifle bullets travel much faster down the barrel and leave heavier deposits of bullet-jacket copper fouling and powder residue. This dirt should be removed after each shooting session. If you are concerned about top accuracy (particularly with a target-grade centerfire), you might take a cleaning rod to the range

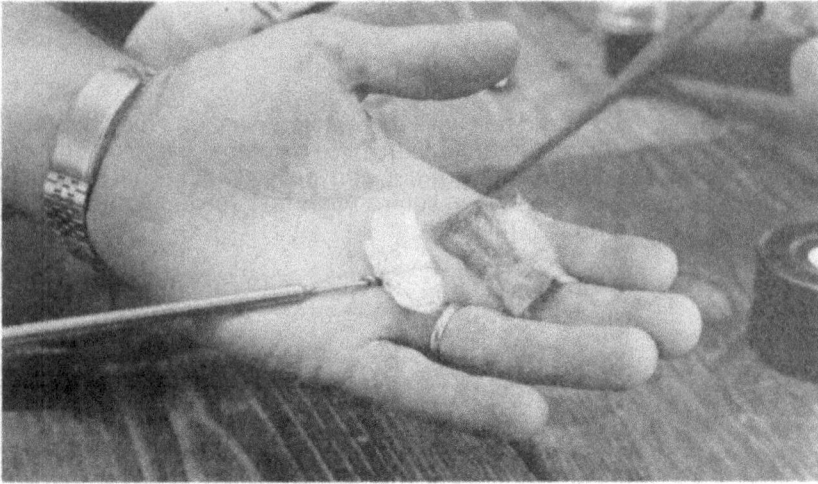

Left — Lastly, run a solvent-soaked patch through the bore followed by two or three clean dry patches. When it's all done the last patch should be relatively clean, not dirty, when it exits the bore.

Below — After you've cleaned the bore be sure to inspect it to be sure you've not left a patch (or portion of a patch) behind. If so, remove it — now!

with you and swab the bore with a solvent-soaked brush and patch after every 15 or 20 shots. Big-bore hunting rifles are not quite as fussy—you can put off the cleaning task until the end of the day.

The procedure for cleaning a centerfire rifle bore is very similar to that followed for removing lead shot deposits from a rimfire bore. Clean from the receiver end only, if at all possible, and start with a solvent soaked patch. Wait 20 minutes, and scrub several times with a brass bristle brush (of the proper caliber size) dipped in solvent. Then swab the bore dry with a couple of clean patches. If the last patch comes out *nearly* clean, well and good. If, on the other hand, it emerges wearing a dark black color or with bright green stains, you should repeat the entire process once again. Be sure to use a cleaning rod that is the right size to fit the bore.

Don't keep cleaning the bore until the patches come out pure white—such dedication isn't necessary and will probably damage the bore. Just make sure you get rid of the dark, green stains. Then pass an oiled patch through the bore. Before you put the rifle away, you might stand it on its muzzle in a corner for 15 minutes to let excess oil drain out (put some newspapers under the muzzle to protect your floor.)

Once again, wipe down the action parts with a lightly oiled cloth and use a toothbrush to clean out the crannies. Then give all exterior metal surfaces a final wiping with the cloth, and you're finished.

This is normal maintenance, and it is all the care your rifle should need—unless it gets soaked in a rain storm, gets dropped in the mud or condenses its own water by being carried from the cold outdoors to a warm room and stuffed into a case. The

Above and below — If your rifle is to be stored after cleaning, get a RIG Rag (RIG grease-impregnated sheepskin wipe) and lightly wipe down all the metal surfaces. You might also get some lens cleaning solvent and tissues from your local camera shop and gently clean the ocular (eye) and objective (front) lenses of the scope if your rifle is so equipped.

business of condensation requires more explanation. Many shooters buy expensive padded gun cases to prevent damage to their rifles. What they do not understand is that these cases are ideal for transportation, but not ideal for storage. The modern hard-side cases, lined with foam, are a particular problem. The foam does a fine job of absorbing moisture from the air and a great job of converting that atmospheric water to gun rust. The old saying "out of sight, out of mind" applies here. The best place to store a rifle is on a rack or in a cabinet where it can be seen by its owner—problems like surface rust may be seen and solved at once.

The best and cheapest gun care tool a shooter can own is a simple oily cotton rag. After a thorough cleaning or after handling your gun at any time, get into the habit of wiping the gun down with the oily cloth. Old British gun makers used a term called "poison hands" to describe people whose fingerprints turn to instant gun rust. All fingerprints are salty, moist and corrosive to some extent—remember to wipe your guns down after each handling.

If you know in advance that you will be carrying your rifle in nasty weather, be sure to wipe it down or spray it before going out the door. Do the same thing as soon as you come back through the door.

If rust should form on your rifle, it should be expunged at once—it will not get better by itself. Small spots may be eradicated by rubbing over them with a typewriter eraser. The best method, at least in this writer's experience, is to use gun oil and 000-steel wool. Many new shooters cringe at the thought of scraping steel wool across a new rifle. Not to worry. It works very well and has little effect on surrounding blue when used sparingly.

Finally, after all is said and done, don't overlook that sling. Mink oil or neat's-foot oil lightly applied will help extend the life of a quality leather sling.

There is one screaming NO-NO when it comes to gun rust: There are jelly-like solvents widely advertised as rust removers. Available from most hardware stores, these products perform exactly as advertised. The only problem here is that the blue on your new rifle is a form of rust. While the jelly goop will certainly remove rust, it will also happily remove all the blue—instantly—wherever you spread it.

A note of caution—don't over-oil your rifle's action. This is a very common mistake most beginners make, reasoning that if a little oil is good, a whole lot might be even better. This isn't the case, as excess oil will only serve to attract dirt and gum up the action. Too, it can damage the stock if it soaks into the wood.

And if you're planning to use your rifle in extremely cold temperatures, be aware that oil can freeze—or at least congeal to the point that operating the action can become difficult or even impossible. To play it safe, when the thermometer's hovering near the zero mark, remove *all* the oil from your gun's bolt, action and trigger assembly with any of the aerosol degreasers, and lubricate with powdered graphite.

Give your rifles the proper care, and they'll last several lifetimes. If an action malfunctions, the proper first aid is a good cleaning to make sure accumulated gunk and powder residue isn't the culprit (remember what I said about using too much oil?). If the problem persists take your ailing firearm to a good gunsmith. In time, there will be some repairs you can learn to do yourself; but while you're still learning to shoot isn't the best time to take up home gunsmithing.

Remember, a 22 rimfire doesn't need a lot of cleaning. You should keep the action free from accumulated powder residue and dirt, but the bore can be neglected for months on end without hurting it (unless you're shooting shot cartridges). Centerfire rifles, on the other hand, are a little more fussy and should be cleaned after every shooting session. In all cases, be careful how you handle that cleaning rod to avoid damaging the lands and grooves of the barrel. One final note to remember—it's a good idea to keep an oiled cloth handy to wipe the surface of your rifles with every time you or an admirer handles them. Fingerprints attract moisture and can cause rust to form on metal surfaces if allowed to remain. Rust *can* be removed, but it's a lot simpler to prevent it in the first place.

chapter 10

CHOOSING YOUR FIRST HANDGUN

PEOPLE BUY handguns for many different reasons. Some want a gun they can use to defend themselves or their home. Before saying any more on this subject, note that armed home defense or personal protection is *not* an activity for the beginner. While we all have a perfect right to defend ourselves, the new shooter must realize that sound instruction preceeds *any* attempt to use a handgun for the very serious business of personal protection. Training video tapes and formal instruction are available through local gun stores, the National Rifle Association and local shooting groups. Whatever the ultimate reasons for owning a handgun might be, be sure to start with a 22 rimfire. Though the gun will not be suitable for some purposes, the real design goal of the first handgun is training and practice. More than a few beginners have started with a 44 Magnum, only to find that recoil, noise and ammo cost make shooting more a chore than a pleasure. Start with the 22, then move up to the bigger hardware.

Most new shooters begin by plinking at tin cans and other informal targets, then move up to honing their skills on more formal paper targets.

As skill levels build, small game hunting with the rimfire may be the next step.

In the past few years, hunters have also learned to take the big-bore handgun seriously for hunting big game. Many states have set aside deer and other hunting seasons for handgun hunters, just as blackpowder hunters and archers have their own reserved dates. In past years, hunters adopted target pistols or police-type handguns to the hunting role, but manufacturers have responded to the new demand by introducing handguns engineered specially for hunting. These guns are generally characterized by long, heavy barrels and provision for scope mounting. The Ruger Super Redhawk is one of the most distinctive of this crop, but Colt, Smith & Wesson and Thompson Center also offer handguns tailored to the hunter's special needs.

Whatever your reason for wanting a handgun, your first revolver or auto pistol should be chambered for the 22 Long Rifle rimfire cartridge. That means it won't really be suitable for big game hunting or serious defense purposes, but if you want to really learn how to use a handgun, you'd better start with a rimfire.

The reason for this is simple: Centerfire handguns — particularly the ones chambered for effec-

Because the recoil and muzzle blast are minimal, the 22 rimfire handgun is far and away the "best buy" when it comes to purchasing that first handgun. Rimfires come in revolver and auto pistol versions. At the top left is a Ruger Single-Six single action; at the bottom left is a Colt Trooper DA; at the top right is a Smith & Wesson Model 41 auto with a Ruger Standard auto at the bottom right.

Centerfire handguns churn up a fair amount of "energy" as well as recoil. Note the exploding can of water and the position of the gun in the shooter's hand — it's still recoiling. Start with a 22, not a 357 Magnum!

tive hunting and defense cartridges — recoil sharply and make a whole bunch of noise when they're fired. Learning to shoot a handgun is hard enough without adding to the problem. If you want to shoot a bigbore centerfire later on, fine. But let's start out with a 22 rimfire. In a short-barreled handgun even the 22 Long Rifle cartridge can be noisy, and while rimfire recoil could hardly be called fierce, you will notice more recoil than a 22 rifleman will feel. This is because the handgun is lighter than a rifle, and what's more, there is no stock to be braced against the shooter's shoulder. Instead, it's held in the hands (you can fire a handgun with one hand, or use both hands to support it) at arm's length away from the body — and that doesn't give the firearm a great deal of firm support. So if there's any recoil produced, you're liable to notice it.

In simple terms, *any* handgun is relatively hard

to control. But the 22 rimfire is the easiest of the lot. Master the 22 first, and you'll be in good shape to move up to the harder-hitting calibers. When you've learned the basics with a rimfire, shooting the centerfires comes fairly easily. But if you try to start out with a hard-kicking magnum or even a 38 Special, it'll take a lot longer to learn the basics of aiming, trigger squeeze and control.

Now that we've arbitrarily narrowed your choices to a single caliber, you still have some decisions to make. For openers, should you buy a revolver or an auto pistol? There are arguments in favor of either selection, and both types of handgun have their good and bad points. Let's list some to see how the two types compare.

22 Rimfire Revolver vs. 22 Autoloader

Let's look at the revolver first. There are two basic types of revolvers — the modern double-action that can be fired by simply pulling on the trigger, and the older single-action design that requires manual cocking between shots. The former is faster to load and shoot, but the classic single-action "six-gun" is favored by a number of gunners who aren't concerned about speed of fire or reloading. The grip shape of many single-action revolvers tends to let the gun rotate upwards in the hand under recoil, and some find these guns more pleasant to shoot because of this. The recoil force isn't directed straight back into the hand, but is felt as more of a torque-like twist.

Generally speaking, the revolver has these basic advantages over the auto pistol (or semi-automatic): 1. It doesn't quit functioning if a cartridge fails to fire (simply squeeze the trigger again [in single-actions simply cock the hammer], and a new round revolves into place); 2. It will function reliably with a larger variety of ammunition (auto pistols are considerably fussier about their diet); 3. It doesn't automatically eject its empty cartridge cases into the surrounding underbrush (a feature that endears the revolver to reloading fans); 4. And it will digest generally heavier, more potent loads than an autoloading pistol will (a plus for hunters).

Finally — and most important for our purposes — the revolver is safer for a beginning shooter to start out with. At least that's my opinion, although others may dispute it. Like a selfloading rifle, an autoloading pistol cocks and loads itself each time it fires. This means it's always ready to go — just

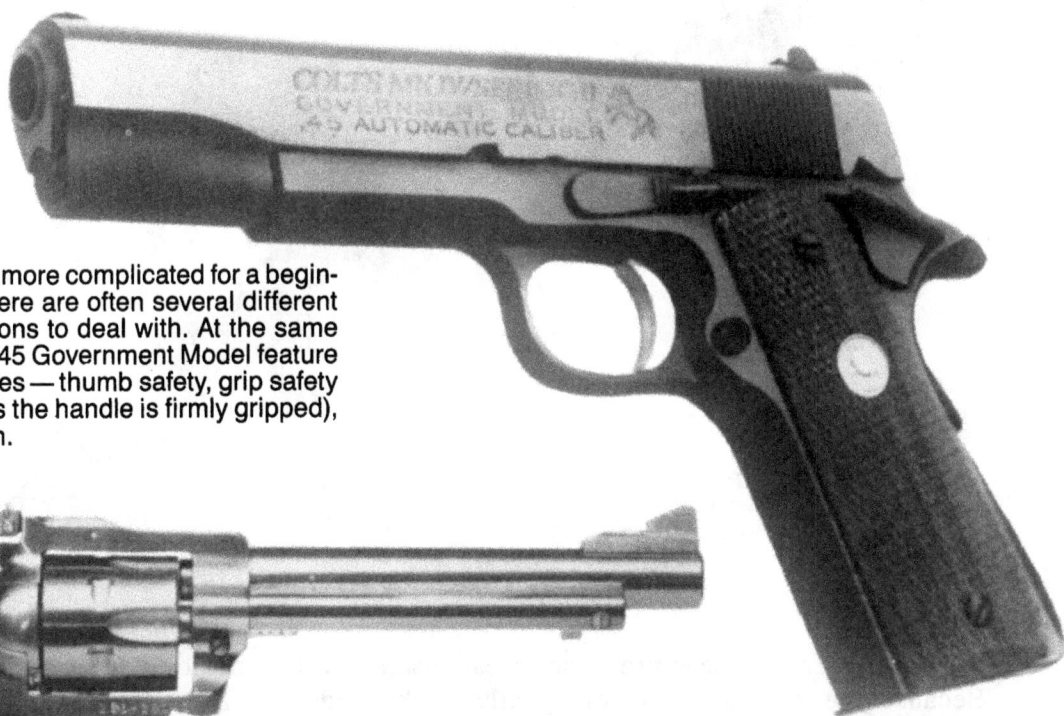

Above—Auto pistols are more complicated for a beginner to understand, as there are often several different levers, catches and buttons to deal with. At the same time, guns like this Colt .45 Government Model feature a number of safety devices — thumb safety, grip safety (the gun won't fire unless the handle is firmly gripped), and rebounding firing pin.

Above — The single-action revolver requires manual cocking of the hammer each time it's fired. The rounded shape of the grip lets the gun rotate upwards in the hand, reducing the effect of recoil.

Right — Smith & Wesson's Model 34, also called the "Kit Gun," is an excellent first handgun. While it is light and compact, it is still a full-size gun with handling characteristics identical to the centerfires.

121

Auto pistols (left) are fed by a removable box magazine, or clip, inserted in the butt, while revolvers hold their ammunition supply in individual firing chambers bored into a revolving cylinder.

trip the trigger and the entire cycle repeats itself. Because the hammer is automatically cocked every time the action cycles, it doesn't take much of a pull on that trigger to make the hammer strike the firing pin. What's more, the only way you can tell whether some auto pistols have a live cartridge chambered is to pull back the slide and take a look. If the chamber was empty but a loaded magazine was left in place, the action of pulling back the slide automatically positions a fresh round for chambering. Let the slide return to its original position, and you *will* have a chamber-loaded pistol in your hand, whether you wanted it or not. A practiced pistoleer can easily avoid these problems, but they present some potentially dangerous situations to the inexperienced handgun shooter.

Auto pistols also have manual safeties to contend with, and some feature more than a single safety device. There are grip safeties, magazine safeties, hammer safeties and slide-mounted safety levers. In addition there's usually a second lever somewhere on the frame that *resembles* the safety lever, but this one merely releases the slide after it's been pulled back and locked in that position. Finally, the auto pistol will have a magazine release button or lever located *somewhere* on the grip or frame. Again, all these mechanical buttons, levers and hammer positions are familiar devices to the experienced gunner but can be awfully confusing to the uninitiated. In contrast, a revolver features a single hammer with — at most — three possible locking positions. And that's it. No safety levers, buttons or slide release mechanisms to worry about. Double-action revolvers *do* have a sliding button usually fastened to the left side of the frame that releases the cylinder so that it can be moved out for loading, but this button will be unique on the gun and can't possibly be mistaken for anything else. Compared to the autoloader, revolvers are marvels of simplicity when it comes to operation. Once you've pulled the trigger to fire the cartridge that's lined up with the bore, the gun won't shoot again until you take a l-o-n-g, hard, double-action pull on the trigger or, for single-actions, manually cock the hammer and *then* pull the trigger. Like I said, I think the revolver is a safer handgun for a beginner to learn with.

On the other hand, revolvers aren't perfect. If you intend to try your hand at organized target competition, you'll probably want an auto pistol. Because auto pistols feature a single, fixed chamber, they're potentially more accurate than a revolver with *its* 5 or 6 individual chambers that have to be rotated into firing position. Because the dimensions of no two revolver chambers will be perfectly identical, and because there may be minor differences in register as the cylinder rotates around its axis, the results on paper targets are less likely to be 100 percent uniform. Too, there's necessarily a gap between the front face of the cylinder and the rear of the barrel, and some gases must leak out the sides of the revolver each time it's fired. That's one more variable to content with.

More important to the serious target shooter, an auto pistol can be fired a little faster and easier than a revolver can. Unless you're satisfied with the heavy double-action pull it takes to rotate a revolver's cylinder into place, you'll have to thumb cock a wheelgun between shots — and that slows you up on the target range. Revolvers take longer to reload, too — particularly since you can carry several pre-loaded magazines with an auto pistol and simply slip a new magazine in when the old one is empty. In addition, even one magazine of an auto pistol usually holds more cartridges than a revolver's cylinder, so you don't have to reload so often.

Because there's no gap between barrel and cylinder, an auto pistol won't shave off bits of lead from each bullet as it enters the barrel. This means better accuracy, and eliminates the "side spitting" (the gaseous expulsion of bits of shaved lead) revolvers are known for. Too, the shape of the grip on certain auto pistols fits the hand better than most revolver stocks, and this is an aid to better shooting.

The above list of features and faults should give you some idea of why handgunners don't all agree on the same type of firearm. Revolvers and autoloaders alike have their fans, and there are excellent guns of each type available.

What to Look for in Buying a Handgun

While I still favor the revolver over the auto pistol for the beginning handgunner, you're the person who must make the final decision of which gun to buy. Both types can be equally safe if you'll take the time to familiarize yourself with them, and follow the safety rules I outlined in Chapter 2.

Regardless of which type you decide on, I'd recommend that you buy the best gun you can afford. There are a number of inexpensive revolvers available, as well as a few low-quality auto pistols. Unfortunately, most of the inferior handguns on the market are in 22 caliber. You can expect to pay $100 and up for a decent revolver or auto pistol. Usually, but not always, anything below $100 is probably an inferior gun.

In looking for a quality 22 handgun, be sure to work the action. The trigger pull of a good quality

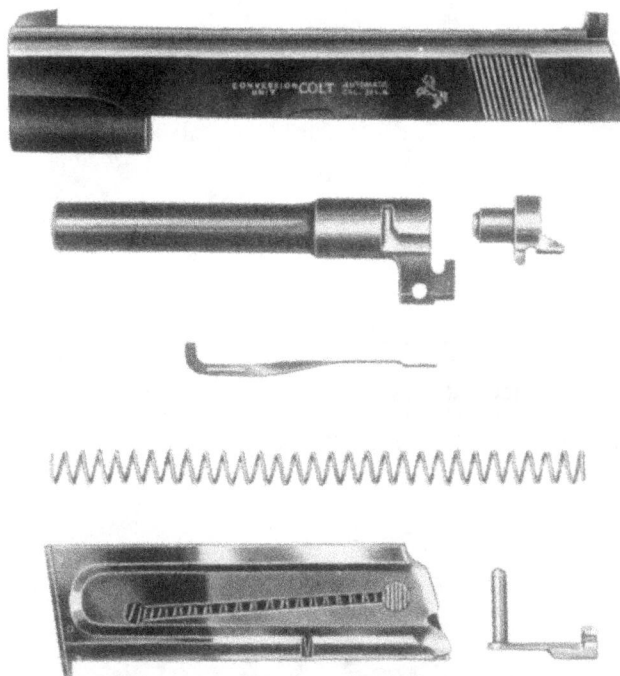

A conversion kit is available from Colt — it allows you to turn the 45 auto into a 22 auto! This approach — buying a Colt 45 and 22 conversion kit — is expensive; however, it enables you to learn the basics on a 22 and instantly progress to the 45 when you feel the time is right. The conversion kit consists of a replacement slide, barrel, ejector, recoil spring, magazine and slide release. No tools are needed, and the "conversion" only takes a couple of minutes.

22 will be smooth, even and crisp, while a poor quality handgun almost always features a hard, gritty trigger pull — and it's next to impossible to learn to shoot well with a trigger you have to drop an anvil on before the gun will fire. Too, sights on cheap handguns are usually crude affairs that lack provision for adjustment, and the overall fit and finish of the parts will be generally substandard. This affects functioning and, in the case of a low-priced revolver, can cause cylinder registration problems that could be annoying or even down-right dangerous. The old saying, "You get what you pay for," certainly applies to handguns.

REVOLVERS

If you decide to purchase a 22 revolver as your first handgun, don't get one with a barrel that's too short. I'd consider 4 inches as the minimum and would be happier with a 6-inch tube on a gun I was going to learn to shoot with. If possible, choose a

The Beretta 22 rimfire pocket auto at top is about the same size as the antique H&R 25 at bottom. While the Beretta is a fine little gun, it is, like the Walther TPH, less than an ideal beginner's choice.

There are some very fine 22 rimfire pocket pistols available, like this American-made stainless steel Walther TPH. Other fine pocket autos come from Beretta and others. These pistols are light, handy and reliable, but are harder for the beginner to learn to shoot well. Leave these guns to more experienced shooters.

To illustrate the complexity facing the beginner, this is an incomplete(!) summary of centerfire handgun cartridges. To make things a bit simpler, the most popular are probably the 38 Special, 357 Magnum, 9mm Luger, 380

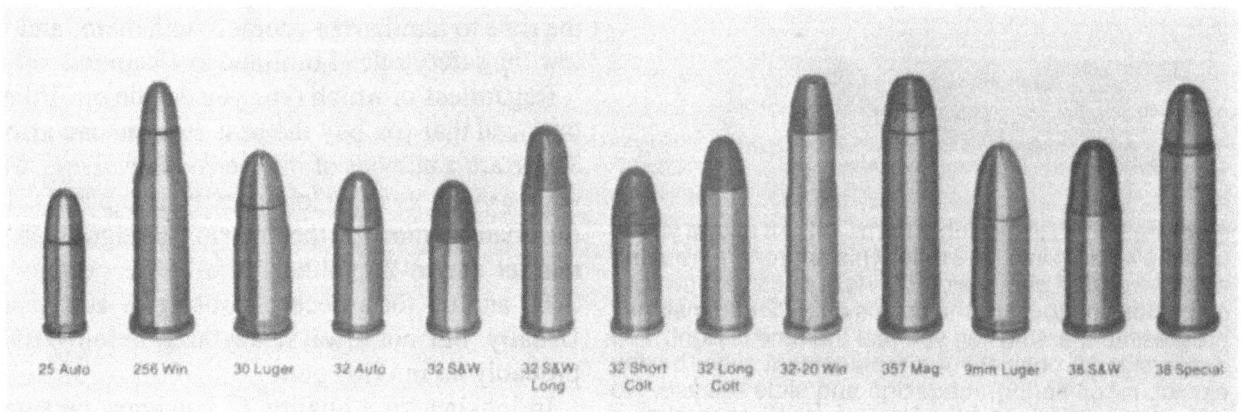

| 25 Auto | 256 Win | 30 Luger | 32 Auto | 32 S&W | 32 S&W Long | 32 Short Colt | 32 Long Colt | 32-20 Win | 357 Mag | 9mm Luger | 38 S&W | 38 Special |

model with adjustable sights. Fixed sights are supposed to be preset to give decent accuracy out to 25 yards or so, but you can't always count on it. If your gun is not shooting to the point-of-aim and you can't adjust the sights, you must learn to compensate by using "Kentucky windage" (aiming far enough off target to make the bullets hit the bullseye).

The grips most revolvers come equipped with are poorly designed to fit the average hand and getting a good pair of custom grips (offered by a number of makers) can be a good investment.

Good quality 22 revolvers are offered by Ruger, Smith & Wesson, Charter Arms, Colt and a number of importers. Some Ruger single-action rimfire revolvers come with an extra cylinder that allows the use of both 22 Long Rifle ammo and the 22 WMR (magnum) rimfire in the same gun. The price premium for the outfit with two cylinders is not great, but the price premium for WMR ammo

is worth mentioning. The new generation of hot 22 Long Rifle ammo — as pioneered by the CCI Stinger in the 1970s — produces nearly the same performance in handguns as the higher priced 22 WMR.

Both single-action and double-action revolvers are available in 22 rimfire chambering. Take your pick — but remember, the single-action guns take longer to load, and must be cocked manually between each shot.

Here is a "beauty and beast" comparison illustrating Smith & Wesson's Model 422 — an excellent beginner's auto pistol at bottom. The upper pistol, a Desert Eagle 44 Magnum, visually resembles the Smith — but the resemblance stops there. The 44 Magnum is not a beginner's handgun.

Auto, 44 Magnum and 45 Auto. The 32 Auto and especially 380 Auto have been seeing a return to popularity. While some of the others can be described as "obsolete," there are still shooters who swear by them.

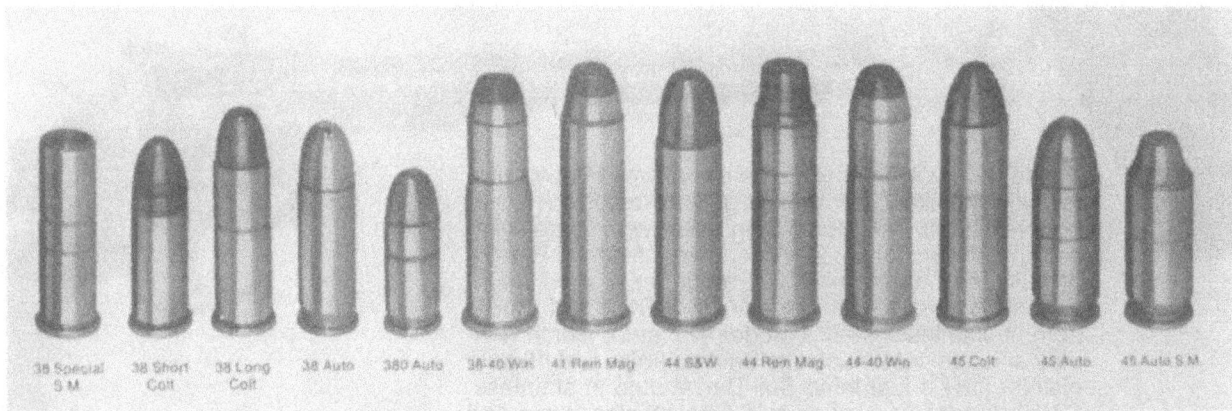

38 Special S.M. — 38 Short Colt — 38 Long Colt — 38 Auto — 380 Auto — 38-40 Win — 41 Rem Mag — 44 S&W — 44 Rem Mag — 44-40 Win — 45 Colt — 45 Auto — 45 Auto S.M.

These 22 revolvers and autos are representative of a few you will likely see in most gun stores. From top to bottom: Ruger New Model Single-Six, with spare cylinder chambered for the 22 Winchester Magnum Rimfire (WMR); the Charter Arms 22 Pathfinder, a good yet inexpensive revolver; Ruger's Mark II Standard Model auto in stainless; S&W's Model 422 auto, which handles and operates much like Smith's 9mm and 45 auto pistols; AMT's Lightning Bull Barrel auto in stainless. S&W's Model 17, built on a 38 Special-size frame and equipped with a long 8³/₈-inch barrel.

AUTO PISTOLS

Rimfire auto pistols are available in several varieties and price ranges, with the heavy-barreled *target models* being the most desirable (and usually the most expensive). The top-of-the-line target guns have excellent sights, crisp-breaking triggers, and a muzzle-heavy feel that really helps you get — and stay — on target. Ruger, Smith & Wesson and Browning all turn out high-quality 22 autoloaders suitable for target shooting, and there are a number of excellent imports on the market, as well. If you're going to buy an auto pistol to learn handgunning with, I'd recommend one intended for target use. Such guns are more accurate than the run-of-the-mill 22, and they're easier to shoot well.

Then there's the *"standard"* auto pistols — these may or may not have adjustable sights, will have barrels either 4 or 6 inches long, as a rule, and some don't mechanically vary too much from the target models that may be offered by the same manufacturer. They're priced accordingly, and you can do a good job of shooting with these slimmed-down models. But unless the price difference is substantial, I'd still recommend a true target pistol.

Next down the line is the pocket pistol. There are some truly excellent pocket pistols available, with offerings from Colt, Walther, Beretta, Iver Johnson and others readily available. These pistols come in several size ranges. Models like the Walther PP and some of the Berettas are only slightly smaller than full-size pistols. Others, like the Walther TPH, the Wilkinson Sherry and the AMT Backup are simply too small for the begin-

Above — One of the reasons for a sportsman to own a handgun is the fact that it can be carried in a holster to leave both hands free. Gun in Lawrence shoulder holster is Charter Arms 22 Pathfinder.

Below — Handgun holsters come in a wide variety of styles to allow guns to be worn on the belt or under the shoulder. Unless you have a permit, some part of the gun must be showing when you wear it.

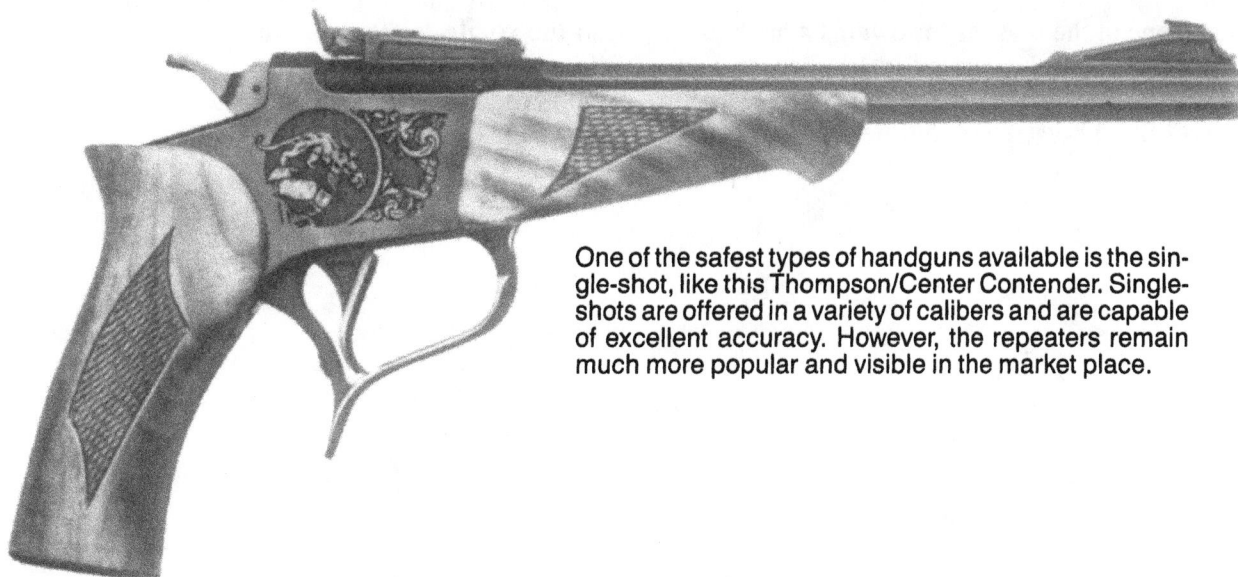

One of the safest types of handguns available is the single-shot, like this Thompson/Center Contender. Single-shots are offered in a variety of calibers and are capable of excellent accuracy. However, the repeaters remain much more popular and visible in the market place.

Padded nylon has made an appearance as holster material in recent years. The nylon is durable, easy to clean and inexpensive. It lacks the traditional appeal of leather, but it works very well.

ner and are nearly impossible for anyone but an expert to shoot well at any range past arm's length. These pistols are designed primarily for concealed carry and personal defense which, as noted earlier, is not a beginner's goal in buying a handgun. Many of them are certainly fine handguns, but they are *not* for the first-time shooter.

Finally comes the *ultra-compact pocket pistol*. These are tiny, super-lightweight affairs that are next to impossible to score well with. The grips on some models are so small that they'll accommodate only part of the fingers on the shooting hand, and nearly all are available only with crude fixed sights. They're small enough to be dangerous — generally to the guy who's doing the shooting — and they present too many problems for the first timer to cope with. *Avoid these guns like the plague*.

Since one of the reasons for owning a handgun is the fact that it can be carried in a holster to leave both hands free, I think it's a good idea to buy a properly fitted holster for your new 22 at the same time you buy the gun. Safariland, Lawrence, Bianchi, and a number of other manufacturers turn out a wide variety of handgun holsters, and these make a fine investment.

Single-Shot Pistols

There is still another category of handgun that is probably the very safest type available anywhere — the single-shot pistol. These guns are excellent for the beginner, provided the gun will digest 22 rimfire ammo. But since they are very slow to operate most shooters quickly become bored with them. The single shot, while undeniably safe, also fails to introduce the beginner to some of the chores associated with safe revolver and auto pistol handling. One primary goal of shooting, as in any sport, is fun. Most beginners simply do not find the single-loaders as much fun as repeaters. While the ultimate decision is made by the shooter writing the check, I hesitate to suggest the single-shot as a learning tool.

chapter 11

HANDGUN SIGHTS AND HOW TO USE THEM

MOST HANDGUN SIGHTS are open iron sights that, at first glance, appear similar to the open sights used on some rifles. There's a front sight blade (that looks like a post when viewed from the rear) and a notched rear blade or leaf. Like open rifle sights, the shooter centers the front "post" in the rear sight notch with the top of the front sight even with the upper surface of the rear blade.

The appearance of these sights will vary. Some handguns come equipped with micro-adjustable target sights that give a very clear, sharply defined sight picture, while other guns have some form of fixed sights. In the case of revolvers, the rear sighting notch may be a shallow groove running nearly the full length of the top strap on the frame. In auto pistols the groove may run the entire length of the top of the slide. The only way this kind of sight can be adjusted is to bend, grind down or build up the front blade. Another disadvantage of fixed sights is that some of them give a very poorly defined sight picture.

Handgun sights differ from rifle sights by appearance (although this difference is sometimes slight) and by the way they are used. Or perhaps I should say by the way a shooter views the target through them. Theoretically, the sight picture should look very much the same — the target is centered just above the front blade or post in the classic 6 o'clock hold, whether the sights being used are on a rifle or

Most handgun sights are similar to the open iron sights supplied on some rifles. The rear sight blade may or may not be adjustable. With "fixed" sights, the only adjustment possible is to bend or grind down the front blade.

129

handgun. But there's one important difference between handgun and rifle sights. On a rifle, the distance between the front and rear iron sights (the sight radius) is relatively long, and the rear sight is much closer to the eye. That means the eye is looking at three widely separated points — the rear sight, the front sight and the target. The sights on a handgun, by way of contrast, are spaced fairly close together. What's more, these sights are held at arm's length away from the eye, and they appear to the shooter to be very nearly in the same visual plane. That means both the front and the rear sight can be held in sharp focus (a virtual impossibility with open rifle sights).

Since both sights can be held in focus at the same time, a handgunner has only two focus points to worry about — the sights (together) and the target. Since the sight radius on a pistol or revolver is necessarily very short (it's limited by the length of the bar-rel), any misalignment whatsoever makes a big change in where the bullets strike the target. Proper sight alignment is much more critical with a handgun than with a rifle, and this means that the handgunner should use his sights differently. A rifleman using iron sights typically focuses on the target and lets the sights blur. When receiver sights are used, the rear sight may be so indistinct as to be practically invisible.

A handgun shooter, on the other hand, should focus on the *sights* and let the *target* blur. This can be difficult at first, particularly for someone who first trained with a rifle. But to become a really good pistol shot or revolver shooter, it'll help to learn this trick. One way to train yourself to focus on the sights rather than the target when shooting a handgun is to sometimes shoot at the center of a plain piece of butcher paper. With no bullseye to concentrate on,

Left — With handgun sights, the proper "picture" is to see the front sight centered in the rear sighting notch, with the top of the front post exactly even with the top of the rear blade. The target is positioned just atop the front sight for a "6 o'clock hold." The sights should be in sharp focus and the target allowed to blur.

Below — Handgun sights are relatively close together, and since they're held away from the eye at arm's length, it's possible to keep both front and rear sight in focus at the same time.

you're free to work on keeping the sights in sharp focus until it becomes habit.

Patridge Sights

The easiest — and best — handgun sights to use are the target-style Patridge sights. This particular sight features sharp, square-faced sighting surfaces (although the outer corners of the rear sight may be rounded slightly to make them less likely to catch on clothing when drawn from a holster). The rear sight notch will be fairly deep, and when the squared-off-post front sight is centered in this notch, twin slots of daylight can be seen on both sides. The amount of light you can see on either side of the front blade varies according to how each sight is designed. Some Patridge-style sights intended strictly for target use may show very thin strips of light — this makes for precise alignment, but such sights may be too slow for hunting or combat use. Most shooters find that they feel more comfortable when they can see a reasonable amount of space on each side of the front blade. You'll develop your own preferences as you gain experience.

Most Patridge-style sights in use today are adjustable for both windage and elevation, and this is a real plus. Most handgun manufacturers make some attempt to zero their guns in before they leave the factory, but this isn't always evident by the results. Besides, most centerfire handgun cartridges are available in a variety of different loadings and bullet weights, and these can't possibly all shoot to the same point-of-aim. So sight adjustment is often necessary, and this is much, much easier when your gun has sights that are designed to allow these changes to be made.

Fixed or Non-Adjustable Sights

Some revolvers and auto pistols are equipped

(Top) — Revolvers, too, are available with Patridge-type adjustable sights. The cost difference between fixed-sight handguns and those with adjustable sights is usually small — get the gun *with* adjustable sights.

(Middle) — These Patridge-type sights have been properly adjusted, and the shooter has done his part in firing the handgun accurately. When you begin to shoot "groups" like this 3-shot cluster, you know you're progressing.

(Right) — Target-style Patridge sights are among the best available. The rear sight features a squared-off blade with a square-cut sighting notch. Sight is adjustable for both windage and elevation.

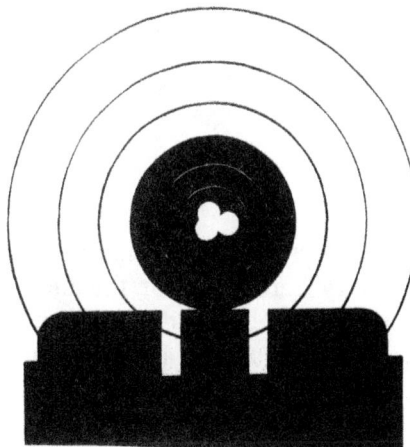

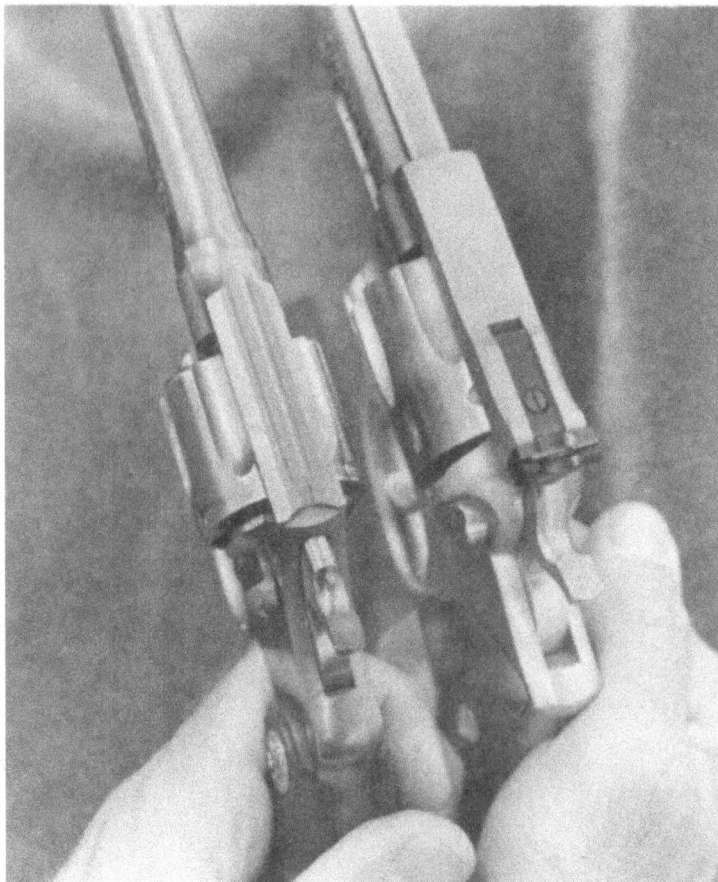

(Left)—The revolver on the right has a rear sight that's screwdriver-adjustable for windage and elevation. Those fully adjustable sights allow the shooter to select any particular brand (or variety) of ammo the handgun is chambered for and sight-in accordingly. The non-adjustable-sighted handgun on the left has been regulated at the factory. The shooter should try two or three (or more) brands of the proper caliber to determine which one(s) prove to be the most "accurate."

Adjustable sights can be moved up and down or sideways to compensate for changing loads. A narrow-bladed screwdriver is usually needed to make these adjustments.

with fixed, or non-adjustable sights. Such sights are theoretically zeroed in for the most popular bullet in that particular caliber of handgun, but again theory and practice are sometimes poles apart. When you find your gun is not sighted-in properly — and the gun shoots 6 inches low and a foot to the left with the load you've settled on — there are three things to do. The easiest way to deal with this problem is to simply adjust your aim to bring the bullets on target. This may mean aiming a half-foot high and several inches to the right over a certain range, and increasing or decreasing this adjustment as the range lengthens or shortens. This is called applying "Kentucky windage." Another easy solution is to try a few different brands of ammunition in your handgun as the bullet's point-of-impact can decidedly shift from brand to brand.

If everything else fails you might want to try to solve the problem by changing the configuration of the sights. This usually isn't easy, and may involve processes your home workshop isn't equipped to handle. It could require the services of a good gunsmith. But if you have access to a good vise (be sure to pad it before clamping your handgun in)

and a set of flat files, you should be able to get your new gun shooting where you point it — at least with one particular load.

Since the rear sight notch of a fixed sight handgun is usually pretty much immobile (it may extend the length of a revolver's topstrap), you'll probably have to work with the front sight. Some auto pistol rear sights are mounted on the slide via a dovetailed notch, and these sights can be driven slightly to one side or the other to make changes in windage. But other than that a "fixed" rear sight doesn't leave you with much to work with.

When working with the *front* sight, you move it the *opposite* direction from the way you want the bullet holes to move on the target. This means that, in order to correct windage, a gun that shoots *left* needs its front blade bent, altered or repositioned to the *left* in order to shift the point of bullet impact *right*. This means that, in order to correct elevation, if a gun is shooting *low* and you want to raise its groups *up* into the target, you file a small

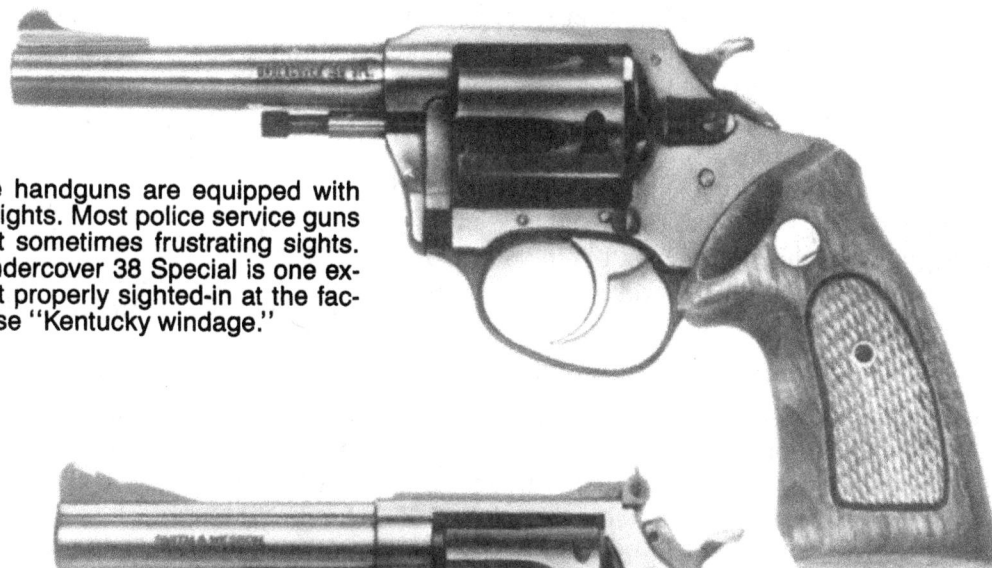

Right, above — Some handguns are equipped with fixed, non-adjustable sights. Most police service guns wear these sturdy, but sometimes frustrating sights. (This Charter Arms Undercover 38 Special is one example.) If the gun isn't properly sighted-in at the factory, you're forced to use "Kentucky windage."

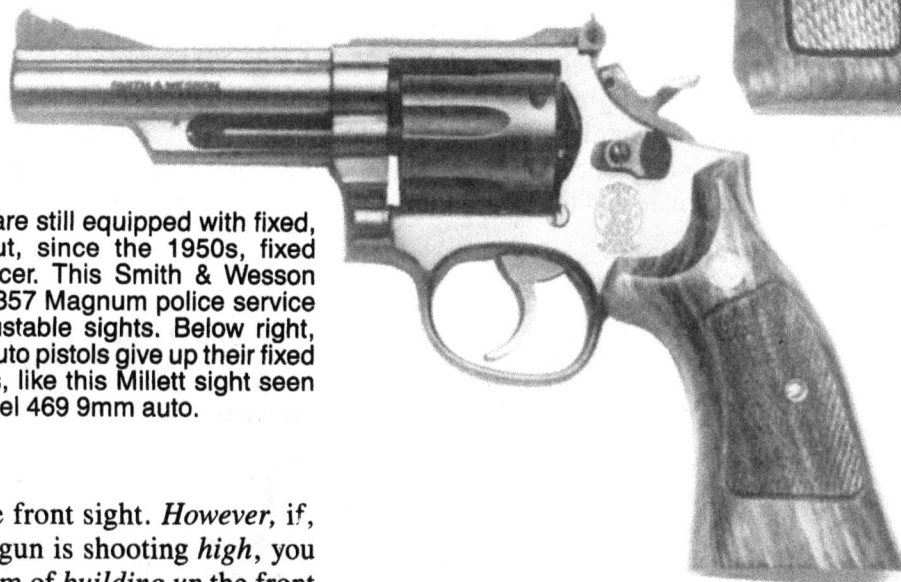

Right — Some handguns are still equipped with fixed, non-adjustable sights, but, since the 1950s, fixed sights have become scarcer. This Smith & Wesson Model 19, a very popular 357 Magnum police service revolver, carries fully adjustable sights. Below right, the same trend has seen auto pistols give up their fixed sights for adjustable types, like this Millett sight seen on a Smith & Wesson Model 469 9mm auto.

amount off the top of the front sight. *However,* if, on the other hand, your gun is shooting *high,* you are faced with the problem of *building up* the front sight by adding metal, or of installing a new sight. If you have the experience and tools to perform either of these operations, I would suggest you make your windage adjustment first, then your elevation correction.

These changes should be made gradually, with three-shot groups fired after every change. (Fire from a steady position — shoot two-handed and rest the bottom hand or wrist against sandbags.) As you can see, bending a front sight, or taking metal off, or putting it on, requires experience — the kind of experience a competent gunsmith possesses. It would be much simpler to farm the job out to a good gunsmith, or better yet, buy a gun with adjustable sights.

Scope Sights

There's still another type of handgun sight that's gradually achieving a measure of popularity — particularly among those who hunt game with revolver or auto pistol. I'm talking about long eye relief scope sights designed specifically for handgun use. These sights feature very little magnifica-

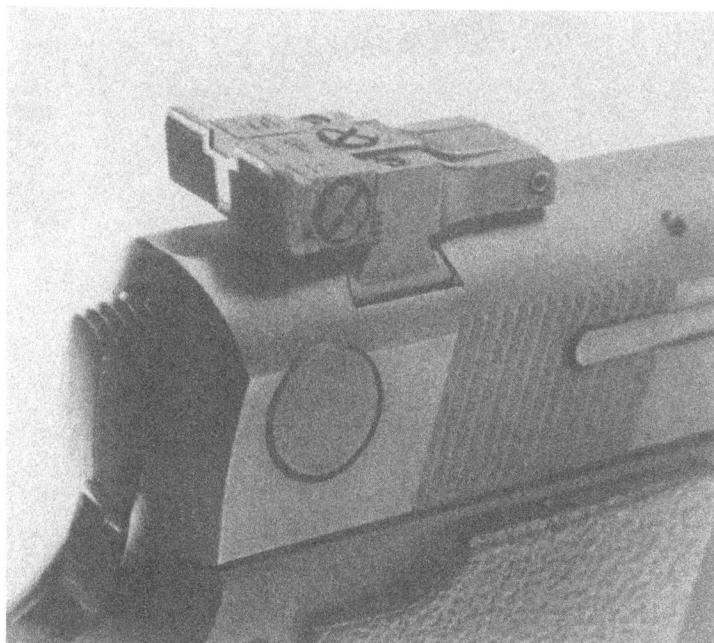

133

tion but offer a sharply defined set of crosshairs that are easy to line up on target. These sights take some getting used to, and they make the guns they're mounted on considerably heavier and bulkier. But long-range varmint hunters and certain other pistol-packing Nimrods have adopted handgun scopes with glee. Such sights aren't for the beginner as they really won't help your scores until you're already well grounded in handgun marksmanship.

Projected Aiming Point Sights

In recent years, a new style of optical sight has been making inroads into the handgun and rifle sighting market — the projected aiming point sight. With some electronic help, this sight projects a transluscent dot of light into the shooter's field of vision (without magnifying the target) to replace the conventional scope reticle. Since the effect is to make the reticle and the target appear in the same focal plane, the sight is theoretically easier and faster to use. It also offers the advantage of not blanking out part of the target as conventional reticles sometimes do.

Handgun Sights and Accessories

Whatever kind of handgun sight you use, you should learn to keep both eyes open. Some marksmen let the eyelid of the eye that isn't directly behind the sights droop slightly, as this works to sharpen sight definition a bit. But don't close the off-eye entirely. You'll be able to see both sights and target better with both eyes operating, and you won't lose your depth perception.

It's a good idea to wear shooting glasses with hardened lenses when firing any kind of handgun, but particularly when you or the guy standing next to you is using an auto pistol. The danger here isn't so much a possible primer or cartridge failure, but the chance of having a hot, empty cartridge case

The long eye relief telescopic sight is designed for handgun use and helps make long-range hunting accuracy possible. Above are Leupold M8-2X scopes mounted on a single-shot Thompson/Center Contender and a Colt Python revolver (inset).

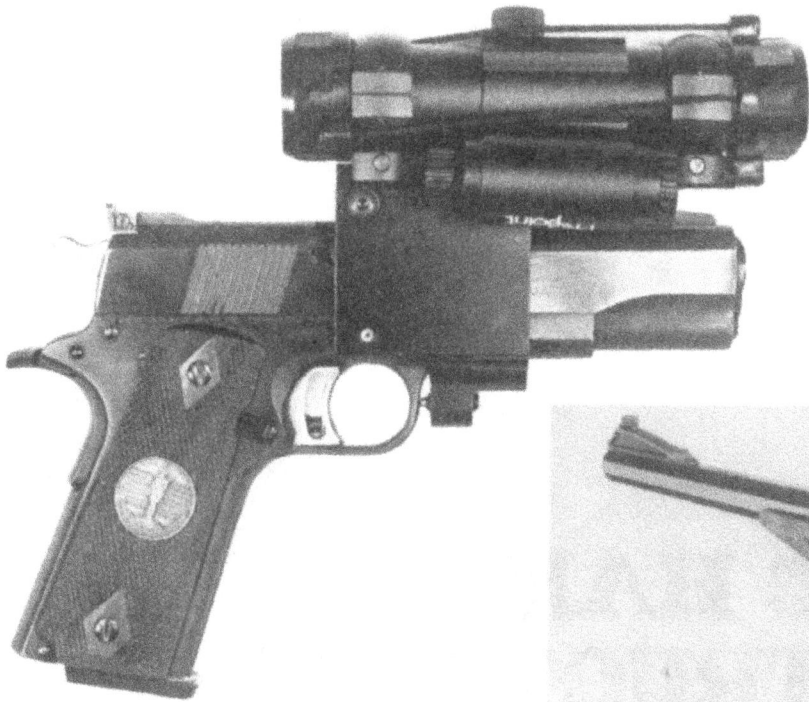

This Pro Point projected aiming point sight, shown here on a Colt Gold Cup 45 auto, is one of the newer type sights on the market. Unlike scope reticles, the translucent dot does not blank out any of the target. Note that a conventional post-type reticle can cover a good deal of deer-sized target at 200 yards.

(Right) — Aperture sights are seldom seen on handguns, but they can be adapted to certain types like this Thompson/Center single-shot.

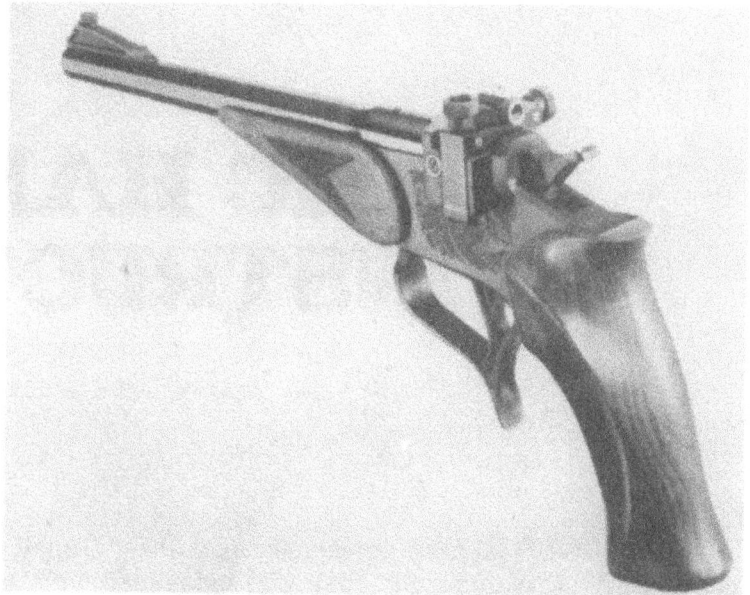

thrown into your eye.

Some serious target shooters attach a Merit lens disc peephole or some other type of aperture to the master eye lens of their shooting glasses. This serves to sharpen the focus and can be of some value in very deliberate, precision shooting. But these attachments require patient and precise arm and head movements on the part of the gunner to get things properly aligned, making them awkward and slow to use. Such extras are for the expert handgunner only, and are mentioned here to simply acquaint you with another "aid" in sighting the handgun.

Some handgun sights feature a front blade with a colored piece of plastic inletted into its rear face to give added contrast. One handgun manufacturer — Dan Wesson — even offers front sights with interchangeable square plastic inserts in a variety of colors to adapt the sight picture to changing light conditions. When shooting in bright sunlight, it's sometimes advantageous to darken the front sight blade by passing it above a candle or match flame.

Once you've learned to use the sights on your handgun effectively, you might try practicing in-stinctive shooting. That's a difficult way to attain skill that involves hitting a target *without* the use of sights. It can be done, but it takes lots and lots of practice and hard work. Don't even attempt this stunt until you've already become a *good* shot when shooting with sights.

One final note about using handgun sights. Until you become able to hit the bullseye on a target regularly and consistently with a handgun at a range of at least 25 yards, don't make the mistake of plinking away at tin cans or other "fun" targets. This destroys discipline and doesn't give you any clear idea of how you're progressing. Holes in a target give you clear, hard facts — misses show up as clearly as hits, and you'll have some idea of what you're doing wrong. (Holes all over the paper indicate that you're simply not concentrating on consistent hold, trigger squeeze or sight picture, while a tight cluster of shots high and to one side could indicate either a consistent flinch, improper trigger control or sights that need adjusting.) But puffs or dirt kicked up around a tin can tell you nothing. Buy a good supply of targets, then use them. Plinking can come later in the game.

chapter 12

BASIC HANDGUN INSTRUCTION

THE HANDGUN IS, in my opinion, the most difficult of all firearms to learn to shoot well and accurately. There's no shoulder stock to help steady the gun or spread recoil forces around the body. Instead, the handgun is held at arm's length, and any recoil that's involved is transmitted directly to the shooting hand.

A beginning shooter will find a handgun a most unsteady implement. Holding a 2- or 3-pound dead weight as far as you can away from your body becomes very tiring in a matter of seconds, and trying to line up a pair of sights and hold them on target at first seems to be a nearly impossible task. The first time you try firing a handgun at a distant mark, you'll probably be convinced that any kind of tack-driving accuracy *is* impossible. But don't be discouraged by your very first performance. Handgunning takes practice, and there's no shortcut I know of that'll turn you into even a passable marksman in an hour or two of shooting.

Actually, things are much better now for the first-time handgunner than they were when *I* started shooting pistols and revolvers. Then, the only acceptable way to shoot a handgun was to grip it in one hand, face about 45 degrees away from the target, and extend the hand, holding the gun fully at arm's

length away from the body in an almost straight line with your shoulders. A right-handed gunner looked at the sights over his right shoulder and did his best to hold the gun steady with his single unsupported hand.

What's more, in my day it was considered "manly" to shoot without ear protection. Some intelligent target shooters did stick wads of cotton in their ears, or maybe a pair of empty cartridge cases in an attempt to muffle the muzzle blast of big-bore pistols. But an equal number just gritted their teeth and did their best to ignore the noise. Talk to a big-bore target shooter who started in the game 20 years ago, and chances are you'll have to raise your voice several decibles. Most of these shooters are now at least partially deaf.

There have been some major improvements made in handgunning methods and techniques since that time, and attitudes toward such safety accessories as ear protectors and shooting glasses now reflect greater intelligence. As a result, today's beginning handgunner has a lot more going for him that his father had to do without. This means he should be able to learn to shoot well quicker than the gunners I started with were able to.

Let's take a look at some of these changes, and see

To load a double action revolver, manipulate the cylinder latch, and gently push the hinged cylinder out of the frame from the off side.

what a beginner needs to do today to get started off on the right foot.

The first thing a fledgling handgunner should do is review all the safety rules (see Chapter 2) and make sure he has them firmly in mind before he even picks up a handgun. Remember, treat all firearms as if they were loaded *at all times*. Watch where that muzzle is pointing — these rules go into effect from the moment you open the box your new handgun came in. That gun should be regarded as *loaded* and potentially dangerous as soon as you lift it from its wrappings, even though you're more than reasonably sure the factory didn't pack it in loaded condition. Observing the safety rules should be — *must be* — a matter of unthinking habit, and now's the time to start developing those automatic reflexes.

The second thing you should do is familiarize yourself with your new handgun. Check the chamber(s) and magazine to make sure they're clear of ammunition (instructions for clearing each type of handgun are found in Chapter 2), and then thoroughly read the instructions that should be packed with your new gun. If you have no instruction manual, you should ask the clerk or whoever sold you the gun to show you how it operates. Most handguns within each type (single-action or double-action revolver or auto pistol) operate in pretty much the same way, but there may be minor differences that should be pointed out at this time by someone who is familiar with the firearm. For instance, all double-action solid-frame revolvers function in *almost* exactly the same way, and the procedure for loading and firing will be nearly identical for each gun falling into this classification. The possible differences that can show up are mostly limited to the location of the cylinder release latch (found at the rear of the cylinder behind the recoil plate on most revolvers, but sometimes located ahead of the cylinder at the top of the crane) and the number of chambers con-

Keeping the gun pointed in a safe direction, load each chamber individually with the *proper* type and caliber of ammo. (Unsure as to the caliber of gun or ammo? See your local gun dealer or reputable gunsmith — he'll set you straight in about 30 seconds, or less.) Lastly, *gently* push the loaded cylinder back into the frame until it audibly locks. The revolver you have in your hands is fully loaded and ready to fire. *Handle with care.*

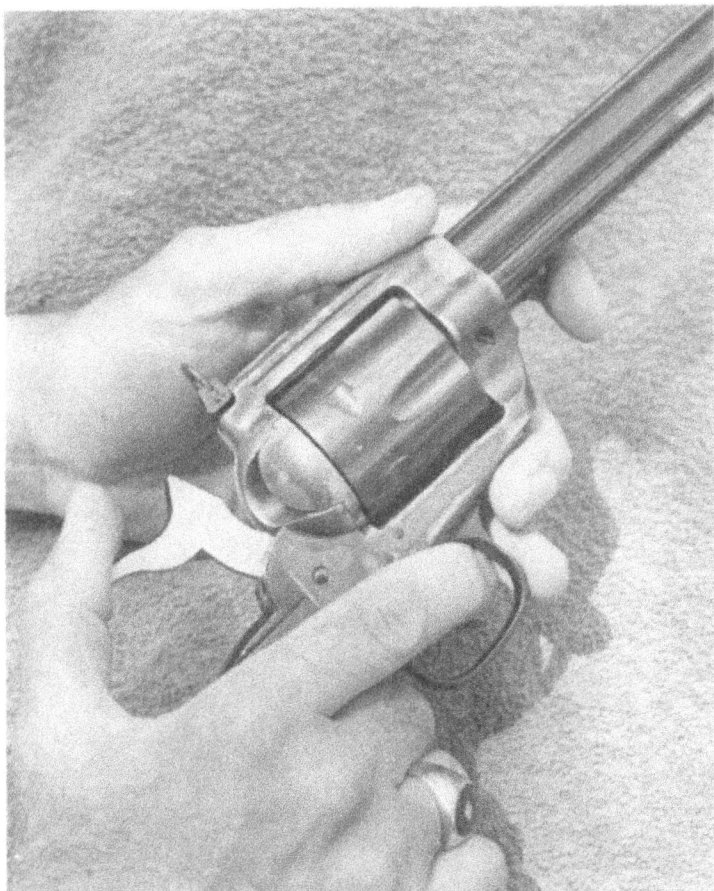

Before you load any single-action revolver, refer to the owner's manual to determine whether or not the revolver is a new "modified" single action, or an older "true" single action. If you're unsure, or don't have a manual, write the manufacturer for one and/or see your local gunshop or reputable gunsmith. To load an original, "true" single action, point the gun in a safe direction and pull the hammer back until you hear *two* audible clicks. Stop. The cylinder should turn freely. Next, open the loading gate and rotate the cylinder making sure all chambers are empty. (If loaded with live ammo or empty brass, it will be instantly apparent. You may remove the ammo or brass by rotating the cylinder 'til the case lines up with the ejection port and then pushing the ejector rod—located under the barrel — fully to the rear.) To load, align an empty chamber with the loading gate and proceed to load only 5 of the 6 chambers. The hammer should be let down gently on the 6th *empty* chamber. Your single action is now loaded and ready to fire by simply pulling the hammer fully to the rear and pulling the trigger. *Handle with care.*

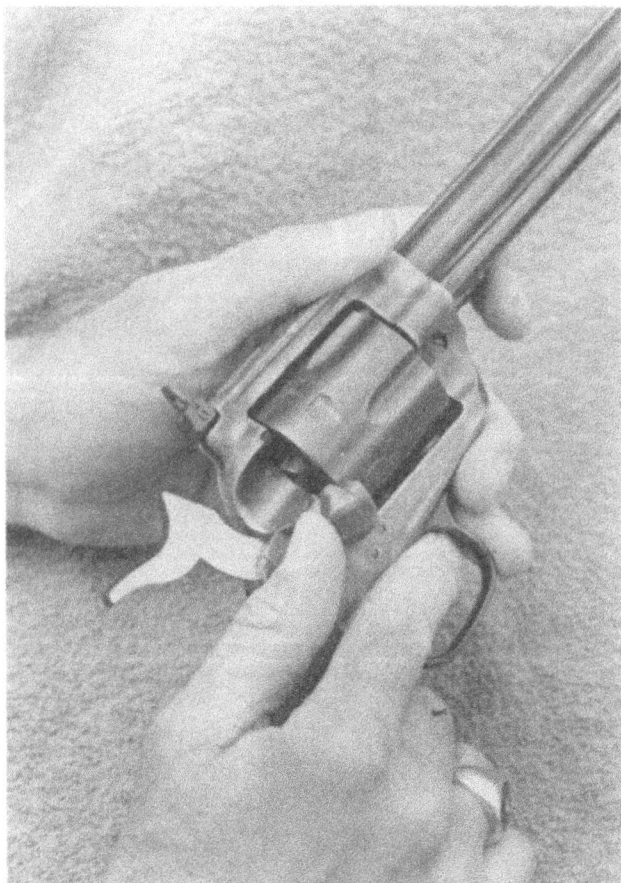

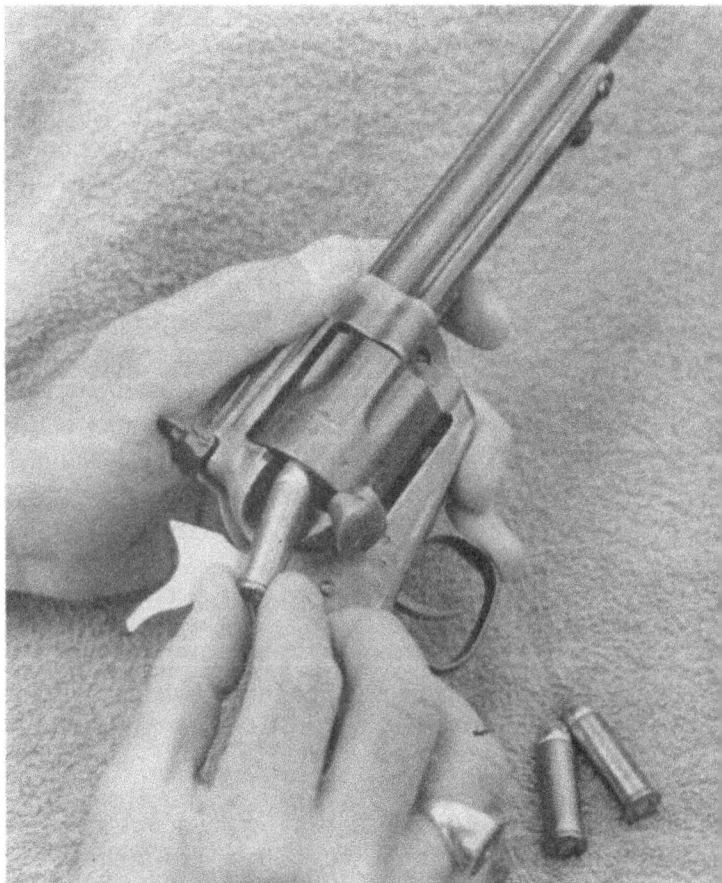

To load a semi-auto pistol, point the muzzle in a safe direction and remove the magazine per manufacturer's (or your dealer/gunsmith's) instructions. Next, pull the slide to the rear and inspect the chamber to make sure it's empty. Setting the pistol aside, take the magazine in your left hand with the flat spine of the magazine against your palm as seen here. The magazine may then be loaded one round at a time with the proper type and caliber of ammo. Once loaded, insert the magazine into the butt of the pistol — with the top cartridge pointed toward the muzzle. A firm shove will produce an audible "click" indicating the magazine is locked into place. To load the chamber you simply pull the slide fully to the rear and release it smartly. The chamber is now loaded, the gun ready to fire. *Handle with care.*

(Ed. Note: Because the design of semi-auto pistols varies tremendously, we urge you to get an owner's manual from the manufacturer and/or visit your gun dealer or gunsmith for hands-on instruction.)

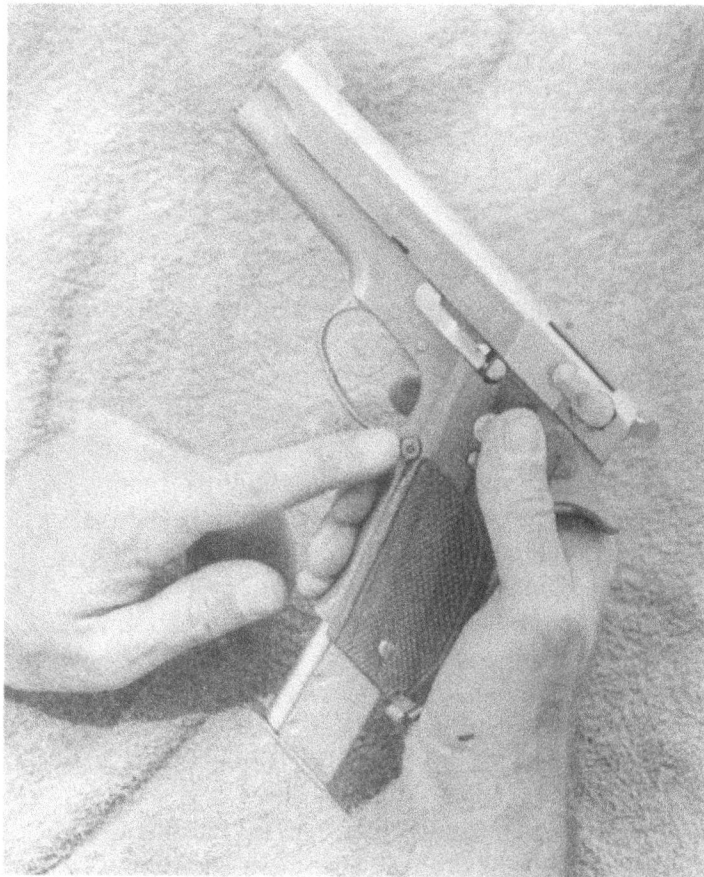

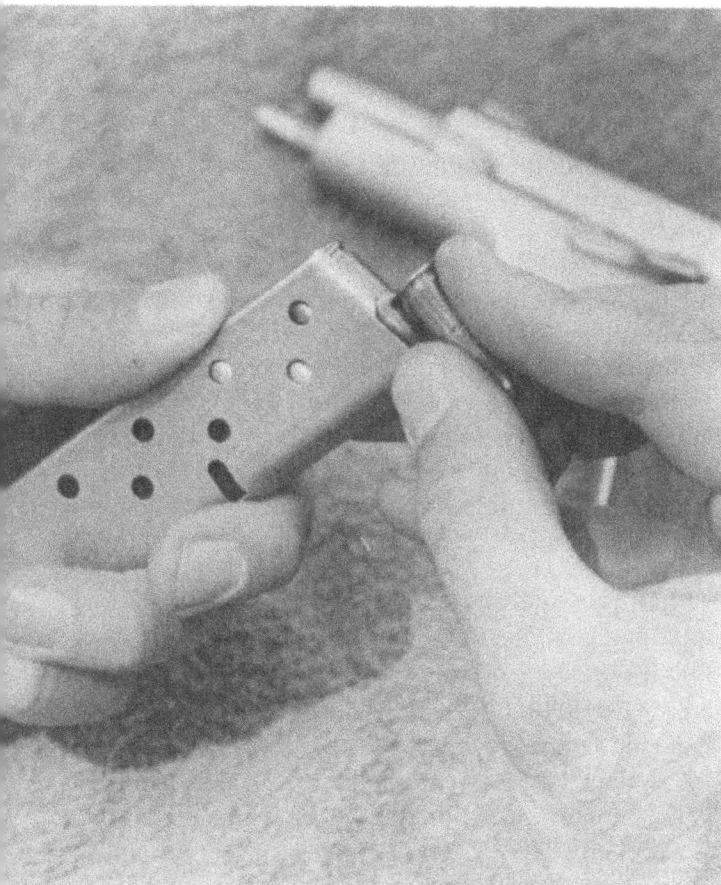

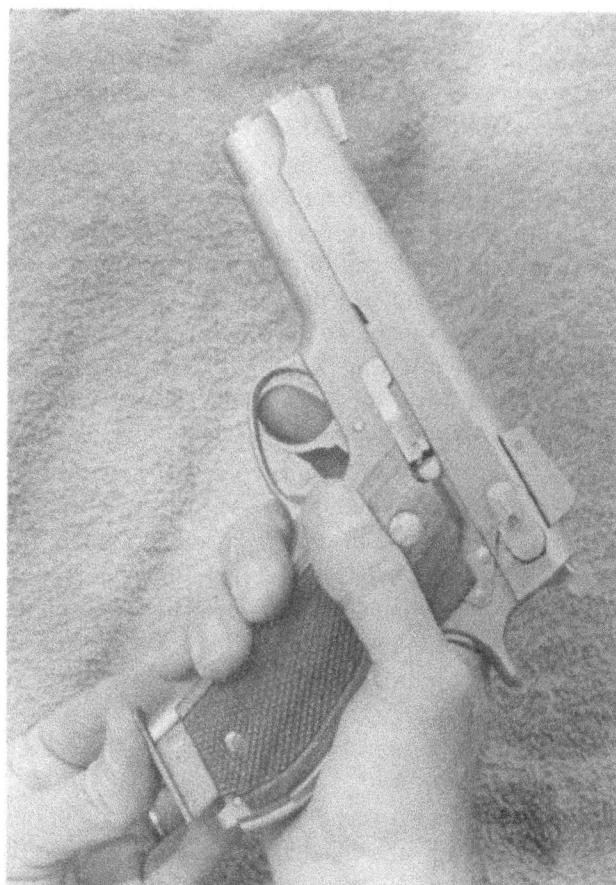

Years ago, this was the only "approved" shooting stance for handgunners, and what's more, ear protection wasn't considered "manly." This is still the stance used in target competition, but the one-hand hold has many disadvantages for the beginning handgunner.

tained in the cylinder (usually six, but varies between five and nine). The direction the cylinder turns can also vary — some models rotate clockwise while others turn in the opposite direction.

There are some exceptions to this rule, as some inexpensive rimfires lack a craning mechanism which moves the cylinder partway out of the frame. Such guns require complete removal of the cylinder (pull the cylinder pin out to free the cylinder) for loading and unloading. There are also some break-top revolvers available, and these open at the top to expose the chambers.

Except for minor differences in the safe way to carry them holstered when loaded (see Chapter 2), all single-action revolvers are loaded and unloaded in an identical manner.

The one type of handgun that can be confusing to a new shooter — and even to some experienced hands who know how to operate a revolver but haven't been around auto pistols much — is the semi-automatic or autoloader. Again, the operating principles are pretty much the same from auto pistol to auto pistol, but the location of the various safety, slide-locking and magazine-releasing buttons and levers can vary considerably between the different models.

An auto pistol clip or magazine can be momentarily confusing to the beginner, who may not be sure exactly how to load cartridges into it. The procedure is simple — the top end of the clip is open to expose a spring-loaded follower. A pair of steel lips or tabs will protrude from the sides of the magazine at one end of the top — that shows you where the rear of the clip is. To load, insert the *base* of a cartridge into the magazine, using enough force to press down the follower, and then slide the *base* or back end of the cartridge back under the retaining lips. Repeat the operation until the magazine is full. Some magazines have a small button or tab extending from one side of the clip — you can compress the follower spring by pressing down on this tab with your thumb while using the free hand to insert the cartridges. This makes loading a bit easier.

Once you've learned how to safely load, unload and operate your new handgun (*don't actually load it during this learning process* — just go through the motions), you're ready for the next step. This is learning the proper grip and stance. Contrary to what most shooters believed a generation ago, there is no *one* shooting stance that all gunners should adhere to. There are several different positions you can take when shooting your handgun, and you should use the one that feels most comfortable and gives you the best results. If you decide to enter formal target competition later on, you'll then have to adopt the one-armed stance I spoke of earlier. But for practical handgunning — doing your best to hit a target or game animal at some distance, and doing everything possible to improve your chances of doing just that — the classic target stance isn't the best possible choice.

First let's consider the grip, or the hold you take on a handgun. We're going to learn to shoot using *both* hands, because that's the most sensible way to hold a handgun. The wrist is the weak point of a handgun support system, and by using both hands the system is greatly strengthened. A two-handed grip is not only steadier than the old one-hand hold, it also helps the shooter absorb recoil comfortably and makes more accurate rapid shooting possible.

While there are minor differences in hand and finger placement, essentially the same hold is used on both double-action revolvers and auto pistols. The gun's grip is grasped solidly by the shooting hand (the right hand for everyone but southpaws), with the heel of the hand making direct contact with the mid-to-lower portions of the backstrap. The hand should be placed as high as possible on the grip — the web between the thumb and forefinger should be forced upward into the curved top portion of the backstrap on auto pistols. High hand placement puts the support as nearly as possible in line with the direction of recoil and aids controllability. The fingers are wrapped around the grip, with the index or trigger finger contacting the trigger at the center of the tip of the index finger. For shooting double-action, you may have to contact the trigger farther back on the finger (somewhere in the area of the first joint) if you don't have enough strength to control the longer, heavier pull with the end of the finger. The thumb should be pointing forward along the left side of the frame (for right-handed shooters). For best results, take a firm, solid grip — don't make the knuckles whiten or grip hard enough to cause strain or fatigue

Here's another variation of the two-handed grip. In this case the shooting hand is supported underneath by the second hand. This offers good support, but doesn't supply the wrist stiffness the "hand-clasping-hand" grip provides.

141

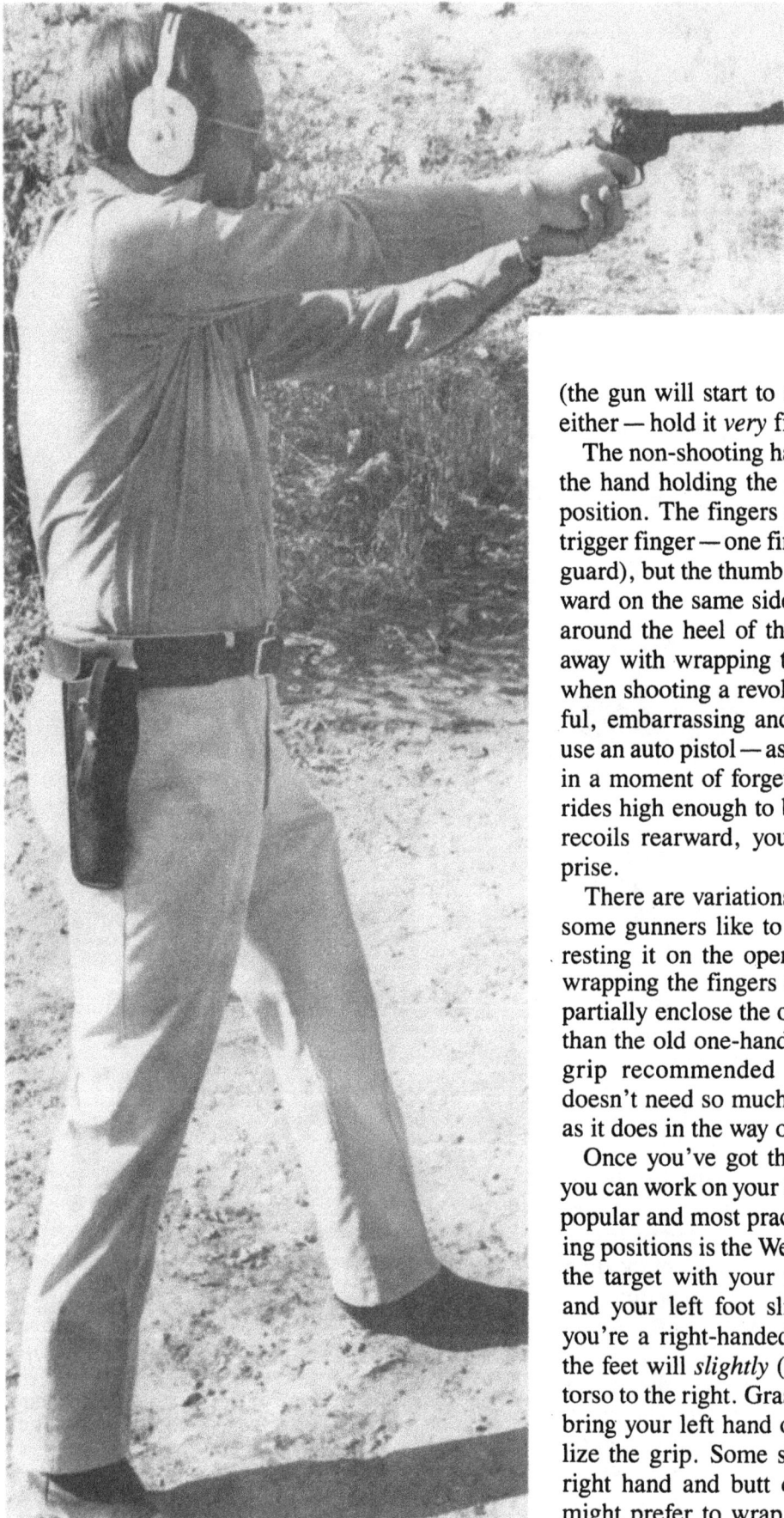

Here's one version of the Weaver stance. Note the shooting arm is fairly straight, while the supporting arm is bent at the elbow. The shooter's left shoulder and foot are moved slightly forward.

(the gun will start to shake), but let's not be dainty either — hold it *very* firmly.

The non-shooting hand should be wrapped around the hand holding the gun in a natural, comfortable position. The fingers should overlap (except for the trigger finger — one finger is plenty inside the trigger guard), but the thumb should be pointing up and forward on the same side of the gun, and not wrapped around the heel of the shooting hand. You can get away with wrapping this thumb all the way around when shooting a revolver, but this hold can be painful, embarrassing and potentially unsafe when you use an auto pistol — as I once demonstrated to myself in a moment of forgetfulness. If that second thumb rides high enough to be in the path of the slide as it recoils rearward, you're in for an unpleasant surprise.

There are variations of this two-handed hold, and some gunners like to support the shooting hand by resting it on the open palm of the other, and then wrapping the fingers of the support hand upward to partially enclose the other. This offers more support than the old one-hand hold, but isn't as rigid as the grip recommended earlier. The shooting hand doesn't need so much in the way of vertical support as it does in the way of reinforced wrist stiffness.

Once you've got the two-handed hold down pat, you can work on your overall stance. One of the most popular and most practical of the two-handed shooting positions is the Weaver stance. To assume it, face the target with your feet spread comfortably apart and your left foot slightly ahead of your right (if you're a right-handed gunner). This positioning of the feet will *slightly* (naturally) "angle" your upper torso to the right. Grasping the gun in the right hand, bring your left hand over to help support and stabilize the grip. Some shooters will prefer to cup the right hand and butt of the gun in the left. Others might prefer to wrap the left hand around the left-front portions of the grip — this is perhaps the better of the two, but it's your choice. Standing erect in this position, bring the gun up and hold it directly in front of you at arm's length. At this point your weight should be evenly distributed. The right arm should be held straight with the left supporting arm bent slightly downward. To fully stabilize the gun, pull your supporting arm slightly backward, applying a stabilizing pressure to your shooting hand and arm. Lean slightly into the gun. (Watch that you don't "cant," or tilt, the gun to one side as this will throw your aim off).

As you bring the gun to eye level your right shoulder will move upward — bring your head down slightly to meet the shoulder and reinforce the stance. The chin/shoulder contact will keep your head, thus your aim, steady. In aiming the gun at the target, focus on the target and, again, bring the gun *up* (not down) to align your sights *with* the target. This is probably the most useful shooting position a handgunner can learn.

Other, more solid positions include the kneeling, sitting and prone stances. The kneeling position is a variation of the Weaver stance in which the shooter kneels on his right knee and rests his left elbow on his upraised left knee. The sitting position is similar to that used by rifleman, except that the arms are supported just behind the elbows by the upraised knees, and the gun is held straight out in front. Shooting from the prone position is done by lying on the stomach and stretching out toward the target, legs spread, with both arms extended in line with the body, elbows touching the ground and the pistol slightly raised. Again, a two-handed hold is used. All of these positions are dependent upon the terrain you're shooting in. If the brush is high, standing may be the only position you can shoot from.

Finally, you should learn how to shoot using external supports. No part of the gun should contact anything but the hands, but the hands can be rested on or against any suitable support. Resting both hands on sandbags, again remembering to use the two-handed hold, gives you a shooting platform stable enough to use for sighting in your gun.

Once you've learned how to hold your handgun and move into a suitable shooting position, or stance, it's time to concentrate on trigger control. At this stage you should be shooting your gun single action — if your revolver or pistol allows you to shoot in either double-action or single-action mode, thumb-cock the hammer before each shot. (Double-action auto pistols may, or may not have to be cocked

Shoot at paper targets until you're able to shrink the size of your groups and bring them on target. Shooting at tin cans early in the game will destroy discipline and won't tell you how you're progressing.

by hand for the first shot — subsequent cycling of the action will automatically cock the hammer for repeat shots.)

With the hammer cocked (remember to keep the muzzle pointing downrange at all times), place the first pad of the index finger on the trigger. Grip the gun firmly into the heel of the hand, take careful aim, and apply steadily increasing pressure to the trigger until the gun fires. Remember to use exactly the same grip, with the same amount of hand pressure, with each and every shot for consistent results. Altering the grip in any way can throw subsequent shots off target.

When firing double action, move the trigger finger forward slightly until the trigger falls at the first joint or slightly farther back. Then pull the trigger straight to the rear (don't exert any sideways motion) as smoothly and evenly as possible until the hammer falls. There will likely be a noticeable build-up of pressure (trigger resistance) through the last stages of this pull, and this can be used to help control the moment of firing.

143

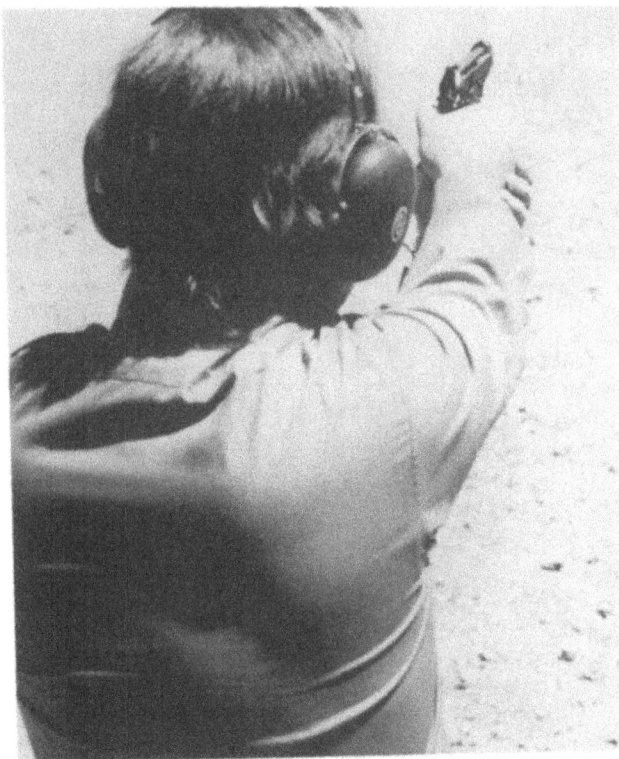

Good ear protection is a must when firing any handgun, but is doubly important when firing magnum centerfires or short-barreled guns.

Like rifle shooting, firing a handgun accurately requires good muscle control. Breathing is important, and most shooters make a habit of taking a deep breath, exhaling, taking another breath and letting it halfway out just before shooting. The "half-breath" is held through the trigger pull.

A great deal has been written about how to rapidly draw and fire a handgun, but the only practical application this skill has is related to combat work. For the beginner, hunter, or almost any sport shooter, a holster should *only* be a means of carrying a handgun conveniently and safely when not in use. A beginner (and many shooters who have owned handguns for several years still classify as a "beginner" in terms of real skill) has no business practicing fast draw, so let's leave it at that.

Learning to shoot a handgun takes a lot of time and practice but, if you go about it the right way, you should eventually be rewarded with a certain amount of skill. Shoot at targets — printed, paper targets, although homemade ones are also acceptable — that will let you see how you're progressing by the size of your bullet groups. Resist plinking indiscriminately at tin cans and other "fun" targets until you're hitting the bullseye with satisfying regularity.

Like rifle shooting, handgun marksmanship and handling is best learned with a rimfire firearm. Recoil and noise are the two big factors to consider here, and cost is also important. You'll learn faster with a mild-kicking handgun, and 22 ammo is inexpensive enough to allow regular practice without breaking the budget. Unless you handload (and few beginners do), buying centerfire cartridges can keep you permanently broke.

Centerfire handguns recoil heavier (some come back at you hard enough to make control impossible until you've mastered the basics), shoot louder, and cost more to operate. Still, anyone who learns how to shoot accurately with a rimfire handgun should have little trouble adapting those skills to centerfire shooting. But I'd advise anyone to start off with a gun chambered for 22 Long Rifle ammunition.

Ear protection is necessary for any handgun shooter, as even rimfire ammunition produces a loud report when fired from a short barrel. And the large centerfires can be downright painful to shoot without something to muffle the sound.

The key to handgunning is to first learn the basics, and then practice, practice, practice. And *never* forget the safety rules. You don't need to go to the range to practice, either. You can work on your stance and trigger squeeze indoors by aiming an *empty* gun and squeezing at some point on a distant wall (but remember to double check the gun for ammunition, and make sure no one is in line to be hurt even if the gun *should* accidentally fire). This is called "dry firing," and it won't hurt most handguns. The exception is some 22 rimfire revolvers, as the offset firing pin can peen into the chamber face if the chamber is empty. In this case, you can cushion the firing pin by inserting *empty* 22 cases into the chambers before practicing.

A ploy you can use (when you're shooting *outdoors*) to tell if you're developing an accuracy-destroying flinch is to let a trusted partner load your gun (revolvers can be loaded with one or two chambers left empty) and hand it to you — then shoot at a target without checking to see whether there's a shell in the chamber or not. If your hand jerks down to pull the sights off target when the hammer falls (by surprise) on an empty chamber, you're flinching!

When shooting at targets, start with the targets set fairly close — say 10 or 15 yards away. Then gradually move the targets farther away, as you become better able to score consistently at the closer ranges. When you can hold all your shots in the bullseye of a "slow-fire" pistol target at 25 yards, you're on your way to becoming a good shot.

chapter 13

HANDGUN CARE AND CLEANING

CLEANING ANY HANDGUN requires careful attention to safety, gun knowledge and patience on the part of the owner. Note that a sad number of accidents occur because someone with more confidence than common sense thought that "the gun wasn't loaded." In the case of handguns, because it is so easy to lose track of the muzzle's direction, accidents can happen when shooters don't think. Especially in the case of the beginner with an auto pistol, be certain that you know *exactly* what every button and lever on your pistol does. Most firearms manufacturers include a detailed assembly and instruction manual with your new handgun. Read the book *carefully.* If you bought a recently-built used handgun and have no manual, see your local gun shop. Your gun dealer can instruct you or tell you how to write the factory for a manual. Most all gun manufacturers and importers are happy to supply manuals at little or no charge. Follow the instructions and you may be assured that your handgun will shoot well and reliably and stay in fine condition for many decades to come. If you disregard this information, you may find yourself with a handgun badly damaged from neglect or — far worse — with a potentially injurious accident on your hands.

The same advice expressed in the rifle cleaning chapter applies here only to a greater extent. Don't overdue the business of disassembly. Ordinary cleaning and care does not require detail stripping. To make matters still worse, the mechanical complexity of many handguns makes reassembly more difficult than with the average rifle. If you feel that you must take your revolver or auto pistol apart, the instruction book will tell you where to start and where to stop. The book will tell you how far *not* to go. Don't become one of those sad folks who formerly owned a handgun and now owns a cigar box full of pistol parts.

When buying powder solvents and cleaning equipment, stay away from the hardware store. Stick with the gun shop. Many hardware store materials will "do the job," but may be far too harsh or less than ideal for the task. Cleaning materials from a number of manufacturers are widely available — ask at your local gun shop to find out what's available in your area.

Years ago, corrosive priming made immediate cleaning an absolute necessity, as discussed in the rifle cleaning chapter. In the rifle chapter, it was pointed out that most non-U.S. military surplus ammo is still primed with corrosive chemicals. In

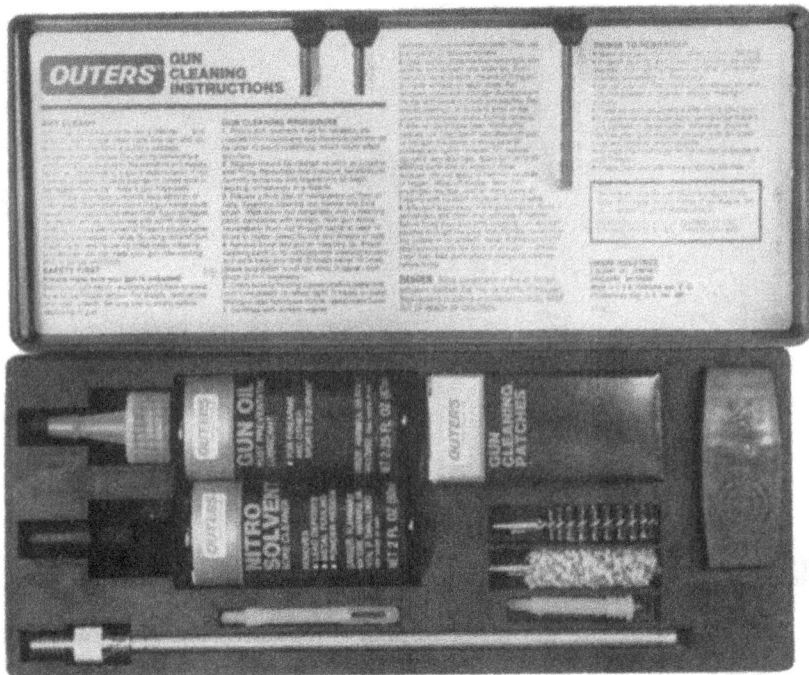

This Outers handgun cleaning kit is typical of kits made for every type of gun. The basic supplies are all here, along with detailed cleaning instructions. Buying the cleaning kit is also usually cheaper than buying supplies one item at a time.

the case of handguns, the problem is a bit simpler, as the number of surplus ammo pistol calibers are so limited, with 9mm by far the most common. As the 9mm Parabellum (Luger) cartridge has become virtually the world standard for good and bad guys, all sorts of 9mm military ammo is turning up at very good prices. As this is being written, Egyptian surplus 9mm ammo is being offered in many gun shops. While this ammo was originally advertised as "non-corrosive," a number of shooters of the Egyptian ammo have seen red grunge grow in their 9mm handgun barrels, which would indicate corrosive properties. As pointed out in the rifle chapter, good practice is to simply assume that *all* military surplus ammo not produced in the U.S. is corrosive. About the only other common handgun caliber ammo offered as military surplus is 45 ACP — and very little of that caliber is available since the 45 has been used by few countries. Little of that 45 ammo has made it back to the U.S.A.

In today's world of non-corrosive priming, immediate cleaning is not the compelling proposition it once was, particularly if your handgun is made of stainless steel. That statement does not mean that neglect is a virtue — it simply means that neglected handguns once very quickly turned to lumps of rust. Today, even if the handgun is made of stainless steel, neglect does not somehow im-

prove the handgun. Stainless guns are a boon to those who live near salt water or who keep a handgun aboard a boat. Still, note that all stainless handguns are not exactly "stainless." Many, particularly less expensive models, use alloys which *will* rust to some extent. Even if the handgun does not rust heavily, fouling and general grunge can very negatively affect reliable function.

If your handgun is a 22 rimfire, it needs only enough cleaning to protect it from corrosion and to keep the moving parts in smooth operation. Modern 22 ammunition is designed to inhibit rust and it generally does a good job of doing so. While some suggest that running a lightly oiled patch through the bore is all that is necessary for rimfire care, I prefer to use a brass or bristle brush and solvent or oil, followed by a patch. Remember that even 22 powder fouling likes to soak up moisture from the air and can make a nasty mess if neglected — temporarily or otherwise.

Cleaning Your Revolver

Since just about all revolvers must be cleaned from the muzzle end of the barrel, you must exercise a certain amount of caution when cleaning these guns. Avoid slam-banging a cleaning rod down the barrel as you may damage the rifling at the muzzle when you do clean your revolver. Those first few inches of the muzzle are critical to

146

Cleaning Your Revolver

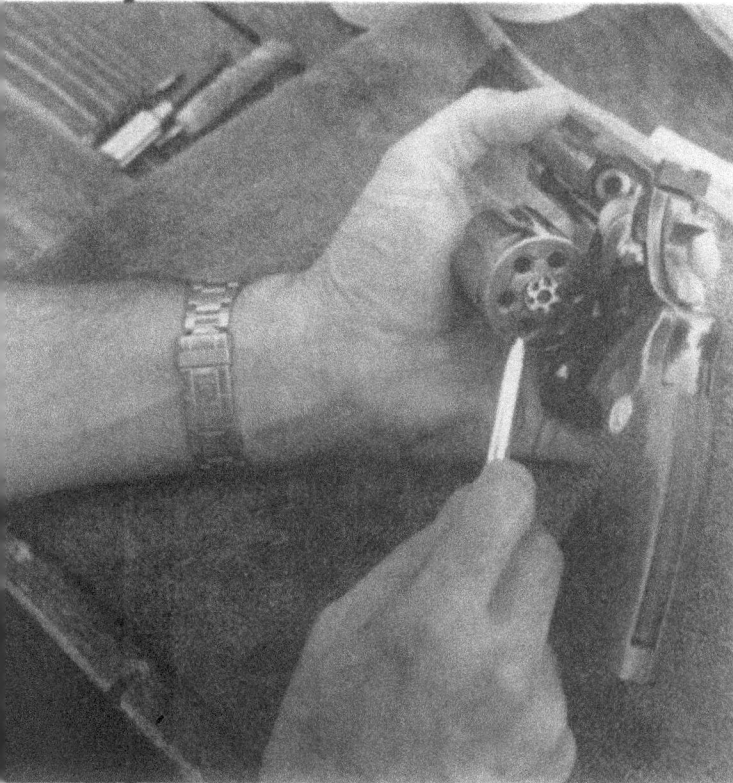

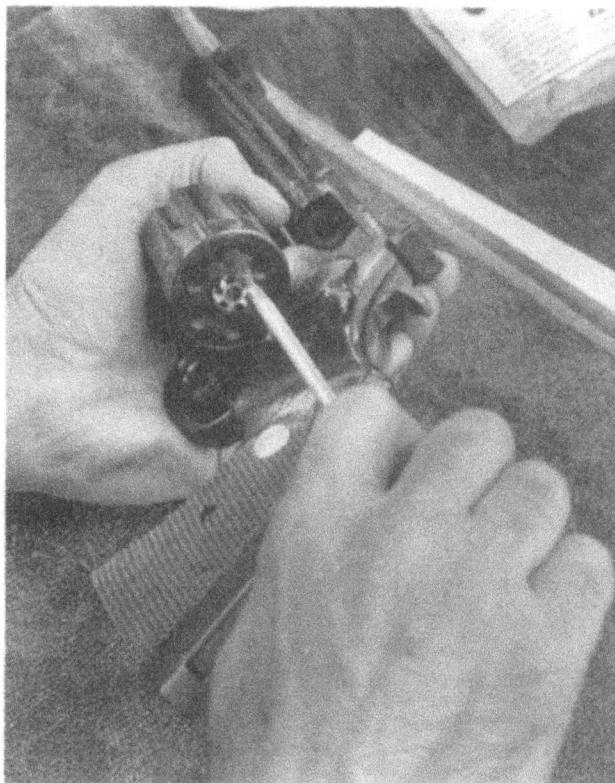

Before you start the cleaning process, point the revolver in a safe direction, swing out the cylinder and make sure each chamber is empty. Next, take a solvent-soaked bore brush of the proper caliber and scrub out each chamber in the cylinder (above). After you've scrubbed out the cylinder, re-moisten the brush with solvent and scrub out the barrel from the muzzle (below, left) being careful not to damage the crown or rifling.

(continued page 148)

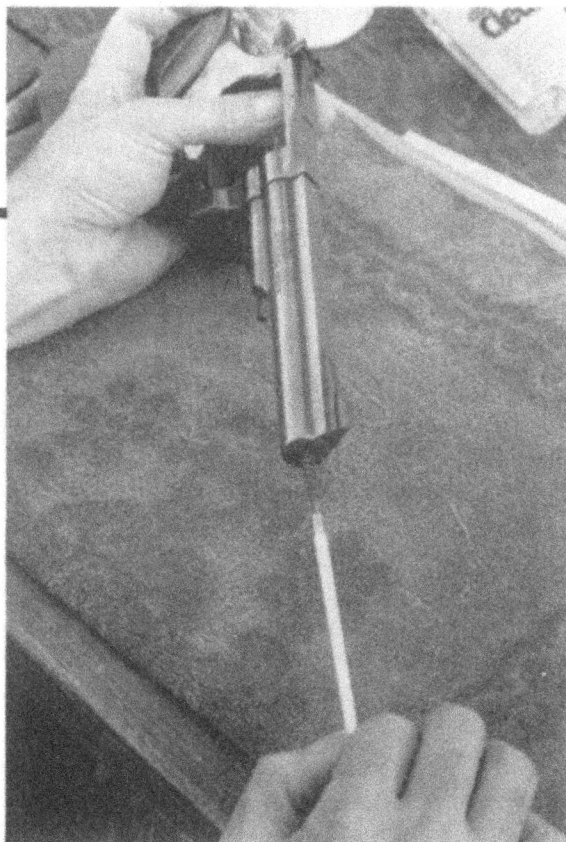

accuracy as they represent the last few inches of barrel the bullet will travel through. A burr or dent in this portion of a barrel — left by a carelessly used cleaning rod — will have a negative effect where accuracy is concerned.

After a heavy shooting session, especially if your gun is a centerfire loaded with jacketed or lead-alloy bullets, you should remove any visible dirt, grime and powder residue. Unload your revolver, check the barrel and, if the rifling appears unsharp or indistinct, run a patch with some solvent on it through the bore *carefully*. Do the same to the chambers of the cylinder. Let the solvent remain in the barrel and the cylinder for a few minutes before you run a clean patch through. The solvent will loosen up powder and lead residue, making the final pass with a clean patch easier — on the bore and you. While you're waiting for the

147

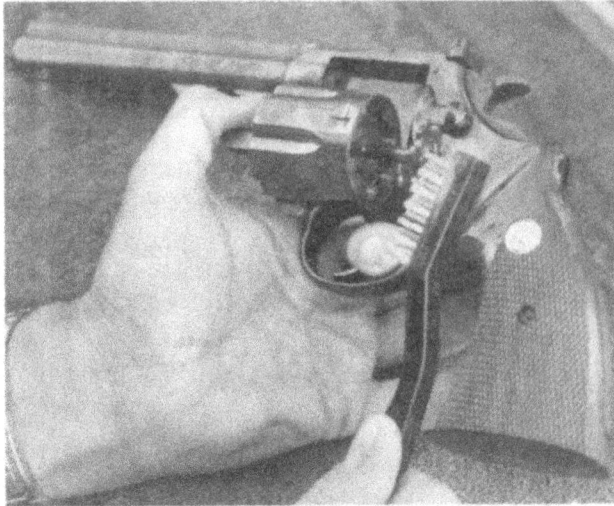

If your dealer has a bronze bristle brush like the one shown here, buy it. Dip those bristles into the solvent and brush off the rear face of the ejector (left) and other areas of powder/lead residue build-up (below).

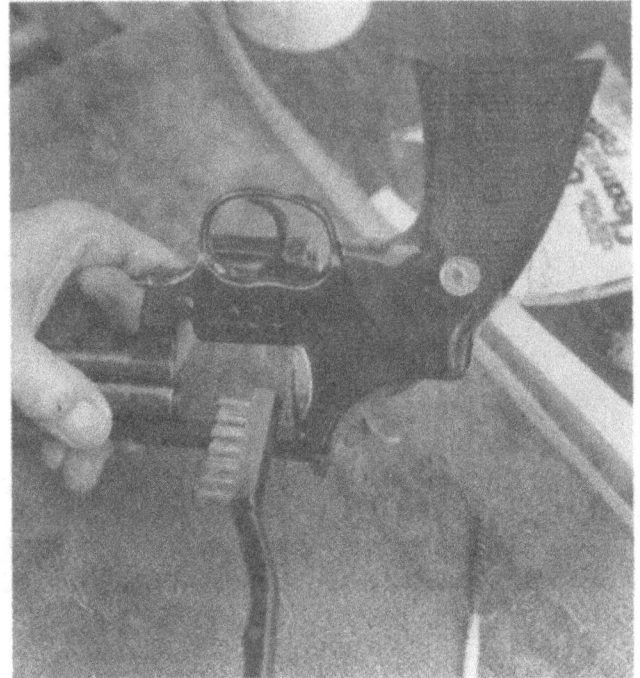

(Left) The face of the cylinder is a prime place for powder residue build up. Brush the face of the cylinder with solvent (or the carburetor cleaner seen here) and finish the job with Birchwood Casey's Gun Scrubber.

solvent to do its work, take a solvent-soaked patch and remove any visible external powder residue from the rear face of the barrel, the inside of the frame and any other area that looks dirty or smudged. A toothbrush can be helpful in removing any accumulated grime that's difficult to remove with the patch.

After you have let the solvent do its work, wipe the gun clean with a piece of flannel rag and carefully run a clean patch or two through the barrel and chambers. If you've done the job correctly, those clean patches and the hunk of flannel rag will be covered with black, oily grime. Make a final inspection of the bore and chambers, and, if you've been using shot cartridges, or the rifling

still appears indistinct (and the chambers look "dirty") you might want to attach a brass bristle brush (of proper caliber size) to your cleaning rod, dip the brush in solvent and make a couple of passes through the bore and chambers. This will remove any powder or lead residue — just run a clean patch through the chambers and bore to finish the job. *Lightly* oil a patch and run it through the bore and chambers and give the external works the same *light* oiling.

If, however, cleaning with a brass brush doesn't do the job (as indicated by heavy blue-green tints on the patches run through the bore), there are other ways to tackle the problem. The Lewis Lead Remover is one gadget designed to remove heavy

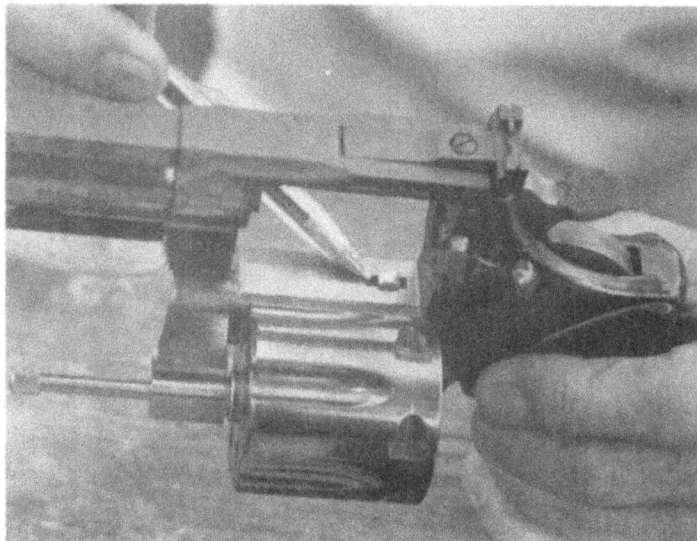

Lubrication: After you've cleaned your D/A revolver, don't forget to lightly oil the various moving parts. Oil the hand which turns the cylinder (bottom right), the cylinder lock (below) and the ejector rod (bottom left).

At this point, the bore, chambers of the cylinder and other areas have been scrubbed. Take a series of clean patches of proper caliber and dip them into the solvent and swab each chamber and the bore. Follow up with clean dry patches.

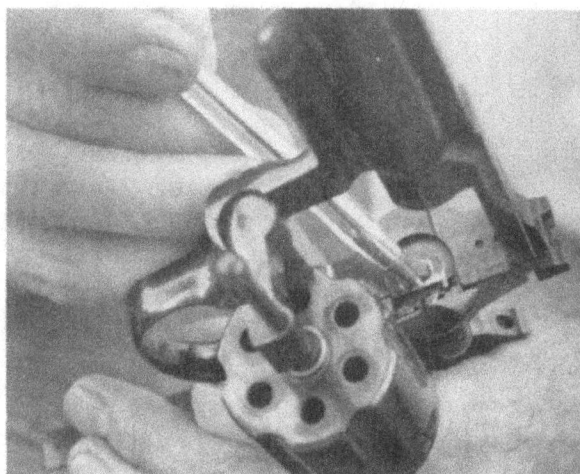

leading from bore, chamber and forcing cone — it uses brass wire screens pulled through the bore by a specially designed pull rod. On the rare times such drastic action is needed, this outfit gets the job done. Now your gun is ready for storage.

When not in use, all handguns I own are wiped down periodically with a lightly oiled cotton cloth to keep the rust demon away. Running a lightly oiled bristle or brass brush through the bore and chambers is also a fine idea — it certainly hurts nothing and can do a great deal of good. If you happen to live in one of the humid areas of the U.S. — which means about any place other than the desert Southwest — you should check your handgun at least once a month for signs of rust. If you happen to keep the handgun on a boat, a weekly or more frequent check is in order. While a good — or even a rather mediocre handgun — will stand up to generations and thousands of rounds of use, note that rust and corrosion will turn a $1,000 handgun to junk as rapidly and gleefully as it will turn a $60 handgun to a paperweight. No matter where you live, always remember to wipe fingerprints off the handgun after handling. Good etiquette dictates that you should handle other people's handguns only by the grips unless you have been told otherwise. Fingerprints from some people can be quite benign, while prints from other folks have about the same effect on polished metal as pickling brine.

Cleaning Your Auto Pistol

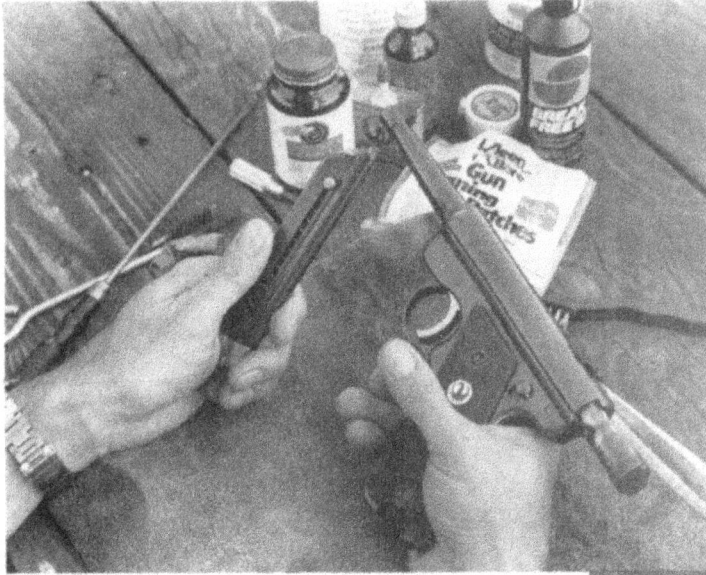

Before you start your cleaning, point the firearm in a safe direction and remove the magazine and lock the bolt open (left) per manufacturer's instruction. Make sure the chamber and magazine are fully unloaded (below) before proceeding.

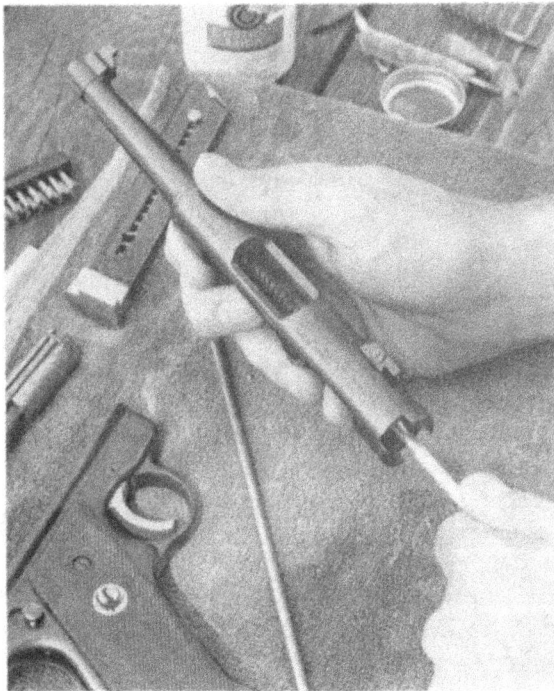

(Left) Next, field-strip the pistol (per manufacturer's instructions) and proceed to scrub out the bore, from the breech, with a solvent-soaked brush of proper caliber.

Cleaning Your Auto Pistol

A rimfire auto pistol is a slightly different matter. An auto pistol should be taken apart and given a good cleaning after every shooting session or it's soon going to stop functioning. Powder and dirt will build up inside until the gun refuses to feed and/or fire — and when that happens, it's a good hint that cleaning may be overdue.

Most auto pistols (again, read your instruction book) are easy to disassemble or take down into their basic component groups — slide, barrel and frame — and that's as far as you need to go to give it a proper cleaning. First making sure the gun is unloaded, use a dry or solvent-soaked toothbrush to get at those hard-to-reach openings, then wipe everything down with an oiled cloth. Don't overdo it with the oil, though, or it'll just collect gunk all

(Right) When most semi-auto pistols are field-stripped, you end up with the frame assembly, the barrel (or frame/barrel) assembly and the slide (or bolt, in the case of this Ruger). At this point you can clean out the various sub-assemblies with solvent, carburetor cleaner or degreaser.

(Below) Pay close attention to the face of the slide/bolt. It rapidly picks up fouling during normal firing. Our suggestion? Take a bronze bristle brush, dip it in Shooter's Choice and thoroughly scrub that area. By doing so you'll help guarantee the reliable function of your autoloader.

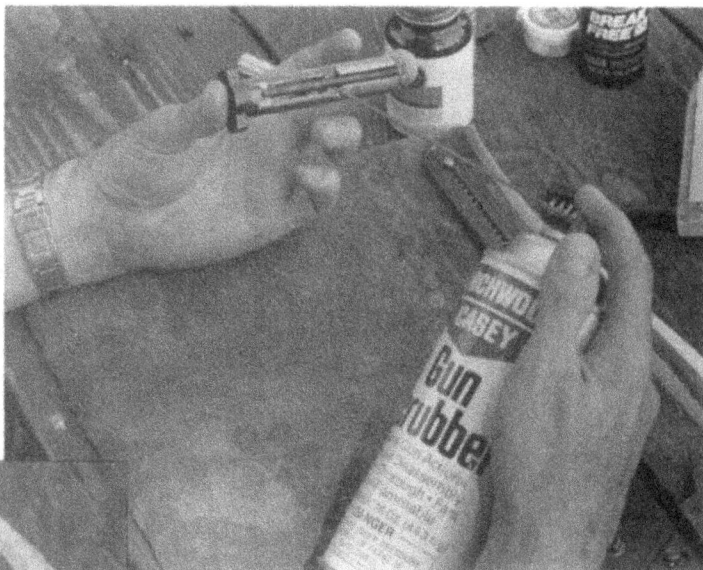

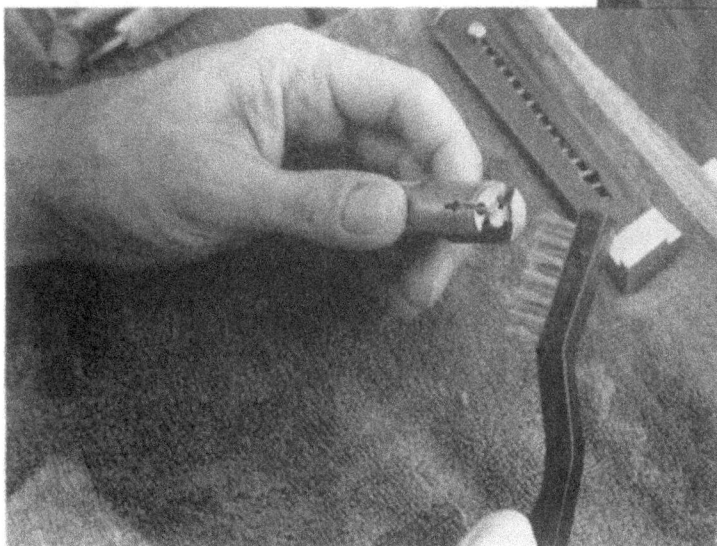

(Right) After you've scrubbed the bore, take a solvent-soaked patch and run it through the bore. Follow up with two or three dry patches to remove the dirty solvent.

(continued page 152)

that much quicker.

A centerfire auto pistol is cleaned the same way a 22 is — take it apart, clean the bore, and get rid of the dirt and residue. Wipe with an oiled cloth, and you're ready to reassemble the gun. Before you do, remove the barrel from its slide (most centerfire auto barrels are removable) and, using a short cleaning rod of the proper size, swab the bore with first a powder solvent-soaked cleaning patch (again of the right size), a dry patch, and a patch lightly covered with oil. (Use of a brass bore brush may be necessary if the gun has been heavily used.) Do your cleaning from the chamber end, as cleaning from the muzzle may damage the lands and grooves and destroy the gun's accuracy.

Be careful in handling the magazines from an auto pistol, as the magazine lips may be readily and permanently damaged. The magazine lips are

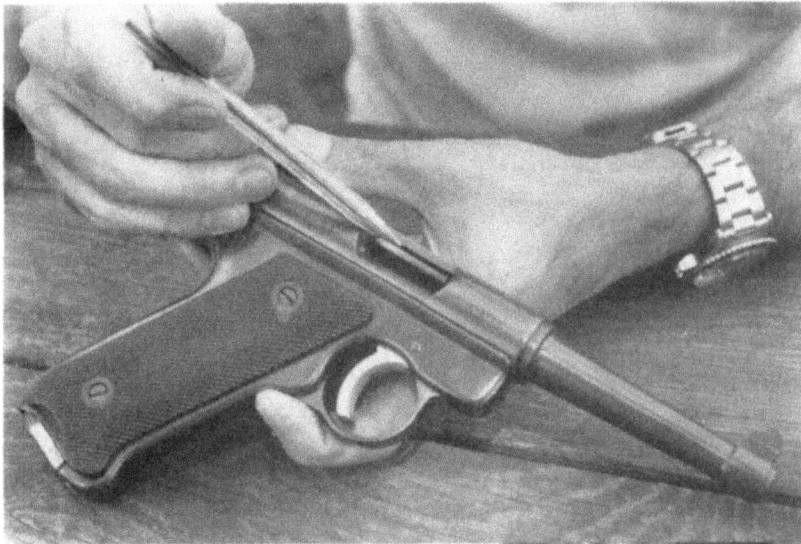

Lubrication: After you've finished your cleaning chore, place only a drop or two of gun oil on the slide rails (bolt guide and bolt on the Ruger seen here — top and center), and on the extractor (bottom). Additional lubrication and reassembly should be per the manufacturer's instruction manual.

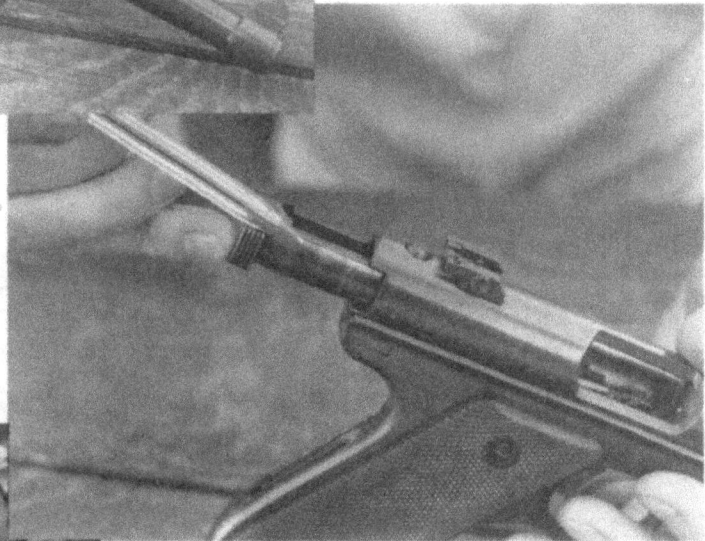

at the very top of the magazine and serve to align the cartridge precisely with the chamber. If they are dented or cracked, chances are that your pistol will not work. If the magazine is damaged — a lot or a little — the result will be that your auto pistol does not function reliably. Note that many auto pistols have been traded off or otherwise disposed of because of magazine damage. The shooter believed that something was wrong with the pistol, when the problem involved only the magazine. Some writers have suggested that magazine lips may be repaired with pliers. Perhaps. In 30-odd years of shooting, I have seen many attempts, but never a really successful plier repair of an auto pistol magazine. Some magazines may be disassembled for a thorough toothbrush cleaning. Whether the magazine comes apart or not, give it at least a careful wipe-down when cleaning the handgun to

Cleaning Your Single Action

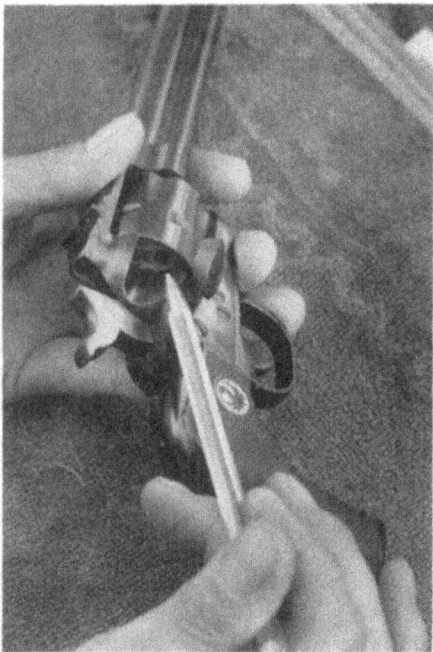

Prior to cleaning, point your SA revolver in a safe direction and check each chamber in the cylinder to make sure the gun is unloaded.

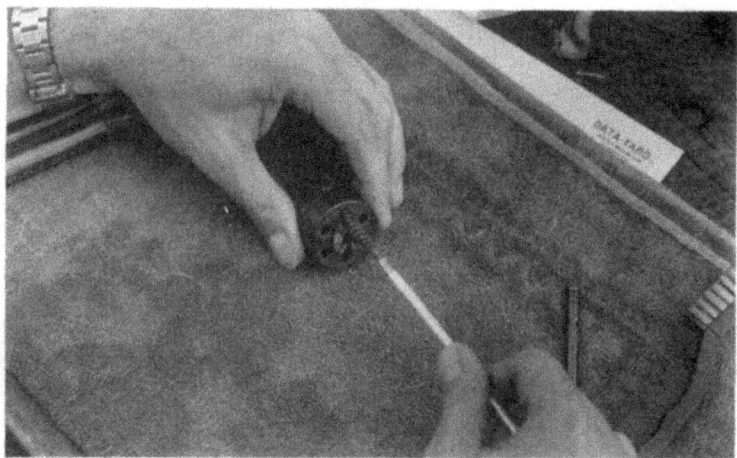

Once you've removed the cylinder (per your manufacturer's instruction manual), take a solvent-soaked brush and clean each chamber in the cylinder (above). Next, re-moisten the brush with solvent and carefully scrub the barrel from the muzzle (right).

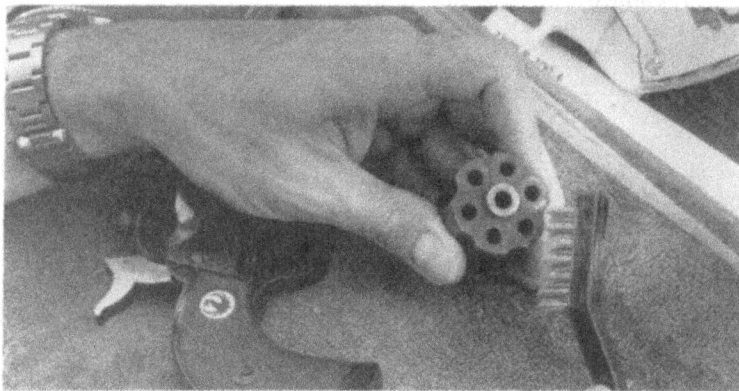

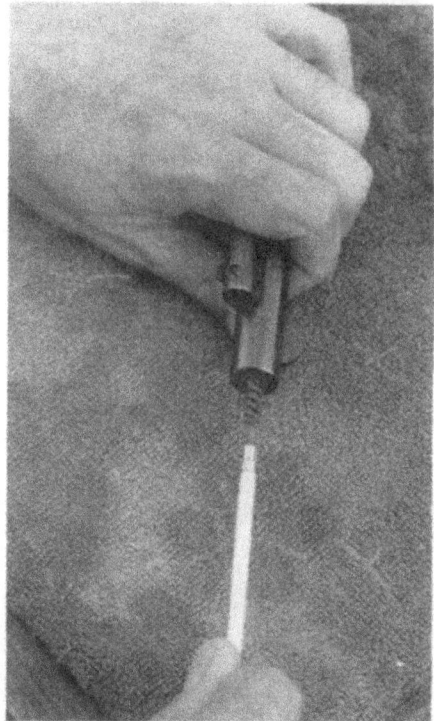

While everything is soaking, take a bronze bristle brush, dip it in solvent and brush out the powder fouling on the interior of the top-strap (right) and on the face of the cylinder (above).

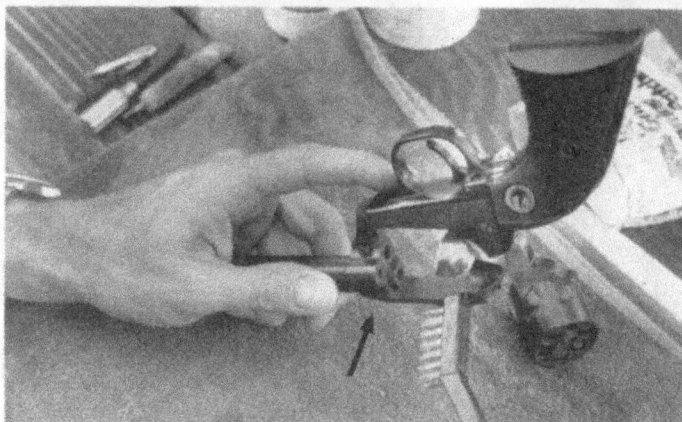

(continued page 154)

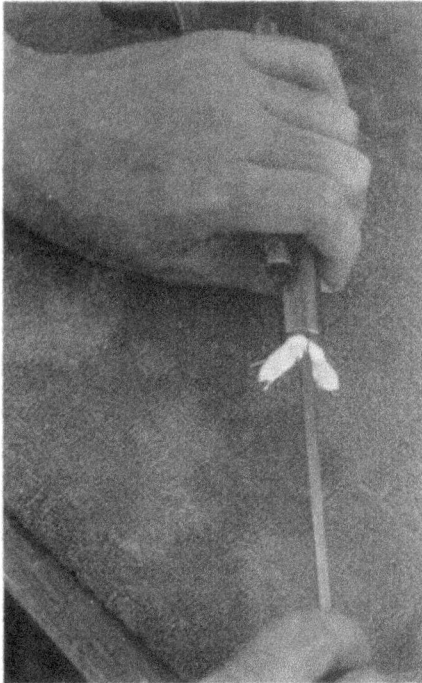

After you've bore-brushed the cylinder and barrel, run a solvent-dampened patch through each chamber in the cylinder and the bore. Lastly, run a series of clean dry patches through the cylinder (below) and the bore (left).

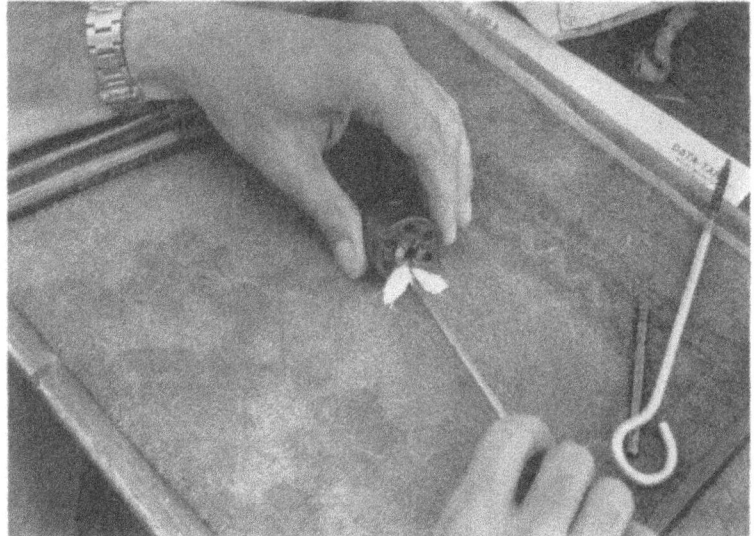

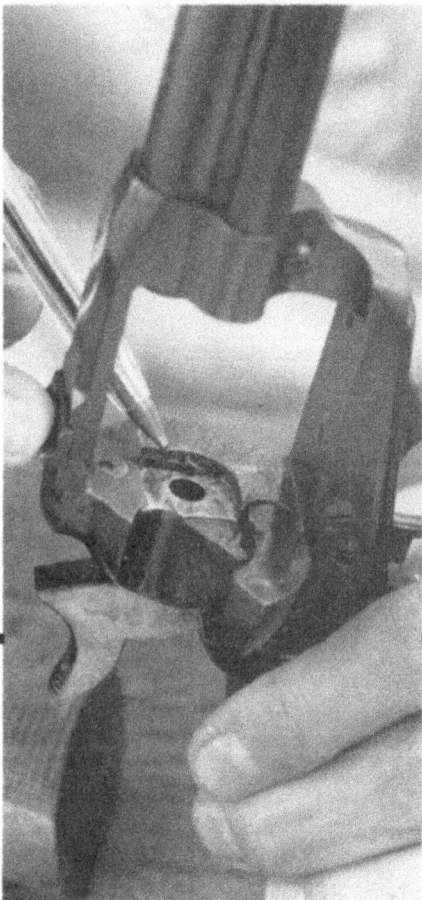

Lubrication: Place a drop or two of gun oil (Break Free is a good choice) on the hand (left), the cylinder lock (below), at the base of the hammer (opposite page, above) and on the ejector rod (opposite page, below). Lastly, reassemble per the manufacturer's instruction manual.

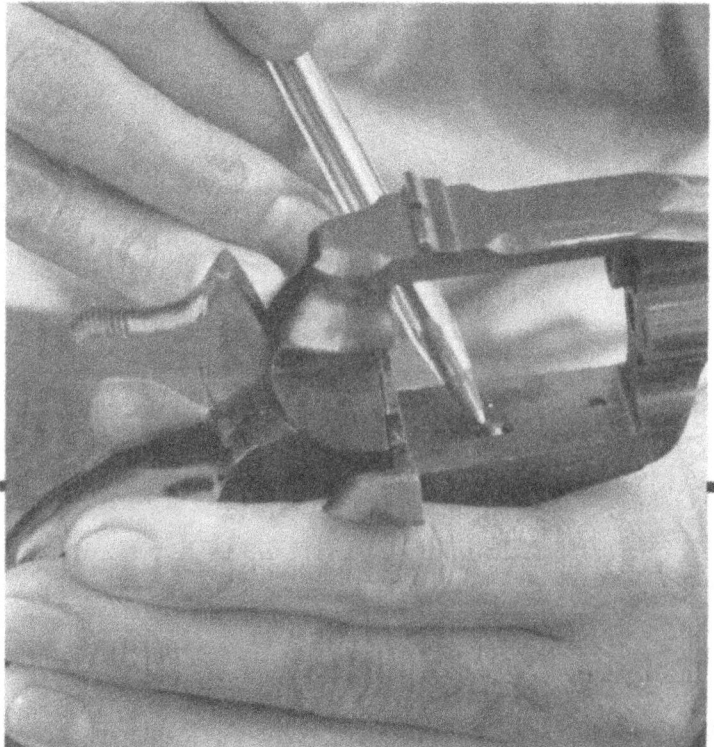

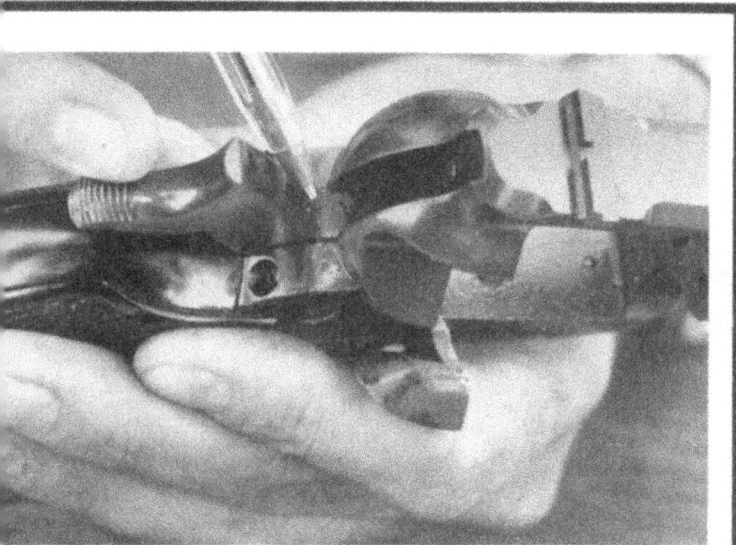

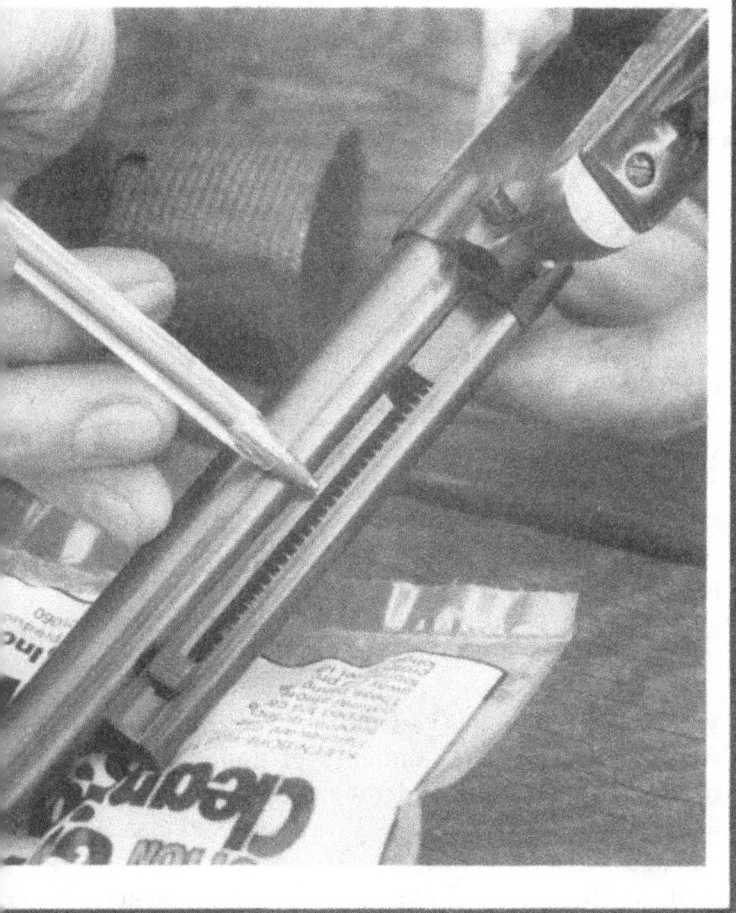

remove bullet lube or fouling.

Remember that holsters are for *carrying* handguns, not storing them. Many shooters believe that the quality and cost of a holster must relate to its storage ability. Don't believe it. Leather gathers moisture from the air and is happy to convert that moisture to nasty, grainy, damaging red rust. Chemicals used to tan the leather in some holsters may also be corrosive. The same applies to the hard-side, foam-lined storage cases mentioned in the rifle chapter. The foam is very efficient in preventing handling, storage and carrying damage. But, even more so than leather, the foam absorbs water from the air and converts it to rust — all over your handgun.

Many people who don't care for the cleaning chore opt for stainless steel handguns. While buying a stainless handgun is hardly a bad decision, don't make that decision for the wrong reason. While rust and corrosion are *less* of a problem with stainless handguns, note that reliable operation of stainless auto pistols and revolvers is very dependent on good care. Failure to provide minimal care and cleaning will create a nearly unrusted handgun — but also one that does not work very well.

Good handguns cost money. To protect your investment and to insure your safety, your handguns must be cleaned and given at least minimal care. Don't be sloppy and uninformed when it comes time to perform the cleaning task.

chapter 14

CHOOSING A SHOTGUN

UNLIKE RIFLES, shotguns are designed to hit moving targets at relatively close ranges and are *pointed* rather than precisely *aimed*. The gunner isn't looking through a set of precision sights, but sees only the barrel of the gun and the target. Since sights aren't used (except for a small bead mounted at the muzzle), the shotgun shooter must depend on holding his gun the same way each time he fires it and maintaining a consistent gun-to-eye relationship when he lines up on a target.

The point I'm trying to make is that a shotgun must fit you — it must feel good in your hands, and when you throw it to your shoulder, it must fit and point naturally. If a shotgun feels at all unnatural or awkward when you pick it up and handle it, it's probably not the gun for you. And in spite of any expert advice or help you seek, only you can make this determination. Choosing a shotgun is a highly personal thing, and can't really be delegated to another individual—no matter how well qualified as a shooter—if you want to find the gun that works best for *you*.

With that bit of wisdom in mind, let's take a look at the different types of shotguns available. Some are better than others for beginners, and budgetary considerations may put certain types at least temporarily out of reach. This book can help you narrow down your choice to a particular type or action style. When you've chosen the type and action style, you can examine several different makes and models of that type and make the final decision based on which one looks, feels and handles best to *you*.

Which Action Type to Choose?

There are six basic action types available in the shotguns sold today: **autoloader, slide action (pump), side-by-side double, over/under** (stack-barrel) **double, bolt,** and **break-top single shot.** They all have their separate advantages and disadvantages, and each type has its share of fans as well as detractors. Let's take a look at the good and bad points of each, in turn.

The **autoloader,** or semi-automatic shotgun is a single barreled gun and has been with us since John Browning patented his famous recoil-operated model at the turn of the century. While that early model is still being manufactured and continues to sell well, there are newer designs that use expanding gas (tapped from the barrel during firing) to work the action. The autoloading shotgun has never been more popular, and one reason is

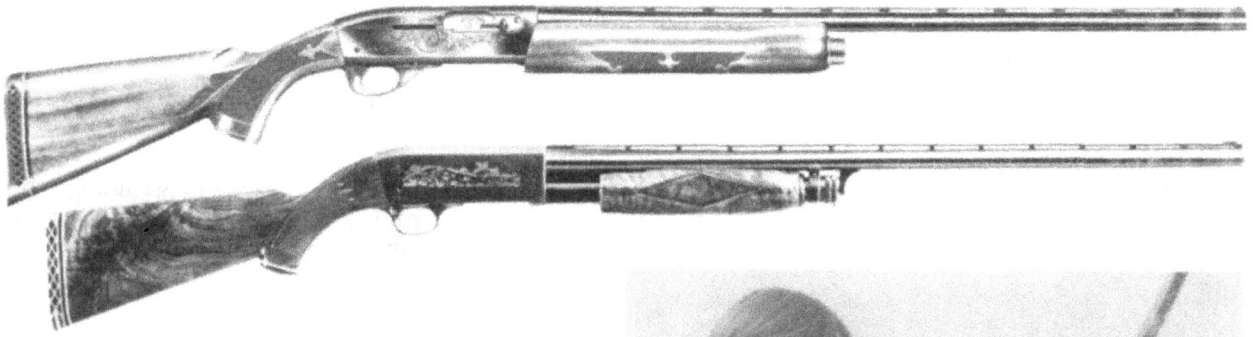

(Top) The autoloader is popular among hunters and target gunners because of its recoil-reduction features. However, this type isn't the best choice for beginners.

(Above) The pump, or slide action, is another popular repeater. Because it's manually operated, it makes a better choice for first-time shooters than the self-loader. Shown is Ithaca Model 37.

(Right) Autoloaders and pumps feature tubular magazines to give three or four shells in reserve, in addition to the load in the chamber.

that today's selfloader is an exceptionally reliable, soft-recoiling firearm. If an auto shotgun is properly cleaned and cared for, it will give long service and malfunction but rarely. At one time autoloaders had a reputation for jamming at critical moments (during a covey rise of quail, for instance), but if this was ever true, it certainly isn't so today.

The big advantage modern autoloading shotguns offer isn't their "just-pull-the-trigger" convenience or speed of fire. The big thing selfloading smoothbores have going for them is the fact that they are relatively pleasant to shoot with even heavy loads. In effect, they "soak up" part of the recoil and make firing these guns easy on the shoulder. While they don't actually absorb a significant amount of recoil (some of the recoil force *is* used up in operating the action, but this amounts to only a small fraction of the rear-directed forces acting on the gun and shooter), autoloaders stretch out the time the recoil force acts upon the shooter. Instead of feeling a short, sharp "kick," recoil comes as more of a drawn-out shove. The effect is more pleasant to the shooter.

This "recoil reduction" effect is important to the duck or goose hunter who wants to use magnum loads to help him bring down high-flying waterfowl and is equally appreciated by Skeet and trap shooters who may fire a hundred or more shells in the course of a single afternoon. The effect of recoil tends to be cumulative, and even light target loads can become unpleasant to shoot after 40 or 50 rounds.

Like most magazine-fed shotguns, the majority of autoloaders being sold today give the shooter four or five shotshells to fire before reloading becomes necessary. Such guns used to hunt waterfowl must have their magazines plugged temporarily* to give them a three-shot capacity, but that's still one more than the owner of a double-barreled shotgun has on tap.

Autoloading shotguns are longer than double-barreled shotguns, too — mainly because the action adds 5 or 6 inches to the overall length of the gun. This tends to make these guns slightly muzzle heavy in the hands—an effect trapshooters and waterfowlers like, because it tends to smooth out the swing.

Older autoloaders were typically a bit longer and bulkier than pumps, but that is hardly a factor today. A look through the *Gun Digest* catalog sec-

---

*By Federal regulation shooters hunting migratory waterfowl must reduce the total firing capacity of their autoloading shotguns to three shots.

(Left) Double-barreled guns are well balanced for quick handling and offer 2 degrees of choke.

(Below) The side-by-side double shotgun gives you two shots, yet is nearly as safe as the break-top single-shot. This Savage-Fox has twin triggers — the front trigger activates the more open choke barrel.

(Bottom) Stackbarreled doubles, or over/under shotguns like this Savage, offer a single sighting plane and good balance, but they can be more costly than the side-by-side doubles.

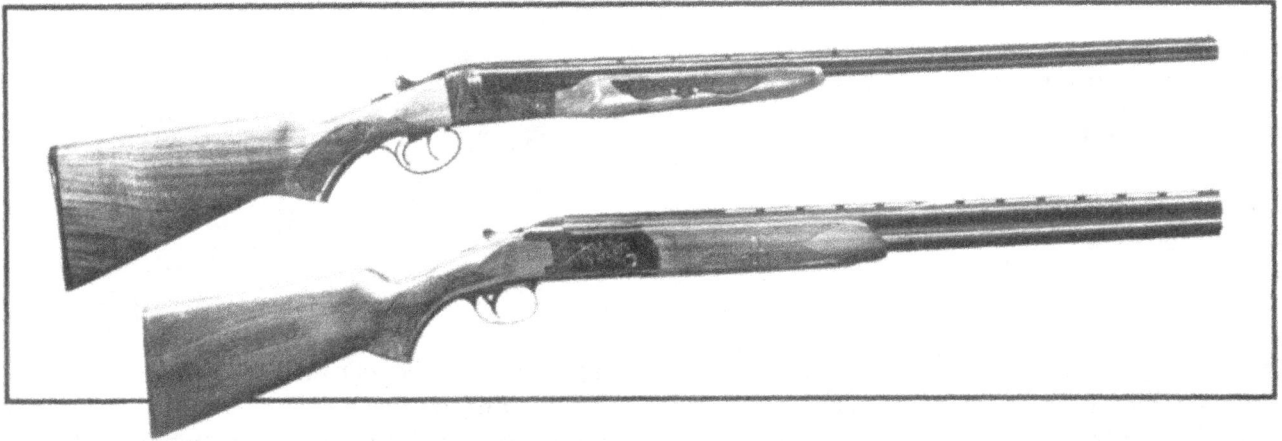

tion shows many light, trim autoloading shotguns. Remember, though, that the lighter guns tend to recoil more than the heavier guns. It comes down to whether you are recoil sensitive or sensitive to carrying weight in the field.

Finally, the autoloader is probably the worst possible choice for a beginner's gun. While it's as safe as *any* shotgun as long as all the safety rules are obeyed, it can be less forgiving of a momentary lapse. Some first-time shotgunners forget that the gun remains fully loaded and ready to go, even after it has once been fired. Too, an autoloader may be too heavy for a young beginner to handle well.

The **slide action,** or pump, a single-barreled gun, is also relatively long and sometimes heavy (although there are exceptions to this rule), and like the autoloader offers three or four shots in reserve. Because it's manually operated, some shooters feel more comfortable with a pump than a selfloader. They feel that the slide action is more dependable — and this may be the case when a variety of loads are used. Most autoloaders function well with *either* magnum or light target loads — but few models will digest *both* with equal ease. Con-

sequently they lack some of the versatility a pump gun offers, as the pump doesn't depend on metered gas flow to work the action.

Because the pump *is* manually operated, it makes a better choice for the beginning gunner. The "trombone" (pump) action is our most popular repeating shotgun, and nearly every manufacturer offers one or more models. Interchangeable barrels are available for both autoloaders and pumpguns to give the shooter a variety of chokes* he can use under different conditions. This makes both the pump and auto guns very versatile. Pump shotguns are, however, usually less expensive than autoloaders of similar quality, and this is another consideration.

There are two different types of **double-barreled shotguns** you can buy — the *side-by-side* double, with the barrels mounted alongside each other in the same horizontal plane, or the *over/under* "stackbarrel" with its vertically aligned twin

---

*"Choke" is the degree of constriction at the muzzle of a shotgun. The *amount* of choke determines how *dense* or *loose* the shot pattern will be at a distance of 40 yards.

tubes. These guns function in much the same way and share pretty much the same advantages — relatively short overall length, good balance and the immediate choice between two different degrees of choke.

This is one of the most important advantages double-barreled guns offer to hunters. Each barrel usually wears a different degree of choke — which is what affects the way the shot charge spreads when it leaves the muzzle — and the shooter can choose the barrel he or she wants to fire first. If a bird flushes at close range the more open-choked barrel is fired (and some double-barreled guns are permanently set in this sequence), while the tighter-choked tube is available for long shots.

Double-barreled guns are one of the best choices for the beginning shooter, because they can be broken open (like a single shot shotgun) and left that way until you're ready to shoot. A double-barreled gun with its action open can't possibly fire, and you can tell at a glance if there is a shotshell in either chamber.

Some double-barreled guns feature a tang-mounted safety that engages every time the action is opened. That means you have to push the safety to "off safe" before the gun can be fired. I personally don't care for this arrangement as the shooter may come to rely on this "automatic safety" feature and fail to develop the proper safety habits himself as a result.

The *side-by-side* double guns are available in a wide variety of price ranges, running from around $250 for the Stevens Model 311 from Savage on up to many thousand dollars for a deluxe British or European import. Their broad twin barrels catch the eye, and some shooters claim they score better with this type of gun for that very reason. A well-balanced double is one of the fastest handling shotguns you can buy.

Over/under shotguns give you a single barrel to look over, and many bird hunters prefer these "stackbarreled" guns to the horizontally aligned side-by-sides for this reason. These gunners feel uncomfortable with the broad sighting plane offered by the traditional side-by-side because of the fear of cross shooting (shooting to one side of the target through an error in sighting). The single sighting plane offered by the over/under barrel arrangement provides more precise eye reference.

The stackbarreled double has only two real drawbacks when compared to the side-by-side: The barrels strike a wider arc when the action is

(Top) Double-barreled guns can be safely carried in the open position, so they're a good choice for beginners. Fancy SKB 20-gauge is one of the author's favorites.

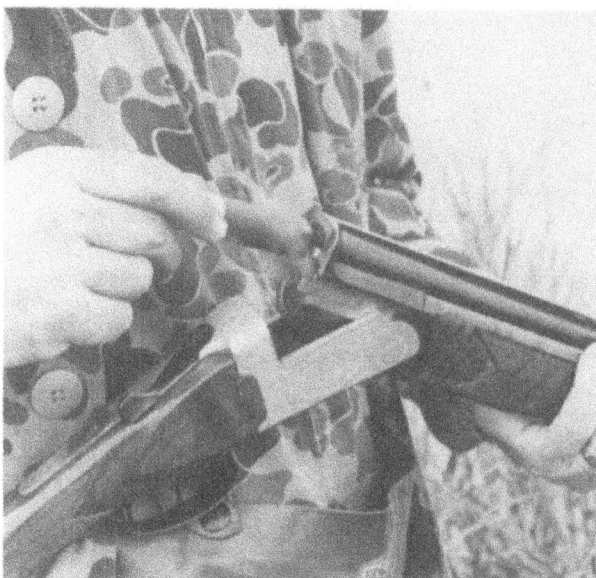

(Above) This over/under has plain extractors. Fired hulls must be removed from the chambers by hand.

opened (which could be a problem in the tight quarters of a duck blind), and guns featuring this barrel arrangement are often more expensive. Over/under shotguns start at around $300, and the majority of these guns sell for nearly twice that amount.

Whichever type of twin-barreled shotgun you

This Franchi over/under features selective ejectors that automatically throw empties clear while leaving live shells chambered. (Note the empty shell near the author's right lapel.) This option can run the price of a gun up, but many hunters like it.

The single shot is the safest type of shotgun made, and has been a traditional beginner's shotgun for over a century. They are also the least expensive way to start shooting the shotgun.

choose (if that turns out to *be* your choice), there are two options that can substantially affect price. These guns are sold with *plain double triggers* (each trigger fires a different barrel); a *non-selective single trigger* (the first pull fires the open-choked barrel, while pulling a second time triggers the other barrel); or a *selective single trigger* (you can choose the firing sequence by flipping a selector lever). The first is the least expensive arrangement, while the last is the most costly.

The other option is often an expensive one, but many shotgunners feel it's worth the extra cost. While nearly all moderately priced doubles come with plain extractors that merely lift the fired shells far enough from their chambers to let you grasp them with your fingers and eject them manually, some guns offer selective automatic ejectors that throw fired hulls clear of the gun while merely raising unfired shells from the chamber when opened.

Double-barrel shotguns offer many of the safety features found on the single-barrel break-tops, yet offer the advantage of an instant second shot. The doubles were on the scene long before the first pumps or autos hit the market. Their popularity is certainly not diminished today.

The **single shot** is the safest type of shotgun on the market, making it a very good selection for the fledgling scattergunner. Like the break-top double, the single shot can be opened by "breaking" it in the middle — and when carried in this condition it simply can't be fired. What's more, the chamber can be easily inspected to make sure it's not loaded at the end of the day.

Another extra safety feature the single shot has that the others usually don't is an external hammer that must be manually cocked before the gun will fire. Yet another plus is the fact that, while most factory-produced shotguns come with stocks of more or less standard dimension, some single shots can be had with a cut down "youth's model" stock. These shortened stocks better fit growing youngsters, and a young shooter will do better with a gun that fits him. You can always cut down a full-sized stock to fit, but a replacement stock will be needed as the shooter grows. Single shot guns are so inexpensive (and there's another plus) that buying a whole new gun in a few years isn't all that painful.

Disadvantages? The only two that I can think of offhand are, first, that single-barreled break-top guns are so lightweight that recoil can be a prob-

160

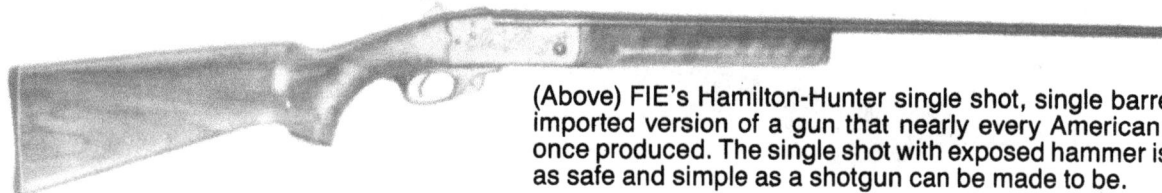

(Above) FIE's Hamilton-Hunter single shot, single barrel is an imported version of a gun that nearly every American maker once produced. The single shot with exposed hammer is about as safe and simple as a shotgun can be made to be.

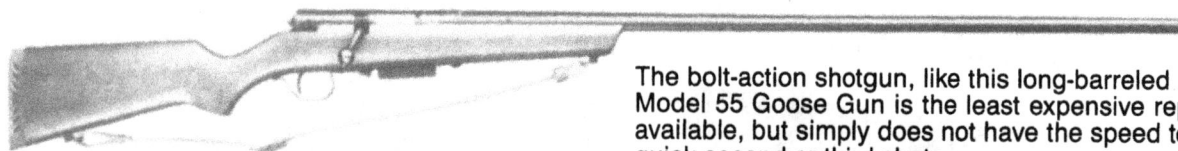

The bolt-action shotgun, like this long-barreled Marlin Model 55 Goose Gun is the least expensive repeater available, but simply does not have the speed to offer quick second or third shots.

SHOTGUN BORE DIMENSIONS

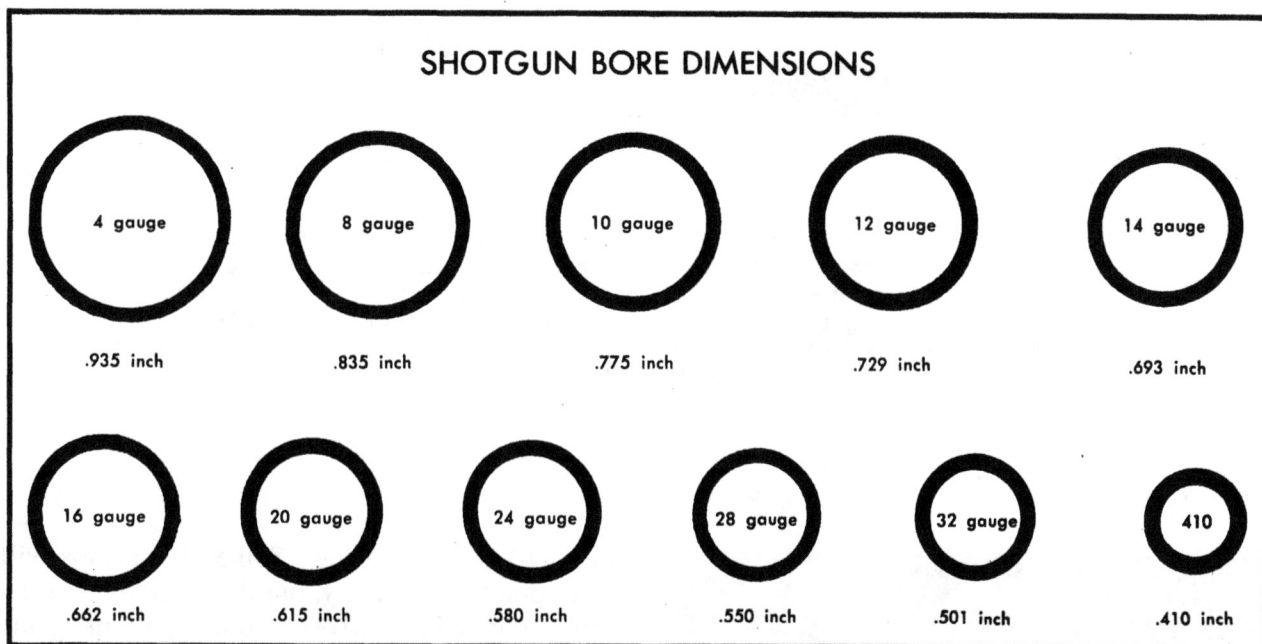

4 gauge	8 gauge	10 gauge	12 gauge	14 gauge
.935 inch	.835 inch	.775 inch	.729 inch	.693 inch

16 gauge	20 gauge	24 gauge	28 gauge	32 gauge	410
.662 inch	.615 inch	.580 inch	.550 inch	.501 inch	.410 inch

The above bore diagrams give you some idea as to the physical size of various shotgun bores. The 10, 12, 16, 20, 28 and 410 are those gauges in common use today.

The other gauges listed above have departed from the shooting scene, but may be found in some limited use, particularly in Europe or the British Isles.

lem, and second, the fact that the shooter will very likely want to move up to something more sophisticated as his skills improve. But for a beginner, a single shot is never a mistake.

That leaves the **bolt action** shotguns. I may be prejudiced against bolt-action smoothbores, but the only good thing I can say about them is that they are the least expensive repeaters available. And I use the term "repeaters" advisedly, as it's difficult to work a bolt action fast enough to give effective repeat shots on either clay targets or flying game. What's more, bolt-action shotguns are homely looking, clumsy handling affairs. I'd far rather have a good single shot.

What Gauge to Choose?

Once you've decided on the general type of gun you want, what about gauge? The 10 gauge is the most "powerful" shotshell, while the 410 is the *least* powerful — the higher the "number" of the gauge, the less it offers in terms of power and vice versa. We can eliminate the 10 gauge right from the start, as this Big Bertha is too heavy and powerful for anyone to learn shooting with. Ammunition is expensive, and the only real use for these mighty boomers is in the turkey woods or when hunting waterfowl.

At the other end of the scale, let's eliminate the

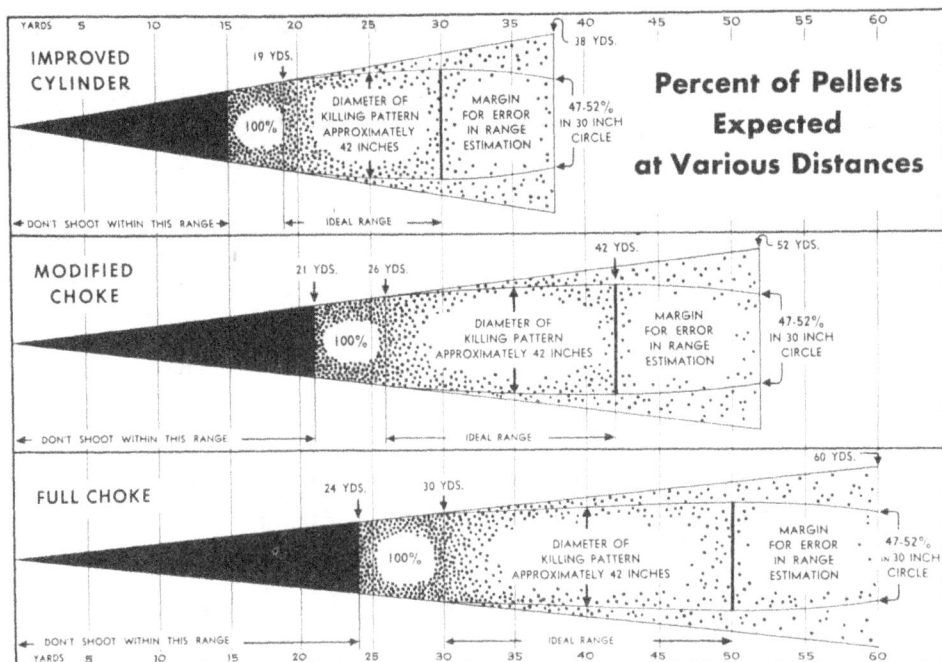

Percent of Pellets Expected at Various Distances

IMPROVED CYLINDER
19 YDS.
100%
DIAMETER OF KILLING PATTERN APPROXIMATELY 42 INCHES
MARGIN FOR ERROR IN RANGE ESTIMATION
47-52% IN 30 INCH CIRCLE
38 YDS.
DON'T SHOOT WITHIN THIS RANGE
IDEAL RANGE

MODIFIED CHOKE
21 YDS. 26 YDS.
100%
DIAMETER OF KILLING PATTERN APPROXIMATELY 42 INCHES
MARGIN FOR ERROR IN RANGE ESTIMATION
42 YDS.
52 YDS.
47-52% IN 30 INCH CIRCLE
DON'T SHOOT WITHIN THIS RANGE
IDEAL RANGE

FULL CHOKE
24 YDS. 30 YDS.
100%
DIAMETER OF KILLING PATTERN APPROXIMATELY 42 INCHES
MARGIN FOR ERROR IN RANGE ESTIMATION
60 YDS.
47-52% IN 30 INCH CIRCLE
DON'T SHOOT WITHIN THIS RANGE
IDEAL RANGE

The most commonly found (and commercially offered) bore chokes are Full, Modified and Improved Cylinder. The Full choke is the "tightest" choke is well suited for waterfowl at longer ranges. For pheasants, grouse and waterfowl at closer ranges (25-45 yards)

the "Modified" choke is excellent. The "Improved Cylinder" choke is superb for close range (15-30 yards) shooting. It's an excellent choice for *close-rising* quail, pheasants and generally any variety of small game to be taken at short range.

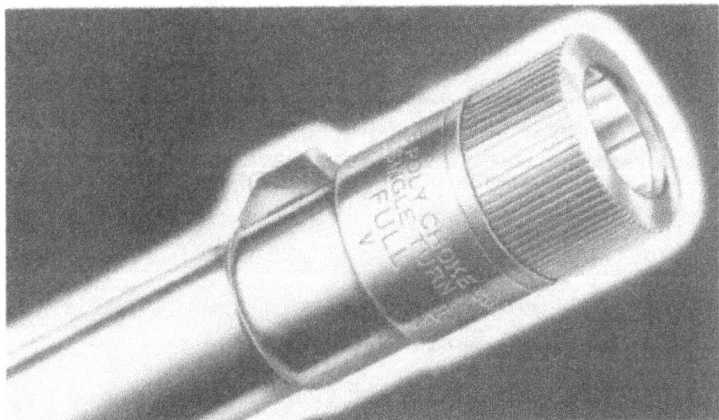

Permanently mounted adjustable chokes allow the shooter to select the choke he wants to use. Choke range is usually from Full on down to Cylinder; and, an adjustable choke like the Poly Choke shown here, is offered as a factory option, or can be mounted by a competent gunsmith. If you are, for instance, shooting waterfowl from a blind and the ducks are flying well out in front (45-50 yards), you can instantly adjust your shot pattern to Full choke. The old Poly Choke, the Cutts Compensator and other similar products offered the shooter a variable choke — but none worked better or looked as good as Ruger's or Browning's screw-in systems.

little 410 bore right now. Some aficionados will look upon this as heresy, but the facts are that the little 410 is so ineffective on both targets and game that a beginning shooter is likely to become discouraged. It's nice to hit something every once in a while, even during those first awkward sessions. Using a 410 cuts down your chances of success — like the 10 gauge, it's a gun best left to the more experienced shooters. The only possible recommendation this gauge has going for it is its light recoil.

That leaves us with four gauges remaining — 12, 16, 20 and 28. If you're a grown man or in your mid-teens, the 12 gauge might be okay. Several different loads are available to fit these guns, and the lightest target loads don't recoil too noticeably. Too, this is probably the most versatile of all shotgun gauges, and is suitable for hunting practically anything that can be hunted with a shotgun. It's an excellent choice unless you're young, of very small stature, or sensitive to recoil.

The 16 gauge is much like the 12 as far as versatility is concerned, and remains very popular in

(Left) Over the years a number of methods have been developed for applying "choke" to the muzzle of a shotgun. This diagram gives you an idea of what the "taper" method of choking looks like. At the muzzle, the inside area of the bore tapers creating a cone effect — this area of the bore is usually called the "forcing cone"; and the degree of choke is determined by the amount of taper. This system is in use today.

Below — Pattern percentages are important, but so is the way the pellets are distributed over the target. Here author checks to see how many target-sized holes appear in pattern that a clay bird or quail could slip through.

Ruger's screw-in choke inserts are one of those good ideas that make you wonder why they take so long to come along. Old-style adjustable chokes were limited to installation on single-barrel pump or autoloaders. Some of them worked well, but no one accused them of making guns look good. Ruger's system allows the choked section of the barrel to simply be screwed out by hand and replaced with another choke in seconds. Browning also sells a system with similar advantages.

Europe. As soon as an American writer suggests the 16 is obsolete, it enjoys a revival. If you like it—shoot it.

The 20 gauge is an excellent choice for a new shotgunner, for several reasons. First, recoil isn't as great (unless you're shooting a featherweight shotgun). And there are a number of different loads available to make this gauge suitable for nearly any kind of shooting. Most 20 gauge guns these days come with 3-inch chambers to accommodate magnum loads that are nearly the equal of regular high-velocity 12 gauge loads. These same guns will also digest standard 2³/4-inch light target loads, making the 20 a very versatile choice.

I'm personally a big fan of the little 28 gauge,

although this number isn't nearly as versatile as the larger gauges mentioned. Recoil is pleasantly light, and this gauge is potent enough for most upland game provided the range isn't stretched too far. Disadvantages include the fact that only a few models are chambered for the 28, and ammunition may be annoyingly hard to find in some areas. This is, however, a far better choice than the 410, which is surprisingly popular in spite of its shortcomings.

How to Select Choke

Now what about choke selection? All modern shotguns come with a certain amount of "choke" or constriction at the muzzle, and this is what determines how fast the shot pattern opens up, or spreads, once it leaves the barrel. You should find the choke marking — *Full, Modified, Improved Cylinder or Skeet* — marked on the left side of the barrel just in front of the receiver, although it may appear elsewhere on some imported guns.

Theoretically, a gun with "Full" choke markings will put between 65 and 75 percent of the pellets in a given load into a 30-inch circle at a range of 40 yards. You can check this on your gun by doing just that — shoot into a sheet of butcher paper at a range of 40 yards (40 long paces). Then use a marking pen attached to a 15-inch length of string to draw a 30-inch diameter circle around the largest concentration of shot holes. Count the number of holes in the circle, and divide by the

163

Left and below — To check shotgun pattern, fire at a large piece of butcher paper at 40 yards, draw a 30-inch circle around the heaviest concentration and then count the holes.

number of pellets contained in the load. (This computation will give you the percentage of shot that hit within the 30-inch circle.) To determine the number of pellets in a shot load, find the number of ounces of shot your shells contain (this information will be printed on the box the shells came in) and multiply by the number of pellets per ounce. Size 9 shot contains 585 pellets per ounce; 8s have 410; 7$\frac{1}{2}$s — 350; 6s — 225; 5s — 170; 4s — 135; and No. 2 shot is 90 to the ounce.

If your gun patterns between 55 and 65 percent, it's Improved Modified; between 45-55 percent, Modified; and between 35 and 45 percent, Improved Cylinder. If the pattern runs between 25 and 35 percent, it's either Skeet or straight cylinder choke. This can vary from gun to gun and load to load, so don't be surprised if your gun patterns differently than the markings on the barrel say it should.

You should match the choke to the range you expect to be shooting at: 35 to 45 yards is about right for Full choke, between 25 and 35 yards for Modified, 20 to 30 yards for Improved Cylinder — this doesn't mean that an Improved Cylinder choked gun won't kill birds or break clay targets beyond 30 yards. It's just that it performs best over the ranges listed.

Choosing Ammunition for Your Shotgun

Once you've decided on a gauge, make sure you buy — and carry with you — *only* ammunition of the *same gauge*. If you mistakenly put a 20-gauge shotshell into a 12 gauge gun, it will fall right through the chamber and hang up in the barrel. If a 12 gauge shell is accidentally inserted behind the 20 and fired, the barrel is guaranteed to burst — and you'd have a better than average chance of ending up in a hospital.

Selecting ammunition for a shotgun is slightly more complicated than simply buying shells of the proper gauge. You also must select between different shot sizes, and between high-velocity and low-velocity loadings. Remember, the diameter of the shot *decreases* as the size number increases. Size 9 shot is tiny and suitable only for clay targets or small vermin, while 2s are too large for anything

No.	12	11	10	9	8	7½	6	5	4	2
DIAMETER IN INCHES	.05	.06	.07	.08	.09	.095	.11	.12	.13	.15
APPROXIMATE NUMBER OF PELLETS TO THE OUNCE	2385	1380	870	585	410	350	225	170	135	90

	Air Rifle	BB	No. 4 Buck	No. 3 Buck	No. 1 Buck	No. 0 Buck	No. 00
DIAMETER IN INCHES	.175	.18	.24	.25	.30	.32	.33
NUMBER TO THE OUNCE / APPROXIMATE NUMBER TO THE POUND	55	50	340	300	175	145	130

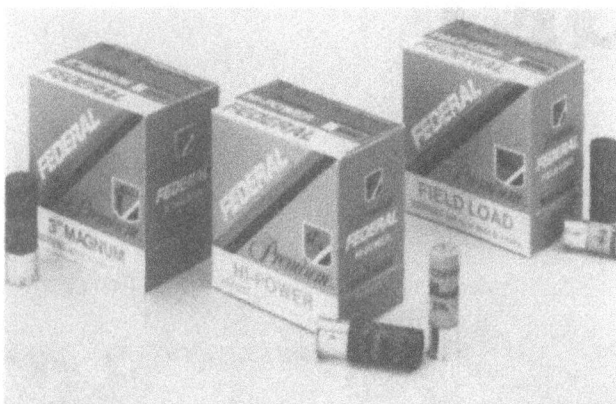

(Top) "Shot" comes in many sizes. The larger the "number" of the shot, the smaller the shot size. Compare No. 12 shot with No. 00 buckshot and you'll see what we mean. (Graphs not actual size.)

Which ammunition should you buy for your new shotgun? You have to know the numbers game to be able to tell the size of shot and the amount you're buying. You must also know the difference between gauges and between magnum, high-power and field loads within each gauge.

but geese and turkey. For clay target work choose 7½s, 8s or 9s. Hunting upland game (rabbits, grouse and pheasants) calls for 5s, 6s or 7½s, while waterfowling (ducks and geese) is best done with 4s, 5s or 6s (reserve 2s for long-range pass shooting at geese). Those aren't hard and fast rules, but they'll get you started until you develop your own shot size preferences. Most waterfowlers choose magnums (if you buy 3-inch magnums, be sure to find out if your gun will digest them first) or high-velocity loads, while target shooters and upland gunners usually prefer "low brass" or

standard velocity ammunition.

The subject of steel shot came up some years ago when environmentalists suggested that wounded waterfowl were suffering lingering deaths from lead poisoning and that lead shot dumped into the environment did no one's health any good. Opponents suggested that steel shot would damage those gun barrels designed to shoot lead shot. Without reviving the emotions of the arguments, suffice it to say that waterfowl is hunted with steel shot, while trap and Skeet shooters use lead. Steel shot does not seem to have injured many gun barrels.

We discuss shot in the context of how many of a particular size fit in a shell. The shotgun slug consists of one bore-size projectile. Slugs, also called "rifled slugs" because of the small stabilizing fins on them, are typically used for deer hunting heavily settled country. While the slug packs a lethal wallop, accuracy does not compare to a good rifle. Recently, new rifled barrels have been offered which are said to impart much greater accuracy to the big lead lump. Considering the number of areas which legally mandate the shotgun for deer hunting, you can count on continuing development.

One final word: Before you buy that first shotgun and stock up on ammunition, make sure the gun fits you as well as possible. Once you've decided on general type and gauge, have the store salesman show you several different models of this action type. Then pick each one up (clearing the action immediately — remember the safety commandments) and throw it to your shoulder. The test is easy — which one feels and handles most naturally? Which one seems to point best? Finally, which one do you really like?

If you're exceptionally tall or have long arms, you can lengthen a gun's stock by simply adding a lace-on recoil pad. Similarly, most gun stores employ a gunsmith who'll be happy to cut an inch or so from the length of a too-long stock. Either of these alterations shouldn't cost you more than a few dollars and can make a big difference in the way you shoot.

When all is said and done, the shotgun you're buying must please only you. There's a large variety of shotgun styles and models in a range of prices to fit every budget. Find one you like that feels right to you, and you probably won't go wrong.

chapter 15

HOW TO HIT A MOVING TARGET

WHILE RIFLES and handguns are precision instruments designed to let you hit a relatively small target at some distance, shotguns are intended to give almost exactly opposite results. The shotgun is a relatively short-range tool, and instead of firing a single projectile, it throws several hundred tiny shot pellets in a pattern that spreads out as the range increases.

What's more, the shotgun is designed primarily for hitting a moving object — a flushing pheasant or a clay target thrown from a mechanical trap. These basic differences have an effect on the way a shotgunner handles his firearm. Where a rifleman or handgunner carefully aims his firearm — and then does his best to hold it perfectly steady until the bullet is on its way — the shotgunner *points* (rather than *aims*) almost by instinct. And unless he swings his gun and *keeps* it swinging even after the trigger has been pulled, chances are he'll miss what he was shooting at.

Because of the way a shotgun is used, the way the stock fits the gunner is important. A rifleman depends on sights that can be precisely aimed at a target regardless of how its stock fits the shooter, and the standardized rifle stock dimensions used by most firearms manufacturers work pretty well

A bird hunter doesn't have time to use sights, so shotguns carry only primitive sighting equipment. Shotguns are instinctively pointed rather than aimed.

166

for nearly everybody. The same rifle can be used by a diminutive housewife and a college basketball center, and both should be able to do a credible job of marksmanship if they have the necessary skills.

The same isn't true in shotgunning. Again, there are more or less standard stock dimensions used on most factory-produced scatterguns, and these work reasonably well for the average-sized adult. But a person much smaller (or larger) than average will have a tough time hitting anything unless he *changes* those stock dimensions. A bird hunter or Skeet shooter doesn't have the time or the inclination to use sights when he shoots. Instead, he relies on his body's ability to point the gun in the right direction while he follows the target with his eyes. If the gun doesn't fit him right, he won't point it right.

Once you have a gun that fits you reasonably well, you'll need a supply of the right ammunition (choose standard velocity loads in a small shot size — 8s or 9s preferably), a hand trap (target thrower), a supply of clay pigeons and a partner who's willing to help. Ideally your partner will know how to shoot a shotgun, and if he's a trap or Skeet competitor, that's even better. However, anyone who can throw targets with the hand trap will do. With these preparations, and a safe area to shoot (the tiny shot pellets won't travel more than a few hundred yards), you should be ready to start your shotgunning career.

The first thing you must learn is the proper stance. If you started shooting with a rifle, you have some habits to unlearn. At first glance, it may look like riflemen and shotgunners have a lot in common as far as stance is concerned, but look closer and you'll see some important differences. In the first place, the supporting arm (the left arm for gunners shooting right-handed) isn't held with the elbow directly underneath the gun. The stiff support this hold gives isn't needed for shooting a shotgun and, in fact, is undesirable. To hit a moving target with a shotgun, the gun needs to move and your body should be free to move with it.

The right arm, or shooting arm, isn't elevated as

Left and Below — Shotguns are intended to hit moving targets like this clay pigeon. Shooting at clay targets thrown from a hand trap is the best way to learn how to hit with a shotgun.

action pose with his legs spread for balance and his knees slightly bent. He stands at about a 45 degree angle to the target area, and leans forward into the gun when he's ready to shoot.

Don't lean *too* far though, or you'll be off balance — and balance is highly important to the shotgunner. The left leg should be forward as you watch for your target, and your body should be loose and relaxed. If your muscles are tense, your movements will be stiff — and you'll probably miss. Practice mounting the gun a few times. Don't just raise it to your shoulder, but push it slightly away as it comes up — and when it's at the right height for shooting pull it straight back into your shoulder.

As the gun comes up, your head should tilt down and slightly to the right. With a little practice your head will be in exactly the right position to let your cheek meet the stock when the butt snugs in place. At this point your eye should be directly behind the receiver, and as you look down the barrel you should see a highly foreshortened tube. If you can't see part of the barrel, and only the front bead is visible the gun is stocked too low. This can be corrected by a lace-on cheek pad, if necessary. But if you see *too* much barrel the gun will shoot high (some wood needs to be removed from the comb).

Bring the gun to shooting position several times to get the feel of how it's done, and ask your friend to critique your movements. Obviously an experienced shotgun target shooter (or even a bird hunter) is more likely to be helpful here, but almost anyone with a critical eye can tell the difference between smooth and unnaturally stiff movements on the part of the shooter. Bend forward slightly from the waist and practice swinging the gun to one side and then the other, tracking an imaginary target. You should move with your whole body — everything from the ankles up. If you merely swing the gun by moving your arms you'll have a tough time hitting thrown targets and an even worse time killing game. Many beginners (particularly those who first trained with a rifle) tend to stiffen up and don't remain flexible. You have to remember to stay loose when you have a shotgun in your hands.

Above — The shotgunner's body should remain flexible to allow movement from the ankles up as the gun is swung on target. Legs are spread apart a comfortable distance, and the shooter leans slightly into the gun.

high when shooting a shotgun, either, for the same reason. You can't let the arm *droop*, but you should hold it only as high as comfort dictates.

While a rifleman stands straight and rigid in the offhand stance, the scattergunner assumes an

After getting the feel of the gun, ask your friend to take a few steps to one side behind you and — on your command — use the hand trap to gently throw a clay target high in front of you. Keep your eye on the target from the minute you first see it. Your body should be moving to bring the gun on target as the gun comes up to your shoulder (remember to push it out, and then pull it into your shoulder). Start swinging from behind the target and as soon as you can see the target directly above the end of your barrel, pull the trigger. Notice I said nothing about a gradual squeeze. A shotgun isn't shot like a rifle, and a plain, quick pull will get the job done. Some shooters practically slap at the trigger when the sight picture looks right. Now the most important thing to remember here is to avoid stopping the swing as soon as the gun goes off. That's the natural inclination — but keep that gun swinging! This is called follow-through, and if you don't keep moving the gun slightly faster than the target — swinging smoothly and naturally — you won't hit it. It's as simple as that.

The theory here is that the gun barrel travels slightly faster than the target, overtakes it from behind, and then continues forward to lead the target. If this is done smoothly (and at the right speed), snapping the trigger at the moment the target appears directly over the shotgun barrel, the moving gun will establish its own lead automatically. This is called "swing-through" shooting, and is the best system I know of to assure consistent results.

Some shotgunners use the sustained lead system — the barrel is pointed at an imaginary mark somewhere ahead of the moving target, and this lead is held or sustained as the gun is fired. This works for some people, but the problem lies in being able to determine the proper lead and then maintaining it.

Finally there's the "point-and-shoot" system, sometimes called instinctive shooting. In this, the gunner simply pokes his barrel toward the target and touches off. He relies on instinct to know exactly where to shoot to have the bird (or target) and shot load arrive in the same space at the same time. This actually works for some people, but is best used at relatively close range. (I've downed birds myself this way, but I couldn't tell you how I did it afterwards.)

If your partner knows something about shotgunning, he can probably tell you what you've doing wrong when you miss the target. Common mistakes include holding the head too high or stopping the swing. Continue shooting at high, easily thrown targets until you get the hang of it and start hitting regularly.

Be sure to shoot with *both* eyes open. The only

The "swing-through" method of wingshooting is the best — it produces results. Above left, the shooter has shouldered his gun, aimed behind the bird, moved into the target and is about to pull the trigger. The trick is to fire as the muzzle passes through the bird — keep that muzzle moving. Don't stop swinging just because you've fired the gun. Keep it moving and you'll hit your bird. As you can see (above right) the shotgun serves as a "paintbrush" when the "swing-through" method is used in wingshooting. Use it to "paint" the bird out of the sky.

"Pointing-out" or a sustained lead, may work for the waterfowler who can take his time as a bird flies on by; however, it may not work on rising game. "Pointing out" consists of placing the muzzle immediately in front of the bird, moving with the flight of the bird, and finally pulling the trigger.

"Point-and-shoot" is a matter of quickly, almost instinctively, "guesstimating" where to aim in front of the bird. The gun is brought quickly to the shoulder, the "guesstimate" made, and the trigger pulled. It's fast, but not necessarily the best method of wingshooting.

time this rule should be ignored is when a right-handed gunner has a left master eye. To tell which is your master eye, hold your finger pointing skyward about a foot away. Then — with both eyes open — look past the finger at some distant object. Focus on the object, not the finger. You should see two images of the finger, offset a couple of inches apart. One image will look solid, while the other will be a "ghost" image. Close the left eye — if the ghost image disappears, leaving the solid one in its place, your right eye is the master eye. To check this, leave your left eye open and close the right. This time the figure will appear to jump from left to right. If you shoot right-handed and your *left* eye proves to be the master one, you'll either have to learn to shoot with the stock of the gun against your left shoulder or close the left eye.

Remember to stay loose and mobile when you're trying to hit a moving target with a shotgun. Your body should become a gun platform, and both torso and legs should be flexible enough to allow sudden weight shifts and considerable movement as you track your bird (clay targets are also called "birds" in trapshooter's parlance).

If at all possible, go to your local Skeet or trap club and ask some experienced shooter to coach you. Most gunners are happy to provide a bit of instruction, and if you've picked a man who knows

what he's doing, he can be of considerable help (one way to do this is to watch the participants and see who posts the best scores during a shoot).

If this is impossible, any willing friend can help you learn. All he really needs to do is throw the targets. It's possible to lob a few targets by yourself, but this really isn't very helpful as you can't throw a clay pigeon high enough or hard enough to do much good unless you use a hand trap — and that puts you too far off balance to recover in time to shoot.

Remember your safety rules, and watch that muzzle at all times. The damage a shotgun load can do at close range has to be seen to be believed. And don't forget your ear protectors and shooting glasses. These should be worn at all times when you're shooting a shotgun, particularly on the target or practice range. Bird hunters seldom wear shooter's ear muffs, but the smart Nimrod will equip himself with hardened eyeglasses and a comfortable pair of compact ear plugs.

As in any sport, practice makes perfect. When you start hitting those lazy high flyers, have your target thrower step up the pace and start changing flight angles. Then you should be ready for the more powerful machine-thrown "birds" at your local trap or Skeet range. Do well here, and upland game birds should be easy to hit on the wing.

chapter 16

TARGET SHOOTING WITH A SHOTGUN

THERE ARE THREE BASIC forms of organized shotgunning competition in this country — trap, Skeet, and Hunter's Clays. Trapshooting dates back to 1793 or before, when it originated in England. By 1831 it was an established sport in the United States. In those early days live pigeons were used as targets, but in 1866 a glass ball was substituted and 14 years later the "clay" pigeon and a trap to throw it with were invented. While these first saucer-shaped targets were made of baked clay, today they're made of pitch and break more easily and reliably when hit by shotgun pellets.

The game of Skeet shooting was developed about 60 years ago in New England by bird hunters who wanted to practice shooting at targets that flew up at odd angles to simulate the flight of grouse. The same "clay" pigeons that are used in trapshooting are thrown at the Skeet range, but the target-throwing machines (traps) are arranged to either side of the gunner instead of directly to his front.

Hunter's Clays were developed in the U.S. in the early 1980s to simulate upland gamebird shooting and waterfowl hunting. Clay targets are thrown from traps hidden behind hedgerows, at treetop level, at the waterline or just about any location so targets simulate the flight of game.

All of these shooting games can be used to sharpen the hunter's shooting skill, but many enthusiasts become so devoted to the game that they shoot only at the frangible clay targets. Competition is nationwide with meets supervised by the Amateur Trapshooting Association, the National Skeet Shooting Association and the U.S. Sporting Clays Association.

Trapshooting

In trapshooting, the gunner stands at least 16 yards behind the trap house, which is usually a concrete bunker projecting only a few feet above the ground, and which houses the mechanism used to throw the clay pigeons. Five shooting lanes are located behind the trap house, spaced 3 yards apart. A squad of five shooters participate at a time, and each shooter takes his turn in each lane. Five shots are fired from each position, and a total of 25 targets make up a "round" of trap.

When each shooter is ready, he calls "pull," "mark," or some other command and the trap operator releases a clay bird. The targets are thrown downrange at random angles, and must be thrown hard enough to travel at least 48 yards if not broken by the gunner. The targets must be between 8 and 12 feet above the ground as they pass 10 yards in front of the trap house.

There are three different events a trapshooter participates in. In the first, known as the "16-yard event," the shooter fires in each lane from the 16-yard mark. In the second, the "handicap event," the shooter stands from 18 to 27 yards behind the trap house according to his ability. The last event is the hardest of all. Known as "doubles," this event features two targets thrown at the same time, but at predetermined angles.

Because trap targets are thrown at a rising angle and are shot at long-range, guns designed for the sport feature fairly tight chokes and a high stock intended to raise the head — this lets the shooter see more barrel and makes the gun shoot slightly high. You can do a credible job of trapshooting with a field gun made for hunting, but nearly all serious competitors eventually acquire a gun designed for sport. Because two of the three events

features targets that are thrown one at a time, single-shot guns are popular. These guns should be heavy enough to help absorb recoil and improve the shooter's swing. Autoloaders are also popular because of their recoil-reducing capability. All guns designed for trapshooting have a wide rib attached to the top of the barrel as an aiming aid. Although there are no minimum gauge restrictions, nearly all competitors use 12-gauge guns (larger gauges are not permitted). Ammunition used for the sport must be loaded with no more than $1\frac{1}{8}$ ounces of shot no larger than No. $7\frac{1}{2}$. Three dram-equivalents is the maximum powder charge.

Skeet Shooting

Skeet shooters take designated positions around a shallow semicircle and fire at targets thrown from traps located at either end of the semicircle. The trap house to the left of the gunner throws its targets from a height of 10 feet and is called the "high house." The trap house to the right throws

Skeet ranges, too, are found throughout the U.S. The word Skeet literally means "shoot" and is Scandinavian in its derivation. The arcing dotted lines at the top of the accompanying illustration show the clay birds' flight path as they are thrown from either (or both) the

"high" house on the left and the "low" house on the right. Unlike trap, the flight of the "birds" remains constant while the shooters move around a semicircle firing from eight different shooting points.

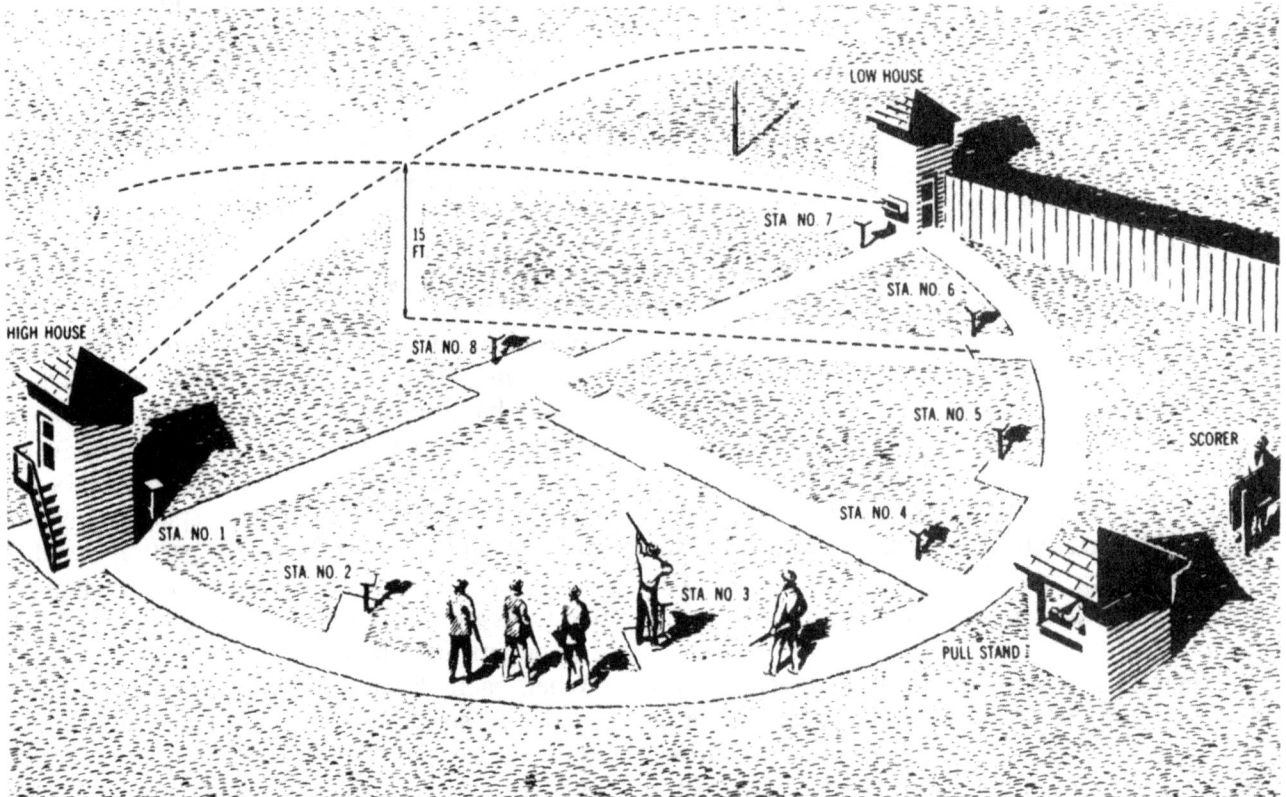

172

its targets from a height of 3½ feet, and this is known as the "low house."

There are seven stations marked out on the semicircle, with an eighth located in the center of a line connecting the two houses. Shooters take turns shooting twice from each station — one shot each at targets thrown from both the high and low houses. Then two targets ("doubles") are thrown simultaneously (one each) from the two houses with the shooter on stations one, two, six and seven. That uses a total of 24 shells. There are 25 shotshells in a box, and the 25th — usually referred to as your "option" shot — is used to repeat the first target missed or it is fired at the end of the round at an extra target. When shooting the first part of the round (at targets thrown individually) the high house "bird" is always thrown first.

Skeet shooting offers four different events — one each for the 12, 20, 28 and 410 gauge. Organized competition is often confined to the spring and summer months, as most Skeet shooters go hunting during the fall. Guns designed for Skeet fea-

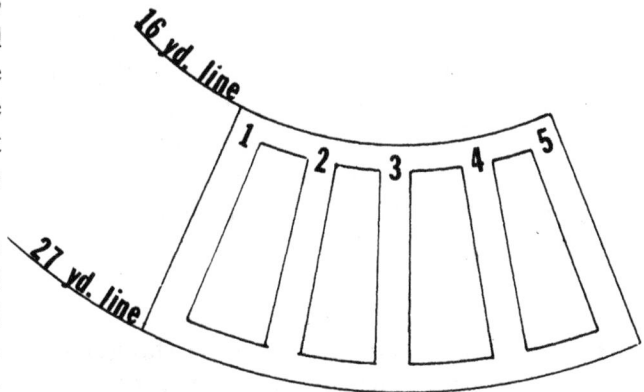

(Above) Trap ranges are found all around the country. This diagram provides some physical idea as to how a trap range is set up. The "birds" are thrown (mechanically) from the trap house while the shooter fires at the "bird" from the positions indicated at the 16-yard line. The shooter rotates from position to position every five rounds and a total of 25 shells are fired in a "round" of trap. The "birds," as they exit the trap house, fly away at different angles making the shooting very challenging indeed.

(Below) Trap and Skeet guns, like this Winchester Diamond Grade oversingle trap gun, have long been known for fine wood and elaborate ornamentation. While trap and Skeet shooters shoot a LOT, they don't drag their guns through the field or duck blind. Special ribs and special trigger arrangements are common on these shotguns for specialized experts.

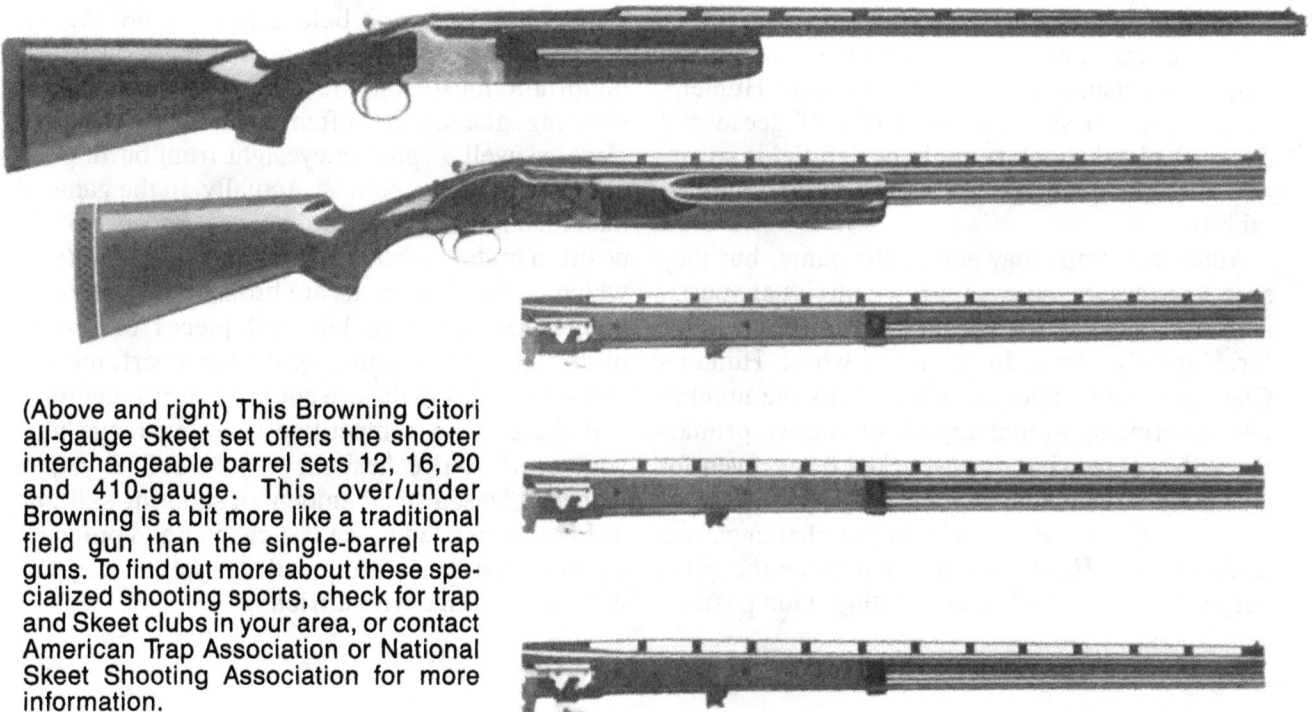

(Above and right) This Browning Citori all-gauge Skeet set offers the shooter interchangeable barrel sets 12, 16, 20 and 410-gauge. This over/under Browning is a bit more like a traditional field gun than the single-barrel trap guns. To find out more about these specialized shooting sports, check for trap and Skeet clubs in your area, or contact American Trap Association or National Skeet Shooting Association for more information.

173

ture open chokes and must be capable of firing two shots in rapid succession. Autoloaders are highly popular, as are pumps. Stackbarreled doubles are considerably more expensive, but have a wide following among those who can afford them — some high-grade models are offered with four sets of barrels to allow the same gun to be used in the four different gauge competitions, and these outfits are anything but cheap.

Shooting either trap or Skeet will do more than any other type of exercise to improve a shotgunner's skill. To locate the nearest shotgun target facility, look in the Yellow Pages of your telephone book under "Trap and Skeet Ranges." The club will be glad to hear from you and will do everything possible to get you started. Shooters compete according to skill levels, so you can have fun right from the beginning.

Hunter's Clays

Hunter's Clays were developed by shotgunners who tired of the regimen of trap and Skeet to simulate upland gamebird shooting and waterfowl hunting under field conditions where the game would actually be found. Regular clay targets are thrown from traps hidden behind hedgerows, at treetop level, at the waterline or just about any location so targets simulate the flight of game.

Fields can be set up so shooters encounter "flushing quail," "springing teal" or "passing mallards." The idea is to use available terrain to create shooting situations that will be both interesting and challenging.

Where waterfowl hunting is so popular, for example, sportsmen might want to design Hunter's Clays fields to simulate the flight of geese and ducks. Upland shooters might design fields so targets simulate quail, grouse, doves or even running rabbits.

Automatic traps may add to the game, but they aren't necessary, so even a few individual sportsmen or a local sportsmen's club can set up fields for Hunter's Clays. In England, where Hunter's Clays is called "Sporting Clays," it's the number one shooting sport and targets are thrown primarily with inexpensive, hand-cocked traps, even for national tournaments.

A variety of features add to the challenge and excitement of Hunter's Clays and make the game very similar to actual field shooting. Gun position is "low gun" with the stock off the shoulder, and there can be up to a 3-second delay after the shooter calls "ready" before the target is thrown. There are "report pairs" where a second target is thrown at the sound of the shooter's firing at the first target. There can also be "simultaneous pairs" where both targets are released at the same time just as in trap and Skeet.

One of the best things about Hunter's Clays is that good field shots can do just as well, sometimes even better, at the game as competitive trap and Skeet shooters. On a difficult Hunter's Clays course, hitting 35 out of 50 targets would be considered a good score. And in competition, a shooter can miss several targets and still be in contention.

More information on Hunter's Clays can be obtained by writing the U.S. Sporting Clays Association, 111 North Post Oak Lane, Suite 130, Houston, TX 77024.

Safety

One final word: Safety rules are rigidly enforced at clubs, and you should learn the conventions practiced at each facility. Guns not actually in use on the shooting line aren't carried about, but are left racked and *unloaded* in a designated area until your turn comes to shoot. Also, you should be aware that most competitive shooters reload their own ammunition to reduce costs — if you decide to do the same, your empty shotshell hulls mustn't be allowed to hit the ground. If they do, they automatically (in most cases) belong to the club. Again, ear protectors and hardened shooting glasses are important musts on the clay bird line. Tinted shooting glasses are often used to cut haze or glare, as well as protect eyesight from burnt powder or ricocheting pellets. Actually, in the game of Skeet, shooting glasses are a *must*. Because of the position of the shooter, in relation to the point(s) at which many of the birds are broken, the shooter is often showered with bits and pieces of broken birds. These shards are, for the most part, moving fast — being hit in the eye could be *very* dangerous.

If there is no competitive shooting range near you, or if you're hesitant to join before you've gained at least a small amount of shooting skill, do as I recommended in the last chapter and purchase an inexpensive hand trap and a supply of clay targets and practice with a friend.

chapter 17

SHOTGUN CARE AND CLEANING

SHOTGUN CARE is reasonably simple for several reasons. Shotgun bores have a smooth finish with no rifling to pick up and hold powder fouling and leading. With the exception of the diminutive 410, shotgun bores are composed of big holes which are easy to see through. Single-barrel break-action shotguns and double-barrels (both over/unders and side-by-sides) are simple and safe in that cleaning from the breech is a snap with no stripping of any sort necessary for routine cleaning. Many of the popular pump and autoloading shotguns also feature readily removable barrels, which give the shooter safe access not only to the gun's bore, but also to the shotgun's action parts. Many newer guns have chrome-plated bores which protect the barrel and make the cleaning task slick and simple. A number of recent shotgun models intended for rugged outdoor use take advantage of the military-type dull Parkerize finish, which goes a long way toward eliminating external rust. The gray Parkerize may not look as spiffy and shiny as traditional bright gun blue looks on the shelf of the gun store, but it does make shotgun care simpler and the shotgun's finish more durable. While stainless steel shotguns are not readily available, a few shotguns are offered with a very durable over-all-satin-chrome finish. But, regardless of finish or trick features, remember that the task of caring for your shotgun may be made simpler but *not* eliminated.

Shotgun cleaning begins with a careful check of the gun for ammunition. If the shotgun is a double or single-barrel break-action, all that is necessary is to open the action and look into the chamber(s). If the shotgun is loaded—and it must *not* be—the shell(s) will be looking at you. With a pump-action or an autoloader, don't forget that a shell may still be in the magazine tube, even if visual inspection of the chamber reveals nothing but daylight and dirt. Double, even triple check, to be sure your gun isn't loaded!

To clean the shotgun—after checking for ammo—you'll need a shotgun cleaning kit. Specialized, packaged cleaning kits are available for all types of shotguns from all full-service gun shops. Hopefully, you bought a cleaning kit when you wrote the check for your new shotgun. The kit will contain a cleaning rod, a fuzzy-fabric bore swab, a brass-bristle bore brush, bore solvent, gun oil, and cloth cleaning patches. Most of the better cleaning kits also feature detailed cleaning instructions printed in the storage box. About all you need

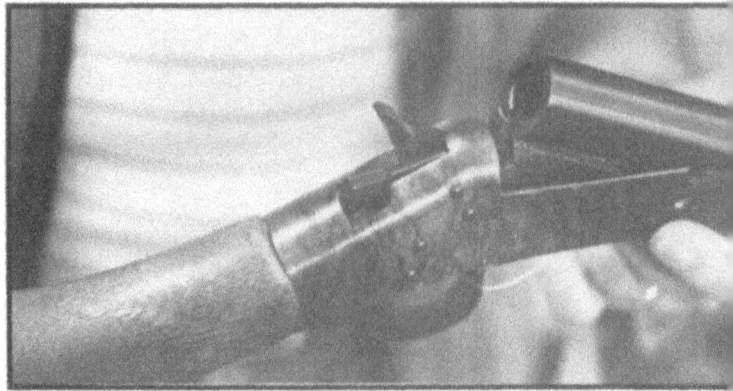

As with ALL firearms, look to make certain that a live round is not in the chamber or in the magazine. Beginners make a common error by opening the action to verify that the chamber is empty, then closing the action and chambering a round forgotten in the magazine. This Mossberg — a sound yet inexpensive beginner's pump shotgun still carries the manufacturer's caution sticker advising the new owner to make certain he knows how the gun works before using it. Some manufacturers — like Ruger — actually engrave the warning on the gun. Be sure you follow the maker's advice and make sure you understand how the gun operates before it's time to do the cleaning task. Note that even with a new gun, it is a good idea to go through the cleaning task before the first shooting session.

The single-shot, break-action shotgun is by far the safest and simplest for the beginner. Cleaning this type of shotgun may be done from the breech. But these days most shooters opt for a pump or other repeater.

that isn't in the kit is some cotton rags for wiping. Old T-shirts work best. In the interests of your happy home, gather some old newspapers to protect your floor from cleaning solvent and gun dirt.

Note that these cleaning kit comments apply equally to rifles and handguns. It is cheaper and simpler for the beginner to buy a basic cleaning kit than to buy the cleaning supplies piecemeal. As skill and experience build, the shooter may supplement the basic kit with more specialized tools and gadgets. Again, the best place for advice is the local gun shop.

Since modern shotshells use plastic collars or wads to protect the shot column as it travels down the bore, barrel leading is a much less serious problem than it was in the days when shotgun wads were merely chunks of cardboard which separated the shot column from the powder charge. Small amounts of leading still occur, but it is easily removed. (It shows up as dull gray stripes when you look down the bore of your *unloaded* shotgun with the other end of the barrel pointed toward a light source.) There will also probably be small amounts of plastic deposits from the wads, but these are also easily scrubbed from the bore with the bore brush. At this piont, it is appropriate to

discuss steel shot. Some years ago, steel shot was mandated for waterfowl hunting. Many very well-informed shooters suggested that steel shot would badly damage shotgun barrels. The controversy went on for some time, with the conclusion that modern barrels would not be damaged. I have no intention of reviving the debate here, or taking one side or the other. Suffice it to say that steel shot has proven to be perfectly acceptable for use in modern shotguns used for hunting waterfowl.

If possible, clean from the breech end of the barrel. Break-action shotguns may be cleaned from the breech simply by opening the action. For more serious cleaning, it is necessary only to unsnap the forend from the barrel(s) and separate the frame, barrel(s) and forend. If you own a pump or autoloader which permits removal of the barrel, be sure to do so. Otherwise, you will be pushing all the dirty solvent and residue back into the working parts of the action. While that gunk can be flushed away with a good aerosol spray, the necessity to do so complicates the cleaning task.

After getting the cleaning implements together and making certain your shotgun isn't loaded, start the real cleaning task by attaching the end of your swab to the cleaning rod. Cover the swab with a solvent-soaked patch. Push this through the bore several times and allow the barrel to stand a few minutes. Go have a cup of coffee or a glass of milk. Allow the solvent to do its work by penetrating the leading and powder residue. Next, replace the swab with the brass bristle brush. Dip the brush in

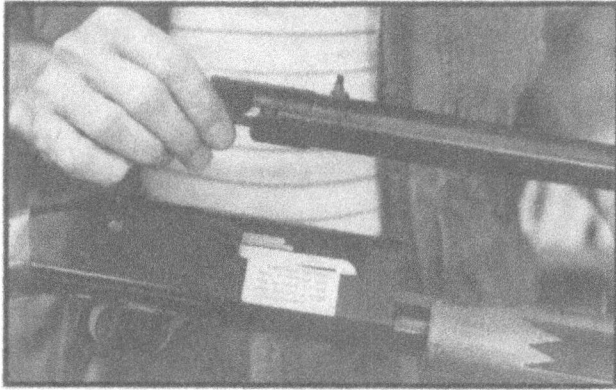

This Mossberg pump, like many other pump and auto-loading shotguns, features a readily removable barrel which simplifies the cleaning chore. Note that this barrel is equipped with rifle-type sights for use with slugs. In other pumps and autoloading shotguns with fixed barrels, cleaning from the muzzle is a necessity.

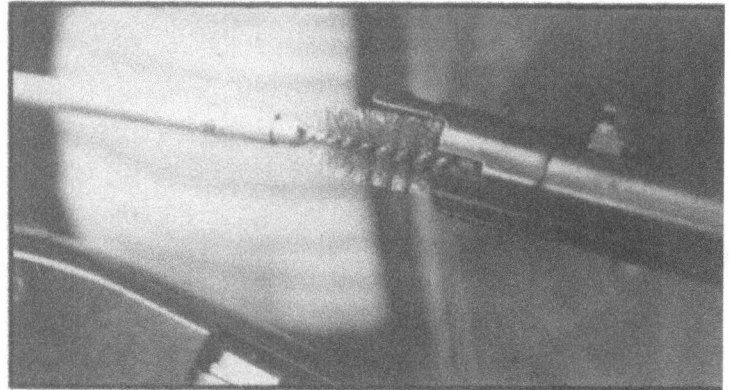

The shotgun barrel, particularly when used with slugs or lead shot, generally requires a bit more effort to clean — hence the use of the brass brush with solvent. For especially stubborn dirt, run the solvent-moistened brush into the barrel, let it stand for a bit and follow up with additional scrubbing. Note that it is common to see a few slightly gray streaks in many otherwise clean barrels. Don't be unduly alarmed if you can't quite remove every streak.

powder solvent and scrub the bore. The grunge should now be softened to the point that the brush will readily dislodge it. After brush scrubbing, put the swab back on and push a few dry patches through the bore. If the third or fourth patch fails to come out clean, or if you can still see streaks of lead or dirt in the bore, repeat the process, but don't overdo it. You might run dozens of patches through a bore which looks perfectly clean, but still see a bit of gray residue. Don't panic. Note that if you rub a perfectly clean cotton cloth over a piece of brilliantly polished, sparkling steel, the result will be some slight gray smudges on the rag. After you are satisfied that the bore is acceptably clean, follow up with a patch moistened with gun oil to protect the bore from atmospheric corrosion.

Most new shooters tend to over-oil their guns and new shotgunners are no exception. While actions should be kept clean, drowning the gun in oil and solvent will do more harm than good. Remember that dripping oil does just that—it drips. Excess oil attracts dust and turns to gum. As the oil volatilizes, in combination with dust, it turns to a hard, gummy varnish. Fine (00 or 000-grade) steel wool and solvent will remove this stuff, but it's best not to create it. The same sort of problem can come from the use of many of the aerosol lubricants. While these products do a fine job of flushing dirt out of complicated mechanical assemblies which need not be disassembled for cleaning, they are *not* a substitute for a light oiling. Like excess oil, the excess aerosols can also

form a gummy varnish.

All new shotguns come packaged with disassembly instructions that should be followed in routine cleaning, or the annual "thorough" cleaning. Disassembly beyond the "field strip" level is hardly necessary every time you fire a half-box of shells, but it should be done if the shotgun is exposed to rain or a mud dunking. Deer hunters can tell you about heavy rain. Duck and goose hunters also know about what happens when they are forced to wade about in the mud to recover a shotgun dropped on the way to the shooting blind. The moisture-displacing aerosols are a great help here. Use the aerosols to flush out the mud and water, then give all accessible parts a wipe with an oiled cloth. If you expect in advance that your gun will get a soaking, give it a spray with a rust-preventative oil before taking it afield.

Before discussing autoloading shotguns, I will make the point that they are not the best bet for the beginner. The skills required to enjoy the autoloader safely and to care for it properly are a bit beyond beginner's basics. But, as surely as I am typing this page, more than a few beginners will decide that they simply can not live without an autoloader. The beginner *can* handle an autoloader—50 years of military experience training beginners tells us so. The autoloader owner must take extra care to keep the shotgun clean. The autoloader action does the work the shooter ordinarily accomplishes with hands and arms. The mechanism which does that work requires special care, how-

177

The toothbrush and Gun Scrubber are an effective way to get dirt out of the action. Unless the gun has been dropped in water — it does happen — disassembly beyond this point should seldom be necessary. The beginner should remember that it is always easier to take things apart than to put them together again.

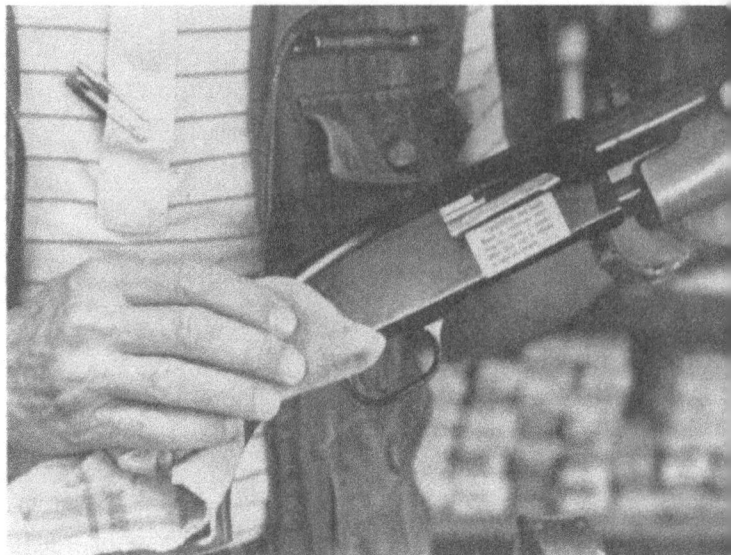

After the action is cleaned, replace the barrel and wipe the gun down. Be sure to wipe any excess solvent off the wood. While modern stock finishes are quite durable, oils, solvents and wood are not a good combination. Remember to get rid of all finger prints, as they often turn to rusty smudges.

ever. With a few exceptions, autoloading shotguns are gas operated. Owners must take extra care to periodically clean the gas piston and the port which channels gas from barrel to the piston chamber. If these areas are neglected, powder fouling will eventually build up to the point that the gun stops working reliably—then stops working. Period. Again, check the factory manual for cleaning and take-down instructions. If the manual is not clear, check with your local gun store people for further explanation. If you don't have a manual, contact the manufacturer/importer to obtain a copy of it. They'll be glad to send one.

Trap and Skeet shooters who fire fast enough to keep their guns hot to the touch take extra care to keep their guns clean. These folks often use gun grease, rather than lighter oils, to lubricate their shotguns. While these heavy-duty measures are a good idea on guns used so heavily, they are not necessary for ordinary hunting arms. The heavy grease could congeal and impede function in very cold weather.

As with any firearm, shotguns should be wiped down with a lightly oiled or silicone-impregnated cloth before being put away. As said previously, fingerprints attract moisture and promote rust—a little or a lot.

Shotguns, even more than handguns, tend to get stuffed into fleece and flannel-lined cases to avoid "closet rash." Closet rash is the minor dinging and scratching that comes from casual storage of the shotgun. A bit of closet battering may be preferable to the rust that may be generated by these moisture-absorbing cases. They are great for transit, but not good for storage. There is no substitute for hands-on, eyes-on inspection at least once a month. A brief inspection and a quick wipe will generally suffice.

chapter 18

AMMUNITION AND BALLISTICS

BALLISTICS. That's a nice big word. Let's see what it means to the beginning shooter. Webster's gives several definitions:

1. The science of the motion of projectiles in flight or the flight characteristics of a projectile
2. The study of the processes within a firearm as it is fired or the firing characteristics of a firearm or cartridge

A basic understanding of ballistics helps the new shooter to choose the right gun for the activity he has in mind. For example, a deer hunter in the Pennsylvania or Oregon woods might choose a very different caliber than a hunter in the high plains of the West. The deer are roughly the same size, but the woods hunter's shots would likely be in heavy cover at short range. The plains hunter would probably be shooting at much longer ranges. The hunter with ballistic knowledge will have a much better chance to put meat in the freezer than the hunter forced to guess what gun and caliber to use.

There are a few terms and concepts to absorb which will take the mystery out of ballistics:

VELOCITY—the speed of a bullet or shot charge as it travels to the target

ENERGY—the force carried by a bullet or shot charge

TRAJECTORY—the curved path of a bullet in flight

Muzzleloaders and early blackpowder cartridge guns generally used large, heavy bullets propelled at relatively low velocity. Modern smokeless powder guns—beginning in the late 1880s—used smaller, lighter bullets propelled at much higher velocities to transmit the same or greater energy to the target than the old blackpowder numbers. The point to be made is that a small, fast bullet generates as much energy, or striking force, as a slow, heavy bullet.

The best of both worlds would be a large, heavy bullet propelled at high velocity. There are such guns—the 458 Winchester and 460 Weatherby Magnums come to mind—but they recoil quite heavily. Used for hunting very large, dangerous game, these heavy calibers are hardly for the beginner.

The new smokeless cartridges also achieved flatter trajectories, which made the shooter's task vastly simpler in that precise range estimation was not nearly so critical. While many people still hunt with muzzleloaders, more patience and shooting

Above — "Muzzle energy" is the result of a number of factors: primarily, the bullet's diameter, weight and speed of flight. Just how much energy is transmitted to a particular target depends upon the expansion characteristics of the projectile, its weight, speed of flight and the composition of target. This 308 caliber bullet is shown in its various stages of expansion (from right to left). The more expansion, the greater the amount of transmitted energy.

Rimmed　　Semi-rimmed　　Rimless　　Rebated　　Belted

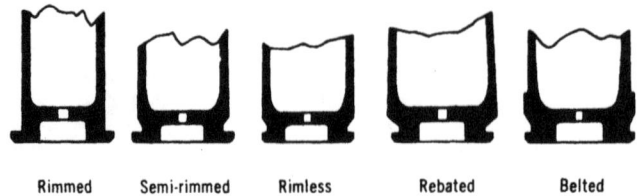

Above — Metallic cartridges have different types of rims as indicated by the accompanying drawing. Most magnum pistol cartridges use rimmed cases (far left) while magnum rifle cartridges use the belted case shown on the far right.

skill is required as compared to hunting with modern guns.

Beginning in the 1890s, hunters discovered the practical advantages early on and swapped guns in calibers like 45-70 and 40-82 for then-new 30-30 and 300 Savage guns. Because the new cartridges were smaller, the guns chambered for them were also lighter and handier.

There is a danger for new shooters called "ballistic hypnosis." This problem surfaces when a new shooter decides that a hotter caliber will solve his inability to hit targets or game animals. With nearly a hundred calibers available, it is easy to be confused. To illustrate the malady, some years ago an acquaintance decided to take up deer hunting. While he had no shooting experience, he selected a good, inexpensive 30-06 as his first rifle. He asked me to help him learn to shoot. I determined that the rifle was sound and shot quite well. The owner, however, could hit almost nothing with the gun. He decided that he did not "have time" to learn basic shooting skills. His solution to the problem was to buy a very expensive, very high-powered new gun. While he could hit almost nothing with the old gun, he could hit absolutely nothing with the new one. He never did learn to shoot, but the gun made a nice conversation piece. The point to remember is that most guns shoot better than most shooters. The beginner should buy new guns because he wants to—not because he believes that the "superior" ballistics of one gun or another will make him a vastly better shot.

Ammunition

Modern ammunition is divided into two basic categories—metallic cartridges, which are used in rifles and handguns, and shotshells. Metallic cartridges are further subdivided by the means of ignition—rimfire or centerfire—and by such identifying characteristics as general cartridge shape (bottlenecked, rimmed, belted, rimless, etc.).

While most shotgun shells are loaded with hundreds of tiny shot pellets and nearly all metallic cartridges hold only a single projectile called a bullet, this isn't always the case. Some shotshells contain a semi-hollow lead projectile known as a shotgun slug, and these loads are used in some areas for hunting deer and other animals. By the same token, there are metallic cartridges available loaded with small shot (called shot cartridges). These are used in handguns and sometimes in rimfire rifles for extremely short-range shooting at very small animals or at breakable targets.

Both rifles and handguns use metallic cartridges, and while it's usually easy to tell the two different types of ammunition apart (handgun cartridges are much shorter than rifle loads, and *usually* have perfectly straight sides) certain revolver loads are also used in lightweight rifles or carbines.

Metallic cartridges are differentiated by a numbering system that can be confusing to the beginner. Some rifle and handgun loads are identified by a pair of numbers—30-30 Winchester, 30-06, 44-40, 30-40 Krag, etc. In all such cases, the first number of the pair designates the bullet or bore diameter in inches (although it may not be an exact measurement—most 30-caliber bullets measure .308 inches in diameter, for instance). The second number in the pair can be confusing because different systems have been used over the years. The second number can stand for the original powder load (in grains) used in the cartridge (this is true of the 30-40 Krag—it fired a 30-caliber bullet with a 40-grain charge of powder). Or this number may

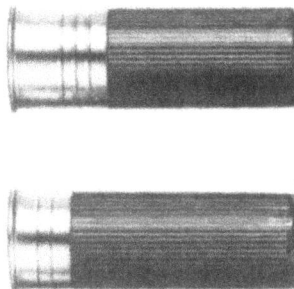

The "high-base" shotshell (upper) is used for long-range hunting. The "low-base" shotshell (lower) is intended for target or close range field hunting. The height of the brass has little to do with the shell strength — it serves to identify the nature of the shell.

From top to bottom: 30-30 Winchester: 30-40 Krag; 30-06 Springfield and the 8mm Mauser. The first set of numbers (in the top three cartridges) indicate caliber; however, the second set of numbers found after each of the above can be confusing. In the 30-30, the second "30" stands for the number of grains of powder used to propel the bullet; the same is true of the 30-40 Krag. The "06" in 30-06 designates that cartridge's year of military adoption—1906. Most European—and some American — cartridges carry a metric designation. The 8mm Mauser is one example, the "8mm" portion of the cartridge designation simply standing for the diameter of the bore (and projectile).

be the date the cartridge was officially adopted (the 30-06 was adopted by the U.S. in 1906).

To make the system even more confusing, most modern cartridges developed in this country now use a single set of numbers, often followed by the name of the company that developed the load (243 Winchester, 222 Remington). Again, the number is an approximation of the caliber, but may not be the exact measurement (the 30 Carbine, 30-06, 300 Savage, 308 Winchester, 300 Winchester Magnum and 308 Norma all use .308-inch bullets, but vary widely in size and performance).

What's more, not all calibers are designated by the English measurement of inches, but wear a metric tag instead. The 6mm Remington and 7mm Mauser are but two examples. The term "magnum" is usually reserved for cartridges that have a belted case and generate more power than another load using the same diameter bullet.

Shotgun shells are identified mainly by gauge, although the length may be tacked on to differentiate certain magnum loads from the shorter non-magnum variety. Still, only two lengths are generally available—2³/4 inches (which is the case length after being fired) and 3 inches. The exceptions to this rule are the 10 gauge (2⁷/8 and 3¹/2 inches long) and 410 hulls (2¹/2 and 3 inches in length). Guns with 3-inch chambers can digest either 2³/4-inch or 3-inch shotshells, but only the standard-length (2³/4-inch) load will fit the shorter chamber.

"High base" and "low base" configurations help identify "express" loads used for long-range hunting from lighter field and target loads, but the height of the brass base has little to do with hull strength.

The type of load contained in the shotshell is identified by another numbering system. The legend "3³/4 1¹/4 6" may appear on either the shell box or the shotshell itself. This means the shell is loaded with 3³/4 drams-equivalent of powder (a throwback to the early days of shooting, as drams are no longer used to measure powder charges), and 1¹/4 ounces of size 6 shot.

Practical shotgun ballistics don't vary a great deal from load to load, with most factory ammunition propelling its shot charge from the muzzle somewhere in the neighborhood of 1100 to 1200 feet per second. Pattern evenness and density are of more interest to the shotgunner, and these factors are determined by firing at a large piece of paper at a range of 40 yards.

Rifle and handgun ballistics are concerned with several different kinds of information. How fast is the bullet moving when it leaves the muzzle? How fast is it still moving at 100, 200 or even 500 yards? How hard will it hit at those ranges? And— very important to the long-range rifleman—how far will the projectile drop over those distances?

For all practical purposes, the main points of concern to riflemen are velocity (measured in feet per second), energy (foot-pounds) and trajectory (drop measured in inches). Handgunners are interested mainly in bullet velocity and energy levels.

chapter 19

WHAT IS HANDLOADING?

ONCE YOU LEARN to shoot, you discover things about ammunition. One discovery is that you use lots of it; another is that it costs a great deal of money. Oh sure, if you stick with 22 rimfire it isn't all that expensive. But if you intend shooting centerfire, unless you go plinking with centerfire calibers available as GI surplus, plan on leaving lots of money at the gun store.

Reloading, if you have the time, can be the answer. The time it takes to reload depends, of course, on the sophistication of the equipment you buy. Less money equals slower work. But even with reasonably inexpensive gear, excellent ammo can be produced. Most serious target shooters and other competitors load their own, as they believe the extra care and time they take will give them consistency impossible to find in factory-loaded ammunition.

Since not all guns are precisely alike, it makes sense that likewise all ammo of the same caliber is not alike and will not perform the same way in every gun. The ability to experiment and determine what your gun likes best is also important.

There was a time, some years ago, when many reloaders tried to outdo one-another in terms of how much power they could stuff into a given car-

Handloading is a matter of *replacing* the expendable components of a cartridge — bullet, powder and primer — in a resized, reusable case. Reloading makes economic sense, is easy to learn and makes a wonderful hobby.

tridge case. The factories had a bit to do with that, claiming some impossible velocities for factory numbers. (Those were the days before chronographs were affordable to the average hobbyist, and it was discovered that some claims were a bit tongue-in-cheek.)

The point of this is to point out that maximum loads are very seldom as accurate or reliable or as comfortable to shoot as more reasonable loads. Maximum loads also are not easy on your expensive guns, any more than driving your car all the time at 120 miles per hour makes your engine last longer.

Safety is also important when reloading. Years ago, I used to cringe when reading the reloading column in a monthly magazine. The author was pictured seated in front of his loading bench happily sucking on his pipe. Reloading and smoking do not mix!

I have also seen people reloading with the TV on, while carrying on conversations with others or while drinking. One of these folks managed to put two maximum powder charges in a 44 Magnum case. It took his gun apart.

He could have as easily failed to put powder in one of these cases which is an equally serious problem. The primer may well manage to push the bullet out of the case and into the bore. In rapid fire, the trigger could be pulled again before the shooter realizes that the first bullet is stuck in the bore.

Smokeless powder is tolerably stable. In the event of combustion, it burns fast and hot, but seldom explodes unless very tightly confined. Remember that a pound or two of powder goes a long way — your loading room need not look like a 16-inch powder handling room aboard the U.S.S. New Jersey.

How do you go about learning to load your own ammo? The procedure is simple and easily learned. When you buy a factory loaded cartridge, you get a precision-formed brass case, a primer, powder and a bullet. When the round is fired, the primer, powder and bullet are used up and can't be recovered. But you still have the empty brass case — and this is the most expensive component. This case is reusable — many times, in fact — and by salvaging it and using it over and over again you're saving the biggest share of the cartridge cost. All you need to do is replace the other three components, and this can be easily (and inexpensively) done. Let's take each of the necessary steps in order.

First, you need to resize the fired case. Firing expands this brass container under the pressure of expanding gas, and when it is extracted from the chamber, it will conform pretty well to the dimensions of that chamber. Too, the case mouth will now be expanded to a size too large to hold a bullet firmly in place. So the case is forced into a sizing die to return it to its original dimensions. If the case is to be fired in the same rifle, only the neck has to be resized — otherwise, the entire case must be reformed to assure proper functioning.

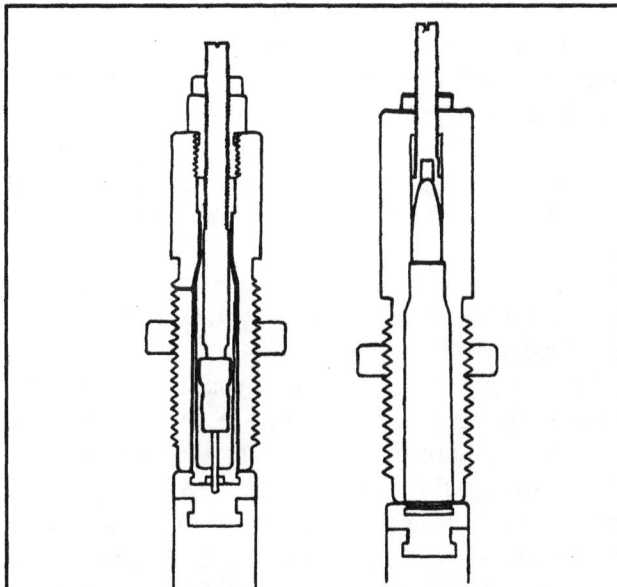

Larger, bench-mounted presses use the type of reloading dies shown here. The drawing on the left shows the first step in reloading a metallic cartridge case. As the fired shell is forced into the die, the dimensions of the case (neck and body) are reduced to more nominal tolerances and the old primer punched out of the bottom of the case. The case may now be primed, refilled with a safe quantity of powder and a new bullet "seated" into the case neck as shown in the drawing on the right.

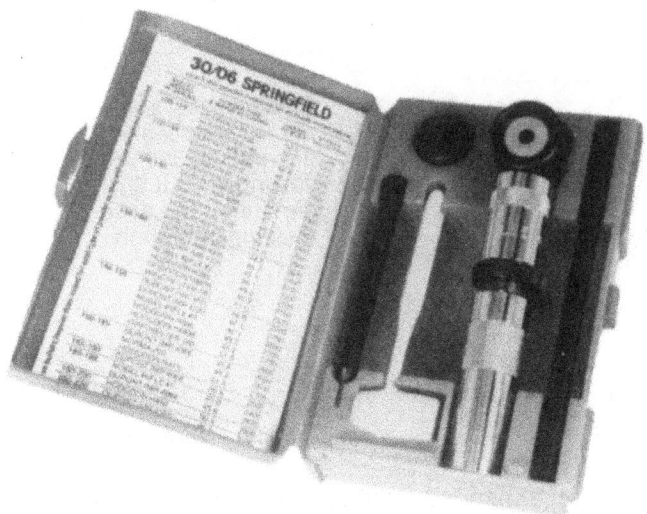

This Lee Loader is about the least expensive way to get started in reloading—and decent ammo can be produced if you have the time. For about $20—you're in business.

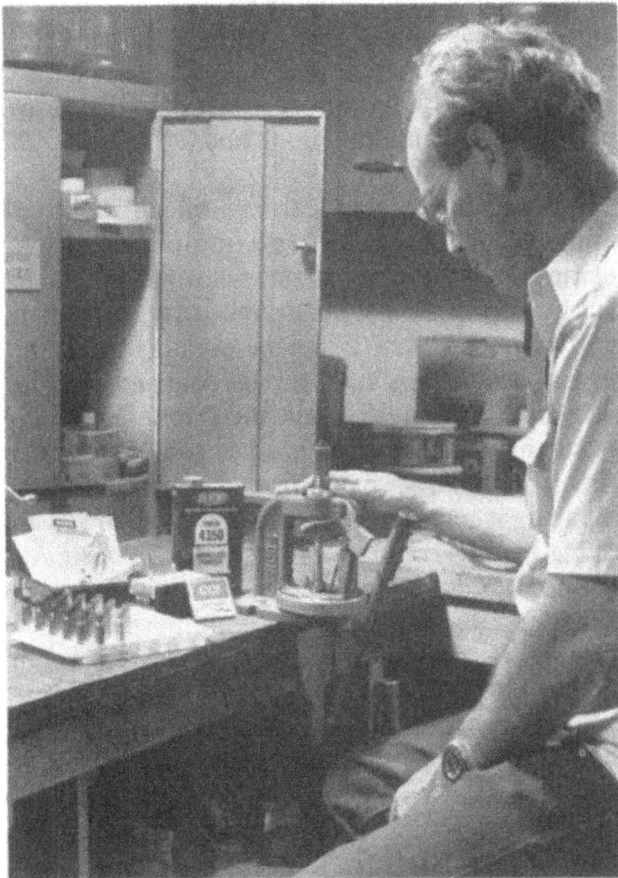

Whenever possible, a reloading press should be securely mounted to a solid workbench. This shooter has a permanent reloading area set up in his basement workshop.

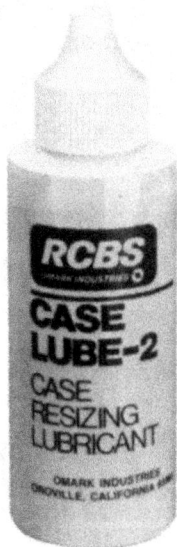

Specialized case lube is necessary to prevent cases from seizing in your sizing die during the case resizing operation.

The expended primer must also be removed from its pocket in the case head. This is called decapping and is performed as part of the resizing operation in most bench-mounted loading tools.

Next, the inside of the case neck must be expanded slightly. Resizing tends to squeeze the neck down to a size too small to allow a new bullet to be inserted, so an expander button is run through the mouth of the case to expand the neck to the proper size. This step is also performed automatically by the resizing die in most cases.

Then a new primer must be pressed into the primer pocket. After priming is completed, a fresh charge of powder is measured into the case. Finally, a new bullet is seated in the case neck. That's basically all there is to it.

These steps can be performed almost equally well with a set of inexpensive hand-operated tools like the Lee Loader or Lyman 310 Tool, or by the more sophisticated bench-mounted equipment dedicated hobbyists prefer. The big differences between the two are the cost of the equipment used and the time it takes to turn out a finished cartridge. The work can be performed at your kitchen table, although it's better to reserve a space in your basement workshop or garage where you can keep your tools and other gear together and work undisturbed. Distractions or interruptions should be avoided when you're working with your loading tools, as you need to keep your mind on what you're doing. Reloading is a safe hobby, but you do have to pay attention to avoid double charging (putting one load of powder on top of another in the case) or other potentially hazardous goofs. *Never smoke or drink during a reloading session.*

These steps can be performed very well — and very slowly — with a set of hand-operated tools like the Lyman 310 or the Lee Loader. Just as the beginner often tires of his single shot rifle, most beginning handloaders quickly tire of their slow hand tools.

The steps just described tell you how to reload a metallic cartridge, but there are some additional details you should perform to get best results. First, you should always examine every empty cartridge case before you load fresh components into it — particularly if the case has been fired more than once. Watch for bulges, cracks in the neck or body and other indications that the case may be unfit for use. Throw these damaged or worn-out cases away.

If the case you're reloading is a new one, or one that has been fired only once before, you should use an inexpensive chamfering tool to remove the sharp edge inside the case mouth. It's also a good idea to clean out the primer pocket — again, low-priced tools are available to handle this chore.

This Hornady powder measure (left) and single-station Lee press (right) will get the job done much faster than the hand tool, but not as quickly as the turret press. With the single-station press, it is normal practice to do all the decapping, change dies, do all the sizing, etc., until the job is done. The turret press turns out complete ammo by moving the tools and the empty brass.

Before you ever try to force a cartridge case into a sizing die, the case must be lightly coated with a special sizing lubricant designed for the purpose. The handiest way to do this is to roll the case across a pad soaked with this lubricant. Failure to properly lubricate a metallic case will cause all sorts of problems — sometimes the case will stick in the sizing die so hard that it can't be removed intact. The inside of the case mouth should also be treated by passing a small, circular brush coated with lubricant in and out of the neck. Don't overdo things, though, as it's possible to coat the cases with too much lubricant. When this happens, the cases sometimes acquire small dents when they pass through the dies. These dents are harmless but unsightly.

You should also make sure to wipe the lubricant from the reloaded cartridge cases with a dry cloth before firing. Slippery cases can cause pressure problems, as they tend to set back — hard — in the chamber when fired.

Once a case has been reloaded a few times, it will gradually become longer. Brass flows under extreme heat and pressure, and the neck will eventually lengthen until cutting or filing is needed to shorten the case to its original length. There are several tools sold to perform this function, and some are very inexpensive. To tell whether or not your fired cases require this treatment (and if you continue to reuse them it's just a matter of time until they will), use a case length gauge to measure them.

If you're using a powder measure to throw charges directly into each cartridge case (most re-loaders use these devices as they can greatly speed loading time), be sure to carefully weigh each fifth or sixth charge to make sure the measure hasn't changed its setting. This is both a safety and a quality control check. Overloads are dangerous, and unless each load is uniformly metered the ammunition produced will give erratic performance.

Be very careful to get just the right *amount* of the *right* powder in those cases, too. Buy a good loading guide (available at most sporting goods stores) and use it. These handy manuals contain reference tables for most popular cartridges and are available from a variety of sources — Lyman, Speer, Nosler, Hodgdon and DBI Books, Inc. all offer good ones, and there are several others published. *Be sure* to start out with one of the milder loads listed and work up gradually. You'll find that the hottest loads shown often give poorer accuracy, and they also shorten case and barrel life noticeably.

I've already mentioned that you can get a usable reloading kit for not much more than you'd pay for a box or two of factory-made cartridges. These work fine and have the added advantage of taking up little room on a storage shelf. Their major disadvantage is that they're slow and sometimes unhandy to use.

The bench-mounted presses are more costly, but if you stick with reloading, you're likely to eventually graduate to one of these heavy-duty tools. They're both faster and easier to operate, and the better ones can be used to let you form *"wildcat"* cases (non-factory-designed ammunition.)

When you load your own cartridges, there are

This Ponsness-Warren Model 375 Du-O-Matic is another time saver for the shotshell loader. The two cylinders atop the machine hold powder and shot. The projections below the cylinders are the dies that do the work of decapping, capping, sizing and charging. A trap or Skeet shooter would find equipment of this sort would pay for itself very quickly.

several important rules to remember. One rule-of-thumb every safety conscious handloader faithfully observes is to always double check each round for possible overcharges before the bullet is seated. The most common cause of overcharging is throwing two charges into the same case — this error is easily detected by a quick visual inspection of a row of charged cases just before the bullets are to be seated. If it occurs, empty the case powder back into the powder hopper, recharge with a single load and continue to check the cases for more double charges.

Another good rule is to never exceed the maximum loads recommended by your reference manual. Many shooters become tempted to see how far they can stretch things, and they're risking damage to both their firearm and themselves. Strangely enough, you can also get into trouble by drastically *under*charging a cartridge. When the volume of powder within a cartridge case is too small, extremely high pressures are sometimes generated — and these can wreck your gun.

Shotgunners also reload, but in this case it's almost always to save money. It's probably possible to improve on the performance of factory loaded shotshells, but this isn't the goal most reloaders have in mind. If you become addicted to either Skeet or trap, chances are you'll be forced to re-load to hold ammo costs at a reasonable level. But if your shotgunning is limited to bird hunting or an occasional session shooting targets from a hand trap, you'll probably be happier buying factory ammunition.

Like metallic cartridges, modern plastic-hulled shotshells can be reused over and over. The reloading process takes eight basic steps with a shotshell: The old primer must be removed (decapping); a new primer inserted (priming); fresh powder added (charging); a felt or plastic wad seated; and a charge of shot poured in. Then the crimp must be started, then finished (crimping is usually a two-stage operation), and finally the shell is run through a sizing die to bring it back to the proper dimensions.

These eight separate operations are usually performed in a series of just five steps or stages. Depending on the kind of loading tool you buy, these steps can be relatively complicated and time-consuming, or performed in seconds by a stroke of a handle. The better (and the most expensive) press-type tools are set up to perform several steps at each stroke. The hulls are set in a rotating platform, and a different operation is automatically performed at each stage.

It's important to settle on a single brand and style of shotshell hull early in the game, and stick with it. Different cases have different internal dimensions, requiring different base wad heights and other changes when it comes time to load them. You can greatly simplify things by standardizing here. And remember to carefully inspect fired hulls before reloading them. Discard those that show split ends, pinholes, or any other kind of external damage, and then check internal dimensions with a length of dowel. If one case has more room inside (greater depth), it's a sign that the inner base wad has been destroyed — once again, discard.

Loading your own ammunition can be both fun and easy, and it can save you a *considerable* amount of money if you plan to shoot a lot. With a little care and attention to detail, you can turn out ammunition that will be equal or even superior to factory made fodder.

More important is the fact that reloading encourages you to shoot more. The fact that each cartridge costs you less helps hold expenses down, and knowing how to load your own ammo makes you a more knowledgeable — and therefore better — shooter.

chapter 20

AIR RIFLES AND HANDGUNS

IT WAS NOT until the industrial revolution was well under way that the average American shooter of any age became aware of the airgun. Until low cost mass-produced products were an everyday fact of life, the airgun was the darling of the monied gentry. Typically, in the 18th and early 19th centuries, fine pneumatic airguns were elaborate handmade arms designed for serious hunting. On the eve of a hunt the lord of the manor would instruct his footman or other manservant to pump up the brass reservoirs of his air rifle. It might take hundreds of pumps to charge each tank.

The arms were well-made muzzleloaders and fully capable of taking a stag or wild boar. And, they had the advantage of being unaffected by rain. Few of these rare weapons found their way across the Atlantic to the New World, although Lewis and Clark carried one on their famous expedition, much to the amazement of the Indian tribes. Imagine, a gun that made no smoke!

The first commonly seen airguns in America were smooth bore spring-piston airguns which appeared in shooting galleries after the Civil War. The airgun really didn't begin to have an impact until the late 1800s when the Plymouth Windmill Co. offered a simple all-metal BB gun as a sales

premium. The windmills didn't sell, but the BB guns did. In due time, the company wisely changed its name to that of the BB guns and Daisy Manufacturing Company was on its way. This writer is of the generation that grew up in the 1930s and cherishes the memory of receiving his first Daisy, a rite of passage more memorable than first long pants or high school graduation.

The value of old airguns of every type has increased as collectors discover this fascinating field. One advantage of the airgun hobby is that there are few restrictions on shipping them. Another is that, like anything new, there are still bargains out there for those who know their airgun history.

Today we are in the "Golden Age of the Airgun," a time when the range of types, styles and prices are wider than at any other period in history. Everything from the simple BB youth-oriented "first guns" to thousand-dollar world-class competition pellet match rifles are readily available. American shooters are becoming more aware that airguns provide a rich and rewarding area of interest well worth their attention. Not to be considered mere firearm substitutes for children, they are high-quality arms delivering high performance

187

Daisy Model 923 — .177 caliber single stroke pneumatic. Moderately priced target rifle. Currently used by U.S. military for training.

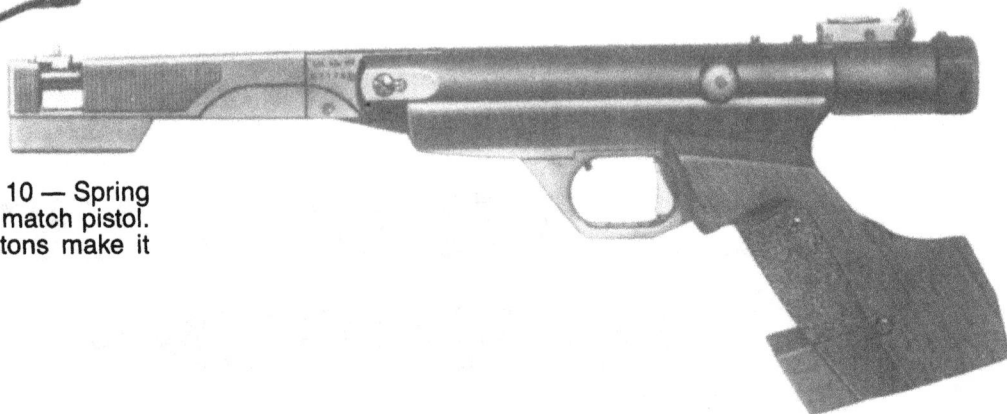

RWS Diana Model 10 — Spring piston .177 caliber match pistol. Counter acting pistons make it virtually recoilless.

according to their design.

Small game hunting with airguns is growing in acceptance and competition is expanding. Field shooting events are the latest interest and look as though they will catch on in a big way. And, as urban areas expand, a place to shoot a firearm is becoming harder to find. The airgun, with its moderate range, practically silent firing noise, and modest power is a welcome solution to those who enjoy shooting but don't have handy access to a safe range. It is always wise to check local ordinances before retiring to a vacant lot or nearby meadow with your trusty airgun, however. You just might be breaking the law.

Airgun Types

Airguns come in three fundamental action types: pneumatic, spring-piston, and CO_2.

PNEUMATIC AIRGUNS

The pneumatic is a true airgun in that air is compressed with a built-in pump and stored in a separate chamber. Pulling the trigger opens the exhaust valve on the storage chamber allowing the expanding air to push the pellet or BB down the barrel. The major parts are the compression cylinder and piston, storage chamber with inlet and exhaust valves, and trigger mechanism. The exception to this setup are those few pneumatics that combine the compression and storage cylinders into one.

The piston forms one end of the storage chamber. They are designed for simplicity and accuracy, not high power, and can only deliver one fixed velocity. Standard pneumatics are variable — the more you pump, the more power/velocity you get. They reach their best performance after six to 10 pumps, depending on make and model. Pumping past the factory designated maximum will only tend to damage the gun and will not result in much improvement in velocity. In other words, if six pumps will go through a tomato can, 12 pumps will not necessarily go through two cans and 100 pumps will definitely not go through a telephone pole! There are advantages to sticking to a conservative number of pumps, especially for practice.

Pneumatics are made in the U.S.A. by the major manufacturers, and a small number are imported from Japan, Spain, and Germany. The latter two countries also produce a number of target arms.

SPRING-PISTON AIRGUNS

Spring-piston airguns are, in my opinion, not true airguns since the power to propel the pellet comes not from air, but from a massive compression spring. Air is merely the medium used to transfer that power to the pellet. The spring is compressed by a hand-operated lever. Most commonly, that lever is the barrel which is connected to the piston by means of a lever or series of levers.

188

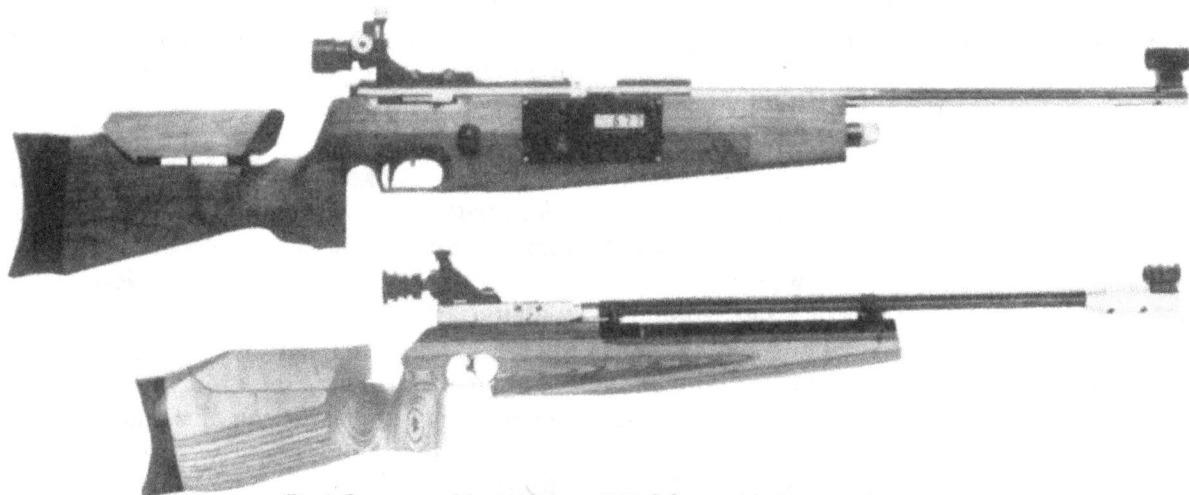

(Top) Crosman Model 84 — .177 CO_2 world class match rifle. The only U.S. made rifle in this class. Gas pressure is electronically controlled.

(Above) Beeman/FWB C60 — CO_2 .177 caliber world class match rifle. State of the art in CO_2.

Some models use a fixed barrel with the cocking lever underneath the barrel or on the right side of the compression tube. The author's beloved first Daisy was a "Red Ryder" spring-piston BB gun with a western-style cocking lever. The spring-piston airguns have evolved into some of the most sophisticated arms of any kind on this planet. At airgun ranges they exceed firearms in accuracy. Testimony to this near perfect precision is the 10-meter international air rifle target. Its 10-ring measures 1 millimeter in diameter, about the size of the period at the end of this sentence. European manufacturers dominate this field, with Germany producing the world champion brands such as Feinwerkbau, RWS, Walther and Anschutz. Daisy imports a moderately priced target rifle from Spain. Prices vary from about $350 for the Daisy to over $1,000 for the top-of-the-line Feinwerkbaus.

CO₂ AIRGUNS

The CO_2 guns were invented by Frenchman Paul Giffard and patented in 1864. His guns and those made later in this country were charged with the compressed gas from special refill cylinders. Crosman pioneered the CO_2 rifle in America and Schimel, a now-defunct company, was the first to manufacture a CO_2 pistol. The Shimel was also significant in that it was the first to use the disposable bulb-type CO_2-charged cylinders now so popular and used in all CO_2 arms. While the power of a pump-up pneumatic can vary with the number of pumps, and the power of a one-stroke cocking spring-piston gun is always the same, the power of CO_2 guns can vary depending on the rapidity of fire. With a modern rapid-firing repeater, shooting several quick shots in succession can turn your CO_2 gun into a miniature refrigerator. If you remember your high school physics, expanding gases absorb heat and lower the temperature. The result is a chilling of the stored compressed gas and a lowering of the power of each shot. So, unlike a firearm that you allow to cool after each shot for maximum accuracy, a CO_2 works best when you let it warm up.

Crosman is the only U.S. firm producing a world-class air rifle. Their Model 84 is a CO_2-powered rifle with electronic trigger and power regulating systems. European match arms that use CO_2 are also available. The prime advantage to using the power of CO_2 is the reduction in muscular effort required of the shooter. Every pump of an airgun lever elevates the heart rate a bit and the goal of a serious target shooter is to eliminate (or minimize) all variables, including those in the heartbeat. After all, if your heart is beating hard, it can cause you to wobble around the target. Until recent years, the problems with CO_2 variations due to changes in temperature and atmospheric pres-

189

sure ruled them out for the top level competition. Modern systems have mastered this.

General Operating Rules

As you can see, there are important characteristics of each type of airgun that should be remembered, and some of these are not obvious. For safe operation and maximum effectiveness, here are some general rules. Of course, airguns of all types should not be cocked, loaded or fired without a thorough reading of the instructions and cautions provided by the manufacturer. And, *eye protection should always be worn while shooting.* Adequate safety glasses are available that cost very little. You can chew with false teeth, you can walk on a wooden leg, but you can't see with a glass eye!

PNEUMATICS

Note the number of pumps recommended by the maker. Do not go below or above that number. Start with the gun unloaded. If you don't know if it is loaded or not, point it in a safe direction (soft dirt, etc.) and fire it. Remember, airguns are real guns and capable of real injury. Follow all the rules of safe gun handling you would observe when handling any firearm. Put on the safety. Pump the desired number of times. Try to be consistent and watch out for "pinch points" that could injure your hands. Open the breech; in most models this cocks the firing mechanism. Load the pellet and close the breech. Disengage the safety only when ready to fire in a safe direction. A pneumatic can be kept loaded for some time with no damage to the gun. It is usually a good idea to leave a couple of pumps in an unloaded pneumatic when it is put away for storage. This keeps moderate pressure on the valve seals.

SPRING-PISTON AIRGUNS

Though simpler in function, spring-piston guns have certain important features that must never be forgotten. Since barrel-cocking airguns are the most common, we'll cover them first. Holding the gun by the pistol grip firmly in the right (or shooting) hand — finger must be off the trigger at all times — slap the barrel smartly near the front sight to break the action open. Some airguns have a breech lock that should be disengaged first. Bracing the butt against the body, bring the barrel down until it "clicks" in the cocked position. The breech is now exposed for loading. Throughout the loading procedure it is imperative that you *keep*

your finger away from the trigger. Many modern spring-piston guns have a built-in trigger block that prevents the gun from firing unless the barrel is closed against the breech, but many do not. If you accidentally pull the trigger when the barrel is down in the cocked position, it will *swing upward with great force.* This could cause considerable injury to you or bystanders. At the least, it could damage the gun. I once saw a fine German air rifle with its barrel bent upwards several degrees by just such a careless act. Always respect the power of that big spring. Once the breech is open, seat the pellet flush with the breech face and close the barrel. The gun is now ready to fire. Some modern airguns have safeties that engage automatically when the gun is cocked. Push off the safety when you are ready to shoot and the gun is pointed in a safe direction.

Unlike a pneumatic, it is not a good idea to keep a spring-piston airgun loaded and cocked. The spring could loose its spring (take a "set") and you have a repair job on your hands. Resist any temptation to disassemble a spring-piston airgun. It is a very dangerous operation for all but the trained expert. Another thing to remember about spring piston airguns: *never dry fire.* Without the resistance of the pellet to push against, the piston will slam with great force against the end of the compression chamber and damage to the gun can result. If you have a cocked airgun with no pellet in the barrel and it is the type that can't be uncocked, press the muzzle against a carpet or piece of leather and fire it. The back pressure will ease the piston on its journey.

There are a few other things to remember, and these are the most important. If you forget them you can hurt yourself, others, or the gun. Maybe all three. Never forget you are handling a device that is perfectly safe when used correctly, but quite dangerous when misused. And don't put all your faith in "safeties." As some wise shooter once said: "The best safety is between your ears."

CO₂ ARMS

Modern CO_2 rifles and pistols are more closely related to firearms in operation. Though the pellets and the CO_2 power bulbs are loaded separately, once loaded they are always ready to fire and should be treated as such. The old adage "all guns are loaded" should never be forgotten. Manufacturer's instructions, no matter how dull, should be read and re-read until the procedures are under-

Beeman Pellet Table

Pellet	Description	Pellet	Description
	Beeman H & N Match—Wadcutter Simply the world's most accurate pellets. The 1986 international award winner. Good penetration and mushrooming effect also favors practical field use.		**Beeman Ram Jet—Round Nose** Extra knock-down power. Popular choice for silhouette. Special lip scores knock-downs even on glancing hits. Great for heavy-duty field use too.
	Beeman Silver Jet—Pointed America's best selling precision field pellet. Deep penetration. Double sealing rings prevent blow-by, increase accuracy and velocity. Never equalled!		**Beeman Kodiak—Round Nose** Beeman's heaviest pellet. Best wind resistance, greatest impact. The best heavy-duty field pellet available. Good for silhouette too.
	Beeman Silver Ace—Round Nose Round nose version of famous Beeman Silver Jet. Super long-range efficiency.		**Beeman Silver Sting—Pointed** Super penetration with outstanding accuracy in the field. Larger waist demonstrates greater stability and better air flow than thinner-waisted copies.
	Beeman Silver Bear—Hollow Point The original and best hollow point. Super expansion for field, silhouette, and target. Special design affords super high velocity, outstanding stability, flat trajectory, and wad-cutting. Great all-around pellet.		**Beeman Laser—Round Nose** The lightest, fastest pellet going! Affords super velocity in magnum airguns. Boosts velocity and accuracy even in domestic airguns.

—Simply the World's Finest Pellets—

Pellets are made in a variety of shapes for special uses: Wadcutters —best for paper targets; Pointed — Airguns kill small game by penetration, not shock. Pointed pellets give best performance; Round-Nose — for all-round use, good for silhouette shooting (note: pellet weight has more influence on velocity than shape); Hollowpoint — for field and silhouette use. Airguns do not generate enough velocity for hollowpoints to open as effectively as firearms. Art courtesy Beeman Precision Arms.

stood. CO_2 guns are generally not as powerful as pneumatic or spring-piston guns, but nonetheless carry dangerous potential when handled carelessly. As mentioned before, shoot deliberately and avoid the temptation to rapid fire these guns. Most CO_2 guns will fire up to 50 or 60 times with each charge, but you will notice a distinct dropping off of power as the bulb loses its pressure. Some CO_2 guns have a feature that dumps the remainder of the pressure when it drops below a useable level. This happens automatically and can be somewhat disturbing if you're unprepared for it. A friend of mine once told me a story of a shooting session with a companion who was unaware of this feature. When the expensive imported CO_2 handgun dumped its charge, the shooter panicked, thought the pistol was about to explode, and threw it as far as he could. The fine finish on the pistol and the quality of the friendship both suffered. The moral to the story: know your guns and educate all who use them.

Airgun Ammunition

Knowing the ammunition you use in your airgun is every bit as important as knowing the gun itself if you are to get the best performance. Ammuni-tion breaks down into lead pellets, steel or lead BBs and darts. Darts are limited to a few smooth-bored airguns and are designed for informal games with special dart board-type targets. Darts can damage guns not made for their use. Lead BBs are rare but can perform adequately in both smooth and rifled barrels. There is even an American-made pneumatic air rifle that fires 22-caliber lead balls. Far and away, steel BBs are the most popular ammunition for the low cost youth spring guns and pneumatics. Many of these guns use magnetic feeding mechanisms so if you come across lead BBs, they won't work except by single-loading. Steel BBs are solid steel spheres that can bounce with a vengeance. That means great care should be taken when choosing targets and backstops. A steel BB can ricochet or bounce back with great force. Wood backstops are notorious offenders. Even soft lead pellets can sometimes rebound from a wooden fence or post. Fairly new on the market are plastic and zinc alloy pellets that feature a low-resistance plastic sabot containing a lighter-than-average sub-caliber pellet. The advantage is higher than average velocity.

Today, lead airgun pellets are being manufac-tured in such varieties of style and quality that a

Beeman Blue Ribbon Airgun Scope — available in 3-9x, 4-12x and fixed powers. Features speed dialing of range and windage, parallax correction.

complete chapter could be written on the subject alone. They are made in this country and in most countries that make airguns. Pellets range in quality from world-class competition grade to low quality backyard plinkers and everything in between. Special designs have been developed for special uses such as hunting and silhouette games. They come in flat-pointed, domed, sharply pointed, hollowpointed and semi-wadcutter shapes. The best advice to the beginner is to try as many different types as possible. It is amazing how any given airgun will respond to different pellets. A gun can shoot badly with one brand or type, and shoot superbly with another. Generally, the better the grade, the better the gun will perform. Cheap brands often contain a high percentage of deformed, irregular pellets that will not shoot consistently. European and Japanese imports have consistently performed better than domestic brands. However, American companies are rising to the challenge and have greatly improved their top grades. Pellets are fragile little lead cups that can be easily crushed, so it is best to keep them in their tins, or in a belt pouch when in the field. Shirt and pants pockets tend to accumulate dust and debris that can raise hob with your barrel. Though pellets can ricochet, they are much safer than steel BBs. They tend to flatten harmlessly when they hit a hard surface. When target shooting, a simple pellet trap can be made of a cardboard box or shopping bag filled with newspapers. There are many fine metal and wood-and-putty traps on the market if you are serious about target shooting with your airgun. A shooting range can be easily set up in a basement, rec room or hallway — something you can't do with a firearm.

Shooting Techniques

Shooting techniques can affect accuracy with airguns, especially with the spring-piston guns. It is often true that where and how well an individual gun shoots depends on how you hold it. A rifle grouped from a rest may shoot to a different point when fired offhand. A rifle gripped and cheeked tightly may shoot to a different point when held lightly. This is believed to be due to the fact that parts of the gun are in motion before the pellet leaves the barrel on its way to the target. Remember, when you pull the trigger you release a massive piston pushed by a large spring. That piston moves forward, compressing air on the way. At a certain point, the compressed air blasts through the vent behind the pellet and drives it down the bore. High-speed studies have shown that the piston and the spring actually bounce before all action stops. So, to hit consistently, shoot and hold the gun consistently.

Airgun Sights

Airgun sights are generally similar to firearm sights. The factories supply a variety of adjustable open sights, and most have click adjustments for elevation and windage. The maximum potential of any air rifle can be achieved only with a scope, and there are many airgun scopes on the market to suit any budget. Air rifles usually have scope mounting grooves or bases. Pneumatics generally require clamp-on adaptors. A spring-piston airgun has a distinctively sharp double snap at the moment of recoil that can shake loose all but the most rigidly installed scopes. Firearm scopes are generally not well adapted to spring-piston airguns for two reasons. First, the lower priced models can't withstand that double snap, and 22 rimfire scopes don't have the parallax setting for the average airgun range. Airguns are usually shot at 25 to 100 feet and airgun scopes have their parallax set to allow for that. Firearm scopes, on the other hand, are set for 50 to 100 yards. Shooting at the shorter ranges with such scopes can result in consistent misses since a slight movement of the eye in relation to the scope can shift the point of impact of the pellet. So, just be sure the scope you are getting is an *airgun* scope. To test this, set the scope on a solid rest, pointed at an object 50 feet away. Without touching the scope, look through it and move your eye back and forth (sideways) in relation to the scope. If the crosshairs move on the target as

Airguns require special care. Specific lubricants must be used in spring-piston guns. Incorrect lubes can cause damage and possible injury. It is essential to check with the manufacturer for recommended products.

you move your head, your scope does not have its parallax set for that range. Some high-quality scopes have parallax adjustments and are worth the extra cost if you are serious about shooting at different ranges, as in silhouette games.

Airgun Maintenance

Care of airguns is a simple matter but there are a few things to remember. It is always a good idea to use *only* the lubricants recommended by the manufacturer or importer. The wrong oils or lubes can damage seals or, in the case of spring-piston guns, can actually damage or destroy the gun. A spring-piston gun is similar to the cylinder of an engine. If the wrong lube finds its way into the compression area at the time of firing, it can explode or "diesel" violently. At the least, you will get a shot of higher than maximum velocity, a sharp crack and a wisp of smoke. This isn't uncommon from the first few shots out of a brand new spring-piston gun. However, if a well-meaning owner squirts ordinary gun oil or household oil down the vent of such a gun, the resultant dieseling could ruin the gun and possibly injure the shooter. Such incidents have been known to blow the barrel assemblies off the receiver. The answer is to only use the low flashpoint chamber lubes recommended by the maker or importer. Spring-piston guns should generally be lubed every 1000 shots or so.

The barrels of all airguns should be cleaned periodically. Felt pellets of the same caliber as your airgun are available and are very easy to use. You just soak one (or two in the case of high-power airguns to create sufficient back pressure) with gun oil and fire them through the bore. Fire a few dry felt pellets until they come out clean, then finish with one that is lightly oiled. Special air rifle and air pistol cleaning rods and kits are available, too. Don't over clean. Pneumatic guns usually have brass or bronze barrels that are much softer than steel and are easily damaged by thoughtless and over-ardent cleaning with rods.

There are so many fine airguns on the market today that it would be impossible to do them all justice in this chapter. For reference and specifications on what is available today, I recommend you read the current edition of *Air Gun Digest*, published by DBI Books, Inc., and the catalogs from Air Shot Corporation, Beeman Precision Arms, Benjamin/Sheridan Air Guns, Crosman Air Guns, Daisy Mfg., Dynamit Nobel of America (RWS), and Marksman Products.

While informal target shooting, backyard or barnyard plinking, and small game and pest control shooting are the major uses of airguns, there is a growing number of formal competitions of many kinds. The National Rifle Association maintains an Airgun Committee and offers brochures, booklets and training films. In addition, they coordinate, plan and develop international and Olympic airgun shooting activities. The major U.S. companies — Daisy Mfg. and Crosman — have marksmanship programs, sponsor and manage shoots, and, in general, support airgun shooting. For years, Daisy has worked with the Jaycees across the country in sponsoring airgun competitions. Crosman has introduced an airgun video and sponsors a new sport for youngsters called "Bikathon" that combines bicycle racing and airgun shooting.

Trends for the future of airguns seem endless. At the time of this writing there are air shotguns, automatic pellet guns, and endless refinements and improvements on all fronts. And, 25-caliber rifles and pellets are making a comeback. Imports from mainland China offer sturdy, well-made adult airguns at affordable prices. There is an excellent new magazine called *American Airgunner* offering timely articles on all areas of airgun shooting. The publisher is actively promoting airgun field target shooting competitions. Shops providing custom airgun services are appearing across the country. Aftermarket products are springing up like mushrooms after the first rain. Yes, the horizon is filled with possibilities. See you there.

chapter 21

Shooting the Muzzleloader

WHILE THE MODERN RIFLE, handgun and shotgun shooter has it all over the shooting sportsmen of the last century in terms of care, convenience and ease of operation, there is a rapidly growing trend toward reviving the old, slow muzzleloaders. As a matter of fact, the blackpowder "frontloaders" have become so popular in recent years that supplies of shootable relics have pretty well been exhausted, and a whole new industry has been created for the manufacture of replicas and kits to satisfy the demand.

What has caused this sudden reversal in attitude? For centuries men have labored to create more efficient firearms that were easier and faster to load. Thousands of unsuccessful attempts were made to build a reliable repeater before the first really practical design evolved. Improvements in firearm ignition systems evolved with equal slowness, and blackpowder itself didn't change appreciably over the course of all but most recent history. When the cleaner burning, better performing smokeless powder finally appeared on the scene, it was so clearly superior that it almost completely replaced blackpowder as a propellant charge in a few short years. New guns were designed for the new smokeless powder cartridges, and the manu-

facture of blackpowder was continued primarily to accommodate shooters who still owned the now-obsolete frontloaders.

So how come shooters are turning to this outdated shooting system in droves these days? Why would anyone want to shoot a gun that takes upwards of a half-minute to painstakingly load and that must be taken apart and thoroughly scrubbed in hot, soapy water before being put away at the end of the day? Who in his right mind would prefer the "charcoal" burning firearms of yesteryear to the modern, convenient and highly accurate sporting arms that are available today?

The answer to that is easy as soon as you understand that most of the people who shoot rifles, handguns and shotguns today do so for fun. *Fun* — pure and simple recreation. Most people who own these guns today aren't concerned with beating off attacks from hostile Indians, and most hunters go to the woods for sport. Blackpowder firearms aren't the life-or-death propositions they once were, and people don't depend on them so much for the serious business of providing groceries and protection as for the fun they have in using them. This change in priorities has also caused a change in attitude among shooters. The change has been slow in coming, but recent generations of shooting sportsmen have embraced the new attitude with fervor. To them, building a ball-and-powder load for a Kentucky rifle isn't a time-consuming task that could be interrupted by the unexpected appearance of a grinning savage. Instead it's regarded as part of the fun — and "doing it yourself" brings more shooter satisfaction than simply buying pre-loaded factory ammunition and feeding it into a multi-shot magazine.

Even keeping a muzzleloader clean and in good working order isn't looked upon as the onerous, filthy chore sportsmen of a century ago faced at the end of the day. The chore itself hasn't changed much since that time (although certain new powders have been developed recently to make the task less messy), but the attitude of those who use blackpowder guns *has*.

Nostalgia also plays a big part in this blackpowder revival. The old muzzleloaders do have a certain romance about them, and large numbers of shooters who turn to using them travel even further along the nostalgia trail by using authentically styled powder horns and other accessories. Period clothing is also worn by the really dedicated participants, and many blackpowder shooting clubs and organizations exist that mirror actual frontiersmen groups and military organizations of days gone by.

Visit one of these unusual shooting groups at one of their periodic full-dress "meets" and it's like stepping into the past. Military uniforms are both brightly colored and historically correct, while the fringed buckskins worn by frontiersmen shooters are almost guaranteed to be the real thing — and were probably crafted from the hide of a deer the wearer shot himself. What's more, you can almost bet that the deer was felled by a blackpowder-propelled lead ball instead of a more streamlined projectile from a modern rifle.

Even if you're not a nostalgia buff, shooting a blackpowder burning firearm can have some real appeal. In the first place, you can exercise greater control over your ammunition — and do it right there on the spot. If a certain load doesn't prove to give great accuracy, you can change it immediately — all it takes is a quick adjustment of your powder measure. You don't need to buy a lot of sophisticated reloading equipment, either — you just pour the components down the bore in the proper order. Shooting muzzleloaders is economical, too — and this is yet another reason it's becoming increasingly more popular.

Some people have turned to muzzleloaders because these obsolete firearms have been largely exempted from recent laws and legislation directed toward limiting free firearms trade. They fear these "gun control laws" may become increasingly more restrictive in future years, and adopt blackpowder firearms as a hedge against this time.

Richard Rees prepares to fire Harrington & Richardson Springfield Stalker, a .45-caliber muzzle-loading rifle. Gun won't be cocked until shooter is actually ready to fire. This H&R blackpowder gun is one of the many modern blackpowder rifles available today.

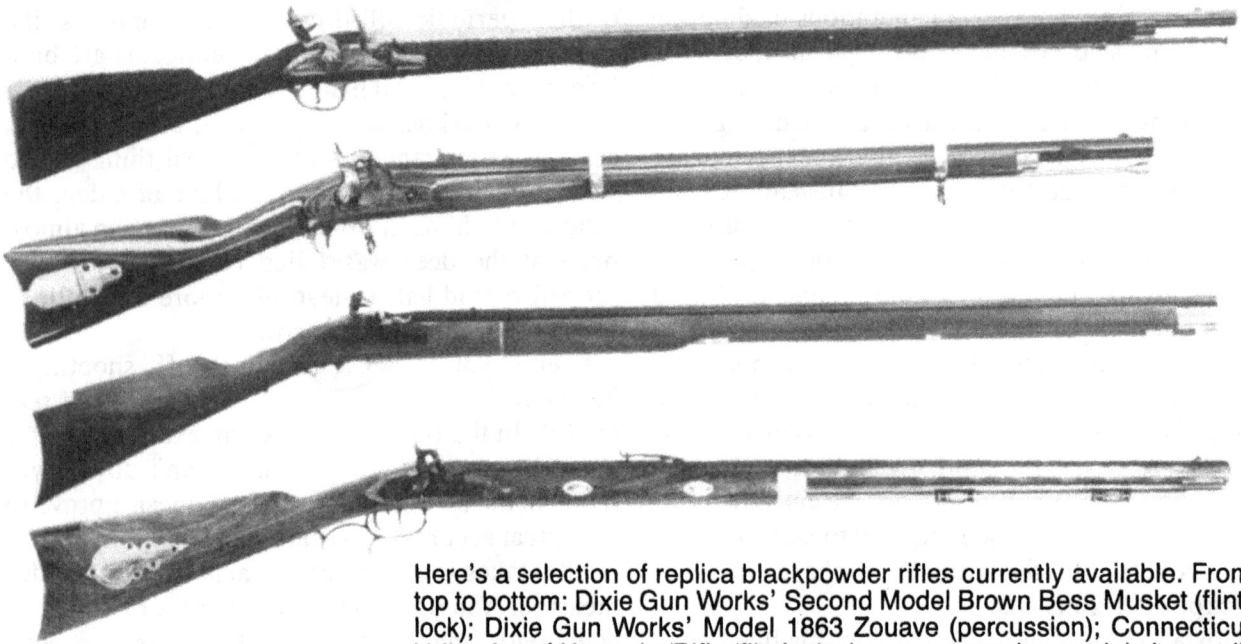

Here's a selection of replica blackpowder rifles currently available. From top to bottom: Dixie Gun Works' Second Model Brown Bess Musket (flintlock); Dixie Gun Works' Model 1863 Zouave (percussion); Connecticut Valley Arms' Kentucky Rifle (flintlock shown, percussion model also available); Connecticut Valley Arms' Mountain Rifle with "set" triggers (percussion).

Whatever the reasons, burning blackpowder in frontloading guns is a highly popular pastime these days, and it's growing more popular each year. Shooting blackpowder guns is a sport that's easy to learn, *provided you've already mastered gun handling and safety with a modern, smokeless powder firearm.* (I don't recommend muzzleloading firearms to any beginning shooter who's not already well grounded in shooting technique and the Ten Commandments of Gun Safety.) At the same time, it's a sport that has its own, special requirements and a brand new set of techniques to master.

Let's take a look at each type of blackpowder gun — rifle, handgun and shotgun — in turn. The basic principles of loading and firing remain the same in all three types, but there are enough differences that each variety deserves individual treatment.

Blackpowder Rifles

The muzzle-loading rifle is undoubtedly the most popular type of blackpowder-burning firearm. A number of these firearms in current use are authentic antiques, carefully restored and usually tested for safety before being fired for the first time. However, the vast majority of blackpowder rifles being shot today are modern replicas manufactured on 20th century assembly lines as complete guns or as do-it-yourself kits. The fact that

these "new-old" firearms take advantage of modern metallurgy often makes them superior to the original guns they were copied from, although in a few cases where quality control is allowed to suffer in the name of lower price tags and higher profits the reverse can also be true.

There are also some thoroughly new designs available, as companies like Harrington & Richardson, Thompson/Center and others are willing to break with tradition to provide a quality gun of more modern concept.

The blackpowder rifles available today — whether original antiques or modern replicas — can be divided into two general categories — flintlock and percussion. There are also a few Civil War era blackpowder breechloaders in use, but these are in the minority. Most blackpowder guns in popular use are front stuffers, and those are the firearms we're most concerned with here. The loading procedure is pretty much the same with both flintlock and percussion lock rifles — up to a point. Once the bore is charged with powder, patch and ball, the flintlock is prepared for firing by pouring a small amount of finely granulated powder into the priming pan, adjusting the flint in the lock to make sure it's positioned to strike a spark, and cocking the action. This means that the flintlock shooter must carry a supply of at least two different granulations of blackpowder and take care to see that the powder in the priming pan

196

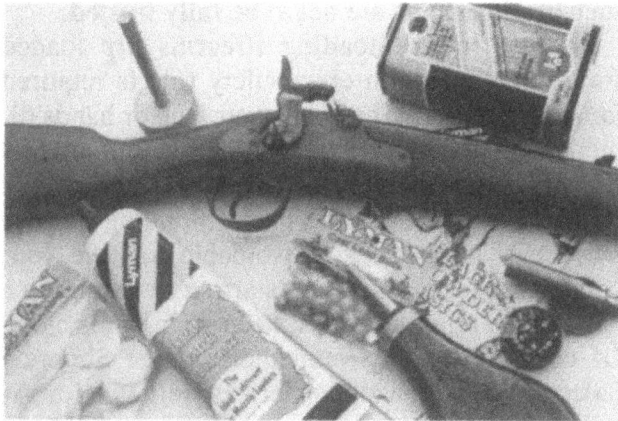

Muzzle-loading rifles require (from upper right) black-powder or the new Pyrodex substitute (shown), powder measure, percussion caps, powder (in powder flask), cast lead balls, lubricant, and cloth patches. Device at upper left is "short starter" used to seat balls in the muzzle.

doesn't get wet. What's more, there is a considerable lag time between the instant the trigger is tripped and the actual moment of firing — this is caused, first, by the relatively slow lock time (when the spring forces the cock forward), the extra time it takes for the spark struck by the flint to ignite the priming compound, and finally by the time needed for the priming powder to burn down through the touchhole to ignite the main charge. This time lag can be very disconcerting to a shooter, who must keep his sights steadied on the target until the ball actually leaves the muzzle.

The percussion rifles are more convenient to use, and in addition they feature a shorter "lock time" than do flintlock firearms. For these reasons, they are more popular among 20th century shooters. All a percussion rifle user needs to do after loading the bore is place a fresh percussion cap on the priming nipple and cock the hammer. The copper cap is fairly weather tight, and a quantity of these caps can be carried more conveniently than loose priming powder.

Before you try to shoot a blackpowder percussion rifle, you should have the following equipment and components on hand. First, you'll need a stout wooden ramrod of the proper diameter — ramrods are usually furnished along with the rifle, so this should be no problem (however, steel or other spark producing metal rods must be avoided). You should also have a bullet starting rod, or "short-starter" — this is basically a short (4-5 inches long) section of ramrod with a broad

palm rest, or handle, attached. You'll also need a powder flask and measure, something to lubricate bullet patches with (commercial lubricants are available for this job, or you can use saliva if you'll be shooting *immediately* after loading), and a container to hold rifle balls and patches. The components you'll need include a supply of blackpowder (FFG granulation works best for rifles between 40- and 58- caliber, while FFFG is used in rifles having a smaller bore), a supply of cast lead bullets of the proper diameter to loosely fit the bore of your rifle, and a supply of precut cloth patches — again, of the proper size.

A properly cared for muzzleloader will have a light protective coating of oil in the bore when it's stored away, and this same protective coating must be removed before the gun is loaded again. Do this by swabbing the full length of the bore with a clean, dry patch, taking particular care to dry out the final 4 or 5 inches near the breech. Then, *with the gun unloaded*, check the vent-hole in the nipple to make sure it's open, place a percussion cap over the nipple, point the gun in a safe direction and fire it. Repeat the operation again with the bore still empty — this will help dry out any oil that remains near the breech end of the bore, and blow any small residue out of the vent. This step is very important before you load for the first shot, incidentally, as any oil left in the bore can dampen the powder and prevent firing. When this happens, you must remove the charge by first making sure to remove the percussion cap from its nipple (do *not* replace it), and then attach a "worm" or "screw" to the ramrod. This is a corkscrew-shaped attachment that is turned to bite into the ball and remove it, and then screwed into the compressed powder charge to loosen it for removal.

Once you're sure the bore and vent are clear and dry, double-check to make certain that there is no percussion cap on the firing nipple. Next, hold the rifle upright with the butt resting on the ground and the muzzle angled *away* from you. Measure an appropriate charge of powder (more on this later) and pour it down the bore. At this stage I usually tap the butt *gently* against the ground a time or two to dislodge any powder that may have adhered to the lands and grooves on the way down, and settle the charge.

Next, the precut cloth patch is lubricated and centered over the bore. Then the lead ball is placed over the patch with the sprue (the tiny protrusion left by the molding process) pointing up. Drive the ball into the bore until it's approximately flush with the muzzle — use the handle of your "short starter" for this. If the patch you're using was cut too large and protrudes noticeably at this point, use a sharp knife to cut it off flush with the muzzle. Next, reverse the "short starter" and use it to drive the ball 3 or 4 inches down the bore.

Finally, use a smooth, firm stroke of your ramrod to seat the ball against the powder charge. Don't continue tamping the ball once it's seated, as this may deform the ball (which will affect accuracy) or crush the powder charge (which may cause a misfire). At the same time, take care that the ball *is* fully seated, and doesn't leave an air space between it and the powder charge — this can create dangerously high pressure when the rifle is fired. You can check the depth of seating by using your thumb to mark the ramrod length when resting against the ball, and then removing the ramrod and measuring it against the outside of the barrel. This will tell you if the ball is fully seated or not.

Now the rifle is loaded, and the only remaining step is to press a percussion cap onto the nipple, cock, and fire. This step should not be taken until you're ready to fire the rifle, though. Most percussion rifles feature a half-cock safety notch, but such arrangements are not to be fully trusted.

Because muzzle-loading firearms *are* loaded from the muzzle, an extra safety rule is required for such guns — never place your face or hands directly over the muzzle during the loading process except when absolutely necessary. It should *never* be necessary to have the barrel pointing at your face or head, but it's unavoidable to have your hands get in the way when starting or ramming a ball. But at all other times, keep your hands clear. Of course the Ten Commandments of Shooting Safety apply to muzzleloaders as well as all modern firearms.

The expanding Minie ball is loaded in much the same way, except that the cloth patch is done away with. Instead, the lubricated Minie ball is placed, hollow-base down — directly over the powder charge.

How do you determine how much powder to use in a blackpowder rifle charge? One good way to come up with a starting charge is to take the caliber of the rifle in question and simply use the same number of grains of the appropriate blackpowder. In other words, a 45-caliber rifle shooting a cast lead ball should do fine with a charge of 45 grains

Author uses a "worm" screwed to the end of a ramrod to pull out a shotgun charge that refused to fire. The corkscrew-shaped device is twisted into the charge (in this case, into the shotgun wad) so that it can be pulled out.

The worm digs into the fiber wad, and the author pulls it out. Then the device will be used to loosen the compacted powder.

of FFG powder. You can then adjust this load up or down to please yourself. There are also blackpowder loading manuals on the market from Lyman and DBI Books, Inc. Just remember, whether using manual or caliber-grain method, don't overload, as unburned powder will only blow out the barrel and be wasted. To see if this is happening, spread out a 6-foot length of butcher's paper on the ground and shoot lengthways across it from the prone position with the muzzle 1½ to 2 feet above the paper. If particles of unburned powder sprinkle the paper after firing, you're overloading. Experienced shooters can tell by the sound of the shot when they've reached the desired loading level. When the rifle produces a sharp "crack" on firing, you know that the ball is moving along faster than the speed of sound, or better than 1080 fps. And that's about right for a muzzleloader — this rule applies *only* to rifles, however.

Extra care should be taken when handling or storing blackpowder, as it is more volatile than modern smokeless powder — smokeless powder tends to burn, blackpowder "explodes." Blackpowder can be detonated by open flame and can also be jarred off by severe concussion. Blackpowder should not be stored in large quantities (over 2

Cleaning blackpowder guns is a chore that must be attended to after every shooting session. Hot, soapy water is the traditional cleaning agent, although there are commercial blackpowder solvents available.

or 3 pounds), and should be treated with real respect. It can be potent stuff.

Blackpowder is also highly corrosive, and guns using this propellant must be thoroughly cleaned after firing. There are modern solvents available now that are specifically designed for this chore, and they do a good job. However, the old traditional method is still regarded as the best one by many experienced shooters.

The traditional way to clean the barrel of a muzzle-loading rifle is to first disassemble it from the stock (if possible) by driving out the metal wedge and/or removing the barrel bands on the forearm. Then lift the barrel clear of the stock and action. (The breech end may have a hook to help hold the barrel in place, but this shouldn't cause any problems during removal).

Next, the barrel should be held muzzle upwards while you pour *hot* soapy water down the bore (when you pour this mixture down the bore, hold the barrel with a towel, pour cautiously, and be sure the breech end is placed in an empty pan as the water will run out the vent-hole.) The water should be boiling when you start, and should be prepared by adding small chips of bar soap — *not* detergent — and stirring until the soap dissolves. Run some patches through the bore to help loosen the powder fouling (water will be forced through the hole in the nipple as you do this). Repeat this process a few times, pour out the suds, and follow with a liberal application of *hot* rinse water. If the water is hot enough, the barrels will dry quickly on their own heat without rusting. Cool water will not evaporate quickly enough, and enough may be left in the bore to cause rusting later on. Then finish off with a light coat of oil swabbed down the bore and also applied to the barrel's exterior.

Some percussion rifles can't be easily disassembled for cleaning, and these firearms can be cared for by placing a cloth patch over the nipple and closing the hammer on it, then carefully pouring hot soapy water down the bore. Stop up the muzzle with a cork and shake the barrel up and down to break loose the fouling. Finally, cock the hammer, remove the cloth patch, and force the water out through the nipple by running a tight patch through the bore from the muzzle end. Take care to keep water out of the lock itself, or you could have rusting problems. Finish off with a hot water rinse, and then oil the bore and barrel. Before oiling, make sure all parts are completely dry. If the rifle isn't to be used for the next few days, it's a good idea to inspect it periodically to make sure no rust is forming.

Blackpowder Handguns

Blackpowder handguns come in two basic varieties — single shot "horse pistols" that are loaded in essentially the same way as muzzle-loading rifles, and percussion cap revolvers. Revolvers require a slightly different loading technique, although the principles involved remain the same with all "muzzle-loading" firearms.

In the first place, a percussion revolver *isn't* loaded through the muzzle. The powder charge and ball are loaded directly into each chamber from the front end of the cylinder, and rammed into place with the help of a levered rammer permanently attached to the handgun's frame.

Before loading a cap and ball revolver, visually check the bore and each chamber by holding the gun to the light and looking through it. Make sure each nipple is clear and clean. Then — *with the chambers still empty* — press a percussion cap onto each nipple from the rear, point the gun in a safe direction, and cock and fire the gun until all the caps have been detonated. (All blackpowder revolvers in use today are single-action designs, and the hammer must be manually cocked each time the gun is fired). This burns off any oil that may have accumulated in the nipples and chambers, and prevents soggy misfire.

Next, hold the revolver in one hand, with the muzzle upright, and pour a measured amount of FFFG powder into each chamber (for safety sake, leave one chamber empty). You'll need to have the hammer set at half-cock to allow you to rotate the cylinder and load each chamber in turn.

Next place a lead ball of the proper diameter (it should be a very snug fit) over each *loaded* chamber in turn and then rotate the cylinder to position that chamber directly under the rammer. Press

Left—Here's what you need to load a cap and ball revolver (clockwise from the top): The gun (in this case a Ruger "Old Army"), commercial lubrication (or vegetable shortening), percussion caps, powder and lead balls of the proper size.

Below, left—The first step in loading a cap and ball revolver is to charge each chamber with a measured amount of blackpowder directly through the front of the cylinder.

Below—The next step is to insert a lead ball in the chamber over the powder. It should be a tight fit.

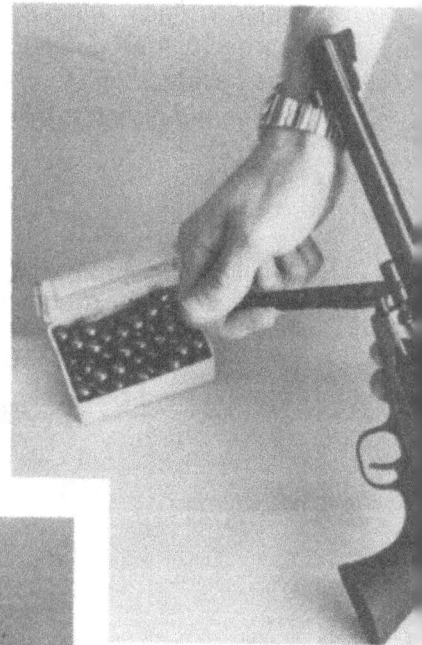

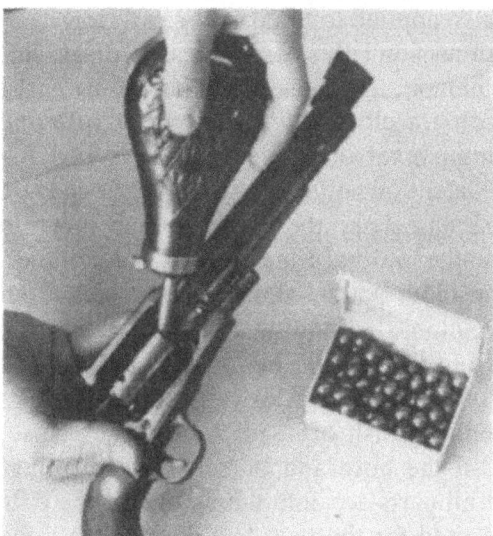

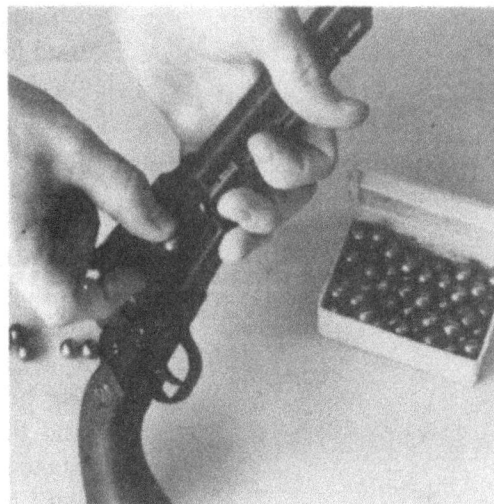

Above—Then the chamber is rotated to the 6 o'clock position so that it falls under the rammer, and the ball is rammed home.

down the rammer lever to seat the ball down over the charge — but be sure you skip the empty chamber, or you'll have some problems getting that ball back out! Finally, fill the opening over each ball with a commercial bullet lubricant like Hodgdon's Spit Ball, or simply a healthy gob of Crisco or some other vegetable shortening. Make sure each chamber is completely filled with this lubricant — leave no gaps for sparks to leak through. Without this lubricant sealing, multiple discharges are possible — in fact, probable — and this can be downright dangerous as well as damaging to the revolver. Such lubrication also softens powder fouling and makes subsequent loading easier.

Finally, cap each loaded chamber (remember, there's one you should skip) — making sure that each cap is firmly in place. Rotate the cylinder once or twice to make sure that it turns free and clear (a cap set too far back on its nipple can jam things up), and then rest the hammer down over the empty chamber.

There are two basic bore sizes or calibers used in blackpowder revolvers — .36- and .44-inch. A good starting load would be 22 or 23 grains of FFFG behind a 375 round ball in .36-caliber revolvers, and 31 to 33 grains of the same powder behind the 451-inch round ball used in 44 revolvers. These charges should be reduced by about 10 grains when using conical bullets.

If a misfire occurs, point the muzzle of your handgun (or rifle) in a safe direction and wait at least 30 seconds — a "hangfire" (delayed firing) is possible in these guns, and they could discharge several seconds after the hammer has fallen. Keep that muzzle pointing downrange! After the waiting period has elapsed, place a fresh cap on the nipple and attempt to fire the chamber once again. If this doesn't do the trick, repeat the waiting period and then run a piece of thin copper wire through the nipple hole to remove possible fouling. Recap and

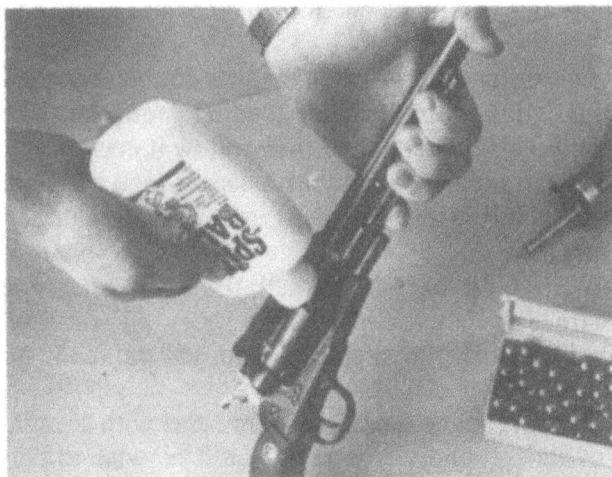

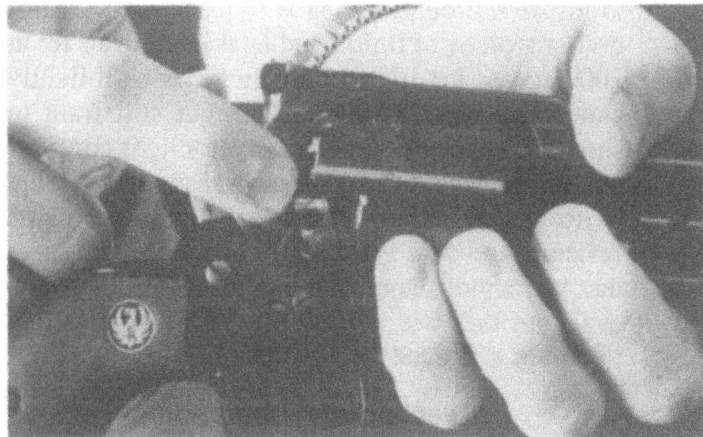

(Above left) Then fill the front part of the chamber with lubricant or vegetable shortening. Be sure to completely fill the space over the seated ball to prevent the possibility of chain fire on other cylinders.

(Above) Finally, place a percussion cap over the nipple behind each loaded chamber (remember to leave one chamber empty for maximum safety). The cap should be pressed firmly in place, and then you should rotate the cylinder by hand to make sure it can turn freely.

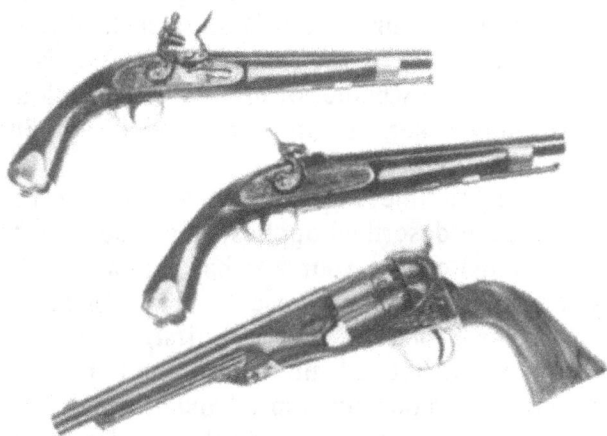

(Left) Replica blackpowder handguns are currently in great demand. Top to bottom: Connecticut Valley Arms' Tower flintlock and percussion pistols; Dixie Gun Works' 1860 Army (percussion) revolver.

try again. If the chamber still refuses to fire (remember to wait for a possible hangfire), fire the other chambers and then remove the ball and powder charge with a screw or worm. Never work on your revolver from the front with a loaded, capped chamber left in the cylinder. This is an invitation to disaster.

Blackpowder revolvers, like rifles, are cleaned in hot, soapy water — but they must be disassembled first. You don't have to take the gun *all* the way apart, baring each spring and screw, but you do need to separate the cylinder from the frame and barrel assembly. This is a relatively simple chore, but it varies slightly from model to model. If you're shooting a newly manufactured replica or one of the modern blackpowder revolvers made by Ruger and others, step-by-step instructions should be packaged with the gun. Otherwise, ask the person you're buying the gun from to show you how to disassemble it at the time of purchase.

Scrub each assembly in the hot suds, then flush with hot rinse water and dry thoroughly. One way to insure thorough drying is to place each assembly on a piece of tinfoil and heat in an oven set at 200 degrees for 10 to 15 minutes. Then oil lightly and reassemble. If you're lazy but still want to shoot blackpowder revolvers, you might consider one of the excellent stainless steel models offered by Ruger and others. These are highly corrosion resistant and won't rust up if not cleaned the same day you shoot them. You can't neglect them forever, but they are more forgiving of procrastination than the blued steel models.

Blackpowder Shotguns

Blackpowder shotgunning may not be as popular as rifle and handgun shooting, but it's still a lot of fun. I particularly enjoy hunting upland game with a muzzle-loading scattergun, and do nearly as well as when I'm toting a more modern, smokeless powder arm.

Here again, the loading principles remain the same, but the details are slightly different from other muzzle-loading firearms. After the usual preliminaries have been attended to (swabbing the bore with a dry patch, then firing a few primers to clear the nipples and dry the unloaded chambers), a charge of FFG powder is poured down the bore. There are many double-barreled blackpowder bird guns available, but each barrel should be *completely* loaded before you move on to the other — this prevents possible double charging and similar accidents. Also, make sure to keep the nipples clear — never allow caps to be placed *until after the loading process is completed.*

Next, you should ram a fiber wad down the barrel over the powder. These over-the-powder wads are the same type used in reloading modern shotgun shells and are available in all gauges. Find one that gives a snug fit in your shotgun's bore (gauge markings may be lacking, and you may have to determine this by trial and error). Then pour a measure of shot down over the powder, and follow up with a thinner, over-the-shot wad of the same diameter. Theoretically, this over-the-shot wad should be thinner than the over-powder wad, but I've used 1/4-inch thick felt fiber wads for both purposes with good results, and this simplifies loading in the field.

These wads can be used dry, or you can moisten the top wad with lubricant or even saliva — don't overdo it though, as the wads aren't all that sturdy to begin with. Moistening the over-the-shot wad will help reduce buildup of caked fouling between shots and will make cleaning a bit easier at the end of the day.

One other thing — when loading double-barreled guns, stick to a particular loading sequence (like "right barrel first") and then follow it *every time.* This can prevent the embarrassment (and possible injury) following the firing of a double-charged barrel.

With both barrels loaded (or with one tube charged in the case of a single shot), all you need to do is cap the nipple(s) and you're ready to shoot.

Because many muzzleloaders come with straight "cylinder" choked barrels, you'd be wise to pattern your blackpowder shotgun before you try it on game or clay targets thrown from a hand trap. You may find that your pattern is so open that you'll pass up targets more than 30 yards out. Now's the time to find out whether or not your gun is a long-range game-taker. It can help avoid disappointment in the field.

At cleaning time, the same hot, soapy water routine already described applies, with one possible exception. Because your gun has smooth (not rifled) bores, you can immerse the breech end of the barrels (take them off the stock first) in the water and pump the hot suds up through the bore by using a tight wad on your ramrod and moving it back and forth like a plunger. I find this is a less messy way of going about this chore, rather than pouring water down the bore.

I've mentioned that FFG powder should be used — this is the case with most blackpowder shotguns, but those with a bore larger than 16 gauge may give better results with coarser FG granulation. What about charges? I generally use about the same amount of shot carried in commercially loaded shotshells of the same gauge (1¼ ounces for 12 gauge, 1-ounce for 20, etc.). I set my dipper or powder scale to meter out this much shot of a particular size, and then use the same measure of powder. (The powder goes down first, of course, but approximately the same volume of powder and shot are used.) This simplifies things in the field, and works fine for my guns. You should experiment a bit to see what works best in *your* muzzle-loading smoothbores, though, as they can be finicky. Overloading generally destroys pattern efficiency, so try not to go this route.

That's basically all there is to shooting the muzzleloaders. Remember the safety rules that apply to *all* firearms, and observe the additional rules that are required to safely load and operate the frontloaders. And treat that blackpowder and those percussion caps with the care they deserve. Do that, and you can enjoy a unique brand of shooting fun safely and economically. The muzzleloaders are a step backward in modern firearms technology, but it can be a rewarding step to take.

In some ways, today's muzzleloader user has it all over his 19th century counterpart. There are new, less corrosive powders now available (Hodgdon's Pyrodex), and stainless steel guns can be had that are less demanding of immediate care and maintenance. What's more, there are many different blackpowder rifles, handguns and shotguns on the market today, and at reasonable prices most shooters can afford. For the do-it-yourselfers, there are an unprecedented number of kits you can buy that let you assemble your own muzzle-loading firearm at home.

Shooting blackpowder firearms is one of the fastest growing pastimes I know of today. These guns are used for hunting, formal target shooting, and simple plinking — and they have a nostalgic appeal that's too hard to ignore.

(Top) This black powder shotgunner is measuring a load of powder into the same measure he will use for the shot charge. A good rule of thumb with muzzle-loading scatterguns is to use equal volumes of powder and shot.

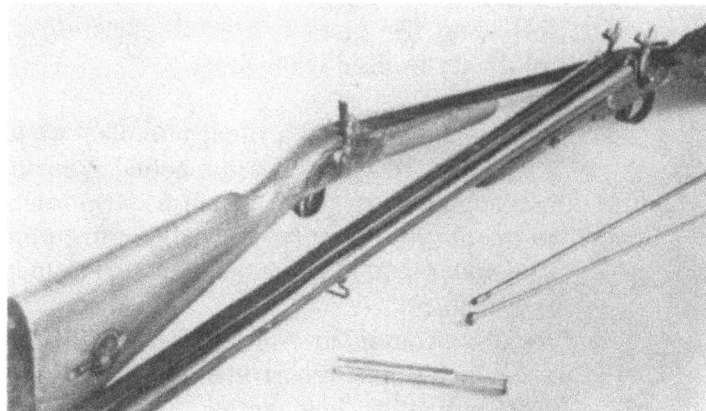

(Above) Muzzle-loading shotguns are fun to shoot, and make good hunting guns for upland game birds and animals. These old guns have practically no choke in the barrel, a fact that showed up only when they were patterned by shooting into butcher paper.

GLOSSARY

A number of specialized terms are used in discussing firearms, and a knowledge of these terms is helpful to anyone who wants to make a serious effort at understanding guns and their functioning. The following list gives a brief explanation of some of the terms used in this book.

Accuracy — the ability of a firearm to place fired projectiles very close to the actual point of aim at a given range, or distance. Also refers to the ability of a rifle or handgun to group successive shots fired at a single target into a small cluster.

ACP — abbreviation for Automatic Colt Pistol. Used to designate cartridges used in some Colt and other autoloading, or semi-automatic pistols (example: 380 ACP, 45 ACP).

Action — The operating mechanism of any firearm. The action of a gun is usually found at the rear, or breech end of the barrel, and consists of the receiver, bolt (or some other form of breechblock), operating handle or action bar, feeding mechanism and trigger mechanism. Actions are usually classified by the means used to operate them, thus there are lever-action, bolt-action, pump-action and semi-automatic (self-loading) rifles and shotguns, as well as falling block, tipping block and break-top types.

Accurize — to improve the accuracy of a rifle or handgun, usually by mechanical improvements or adjustments.

Adapter — a device that allows a smaller cartridge to be fired in a gun than the gun was originally designed to use.

Adjustable Choke — a device attached to the muzzle end of a single-barreled shotgun to allow the shooter to vary the size and spread of the shot pattern fired.

Airgun — a gun that uses compressed air, a spring or gas as a propellant. Since these guns don't use burning gunpowder to fire their projectiles, they aren't classified as true firearms. (Airguns can be dangerous if carelessly handled, however, and deserve the same safety precautions used in operating firearms.)

Air Resistance — one force that affects a bullet in flight, causing it to slow down as it travels along its trajectory, or flight path.

Air Space — a term used by shooters who load their own ammunition (handloaders). It refers to the space within the cartridge case not taken up by either powder or bullet.

Ammunition — a cartridge case loaded with all necessary components — primer, powder and projectile — for firing in a gun. *Ammo* is one popular abbreviation for this term.

Anvil — that part of the primer that provides resistance to enable the firing pin to crush the priming compound when it moves forward under spring pressure. In the snaphaunce and flintlock the anvil was the frizzen.

Aperture — the hole in a peep sight, through which the shooter looks when aligning the front sight with the target.

Aperture Sight — Commonly known and referred

to as the "peep sight," you'll find the aperture sight mounted on a firearm's receiver. It features a windage and elevation adjustable aperture. Its use is generally associated with precision target shooting.

Arm — abbreviation of *firearm*

Arsenal — a large collection of firearms, usually applied to police or military installations.

Assault Rifle — a military or police rifle usually capable of full-automatic fire. Also used to describe military-style civilian rifles capable of semi-automatic fire only.

Assembly — a group of parts that function together to perform a particular operation.

Autoloading — semi-automatic, or self-loading operation. Autoloading guns use recoil or expanding gases from the burning propellant to operate the action.

Automatic — a term sometimes used to describe an autoloading pistol. A true automatic firearm is one that continues to fire as long as the trigger is held down and the ammunition supply holds out.

Automatic Safety — a mechanism used to prevent accidental firing of a gun or rifle, that engages automatically every time that the action is opened.

Automatic Selective Ejector — a device found on break-open shotguns (and, rarely, on rifles) that throws fired cartridge cases from the action, but leaves unfired cartridges chambered whenever the gun is opened.

Backlash — the continued rearward motion of a trigger after the hammer has been released.

Backstop — any material placed behind a target to stop or absorb bullets. A high hillside is often the safest possible backstop, provided it's large enough to capture even bullets fired accidentally downrange.

Backstrap — the metal portion of a handgun frame found at the rear of the grip.

Balance Point — that point along the length of a rifle or shotgun at which the weight is evenly balanced.

Ball — once used only to describe round rifle balls cast from lead and fired from muzzleloaders. Now also used by the military to describe jacketed rifle ammunition.

Ballistics — the study of bullets, or firearms projectiles, in motion. Interior ballistics deals with the ignition and burning of gunpowder within the firearm and the forces that act on the bullet until it leaves the bore of the gun. Exterior ballistics deals with the flight of the projectile from the time it exits the muzzle to the time it comes to rest.

Ball Powder® — gunpowder used by both ammunition manufacturers and handloaders, featuring grains spherical in shape.

Bandoleer — Generally, a canvas or leather strap complete with loops for carrying a number of metallic cartridges or shotshells. Commonly worn over the shoulder and across the chest. Simple cartridge belts (of the same style) are often worn over-the-shoulder, in bandoleer fashion, rather than around the waist.

Barrel — the tube of a firearm or airgun through which a bullet, pellet or shot charge is propelled when the gun is fired. The hollow interior of the barrel is called the bore.

Barrel Band — a metal band that encircles the barrel and fastens it to the forend or magazine tube of a rifle.

Barrel Channel — a groove cut into a gunstock that the barrel rests in when the gun is assembled.

Barrel Lug — a projecting metal lug fastened solidly to the barrel on its underside. This fits into a mating hole in the gunstock, and serves to help transfer recoil forces to the stock.

Base Wad — a fiber, compressed paper, or plastic plug permanently attached to strengthen, and serve as the inside base of, a shotgun shell. If this wad becomes damaged or works its way loose from the hull or shell, the hull is no longer useable for reloading purposes.

Bases — this term usually refers to the pieces of metal attached to a rifle's receiver to make mounting of a telescopic sight possible. The scope is held in metal mounts, and the mounts are clamped to the bases.

Battery — When a firearm's breechblock, bolt or slide is fully locked into firing position.

BB — used either to describe a particular size of lead shotgun pellet, or copper-plated shot used in air rifles. While the term is used in both cases to refer to a spherical projectile, BB-sized shotgun pellets are slightly larger than BB air rifle shot.

BB Cap — Bulleted Breech Cap cartridge, consisting of a short 22-caliber rimfire loaded with priming compound only. The BB Cap was the first self-contained cartridge to feature integral priming.

Bead — a round or spherical ball, normally used at the muzzle of a shotgun as a sighting aid.

Beavertail — a term used to describe a particularly wide, handfilling forend, or front part of a stock used on a shotgun or rifle.

Bedding — the fitting of the metal action and barrel surfaces to the stock.

Belt or Belted — refers to a small, steel belt encircling the base of a rifle case.

Bench — a structure used to solidly support a rifle while the shooter fires at a target. The rifle is usually rested on sandbags, while the shooter is seated with both elbows resting upon the bench.

Berdan Primer — a European primer not commonly used in the United States. Cartridge cases designed to use Berdan primers will not accept the U.S. Boxer primer unless the primer pocket is first reamed out or enlarged.

Big Bore — a term used to designate high-powered rifle cartridges of 30-caliber or greater, and handgun cartridges larger than 38-caliber.

Bipod — a two-legged device used to support the forend, or front part of a rifle during shooting.

Blackpowder — a form of the earliest known gunpowder, this powder is still manufactured today for muzzle-loading firearms. Blackpowder is a more volatile propellant than modern smokeless powder, and requires greater care in handling. This powder is currently offered in four different grades, or granulations: Fg, FFg, FFFg, and FFFFg. Fg blackpowder is relatively coarse, while the FFFFg granulation is the finest available.

Blade Sight — refers to the front sight used on rifles and handguns.

Blank — a cartridge loaded with only primer and powder, designed to produce a loud report when fired. The powder is held in place with a cardboard wad, and no bullet is used.

Blowback — a type of action used in semi-automatic or full-automatic guns, in which the breechbolt is not locked in place during firing.

Blueing — a blue-black colored finish applied to steel actions and gun barrels through a process of chemical oxidation.

Boattail — the shape of a bullet with a tapered base, designed to reduce air drag for long-range accuracy.

Bolt — a breechblock that travels back and forth to open or close the action. Several different types of firearms use some variety of bolt, including automatic rifles, machine guns, autoloading rifles and shotguns, slide-action firearms and even some handguns; however there is a distinct action type known as the *bolt action* (see below).

Bolt Action — an action type popular among many riflemen, although it is also used in some shotguns. This type of action has a tubular receiver that contains a manually operated bolt. A knobbed handle extends from the side of the bolt, and in operation this is grasped in the hand, lifted up and pulled back to open the action. This pulls the expended cartridge case from the chamber, and when the action is fully open a mechanical ejector ejects the case. Pushing forward and down on the bolt handle strips a fresh cartridge from the magazine, chambers it, and locks the bolt in place in preparation for firing. A series of lugs projecting at right angles from the body of the bolt cam into matching recesses in the receiver to lock the bolt in the firing position.

Bolt Face — the forward end of the bolt, which may or may not be recessed to enclose the head of the cartridge case.

Bolt Handle — the operating handle that projects from a gun or rifle bolt.

Bolt Guide — a groove or narrow rib running along the inside of the receiver, designed to keep the bolt in proper alignment as it travels back and forth.

Bolt Knob — the rounded knob at the end of the bolt handle.

Bolt Release — the mechanism, usually operated by a lever or by pulling on the trigger and back on the bolt simultaneously, that allows the bolt to be removed from the action. The term also applies to the lever or button that allows the bolt of an autoloading firearm to move forward from a locked-open position.

Bolt Stop — a metal projection or hook that halts the rearward travel of the bolt. This device sometimes also serves as the cartridge ejector.

Bore — the hollow interior of the barrel. The bore may be rifled or smooth, depending upon the type of ammunition the gun is designed to fire.

Bore Diameter (Caliber) — usually measured across from land to land in a rifled barrel. This measurement, in hundreths of an inch, is the rifle's caliber (some American calibers and most European calibers are measured in millimeters).

Bore Sighting — a rough means of checking the alignment of a rifle's sights by removing the bolt, placing the rifle in a solid rest, and then looking through both the bore and the sights at a distant fixed object. Adjusting the sights so that they are in line with the target as seen through the bore is only the first step to sighting-in a rifle properly, as the method is too crude to assure reliable accuracy when shooting.

Boxer Primer — a metallic primer used in nearly all American-made centerfire cartridges.

Box Lock — a type of action lock used in most moderately priced double- and single-barreled break-top shotguns.

Box Magazine — a magazine which contains cartridges in a stacked arrangement. When this magazine is a removable one, it is sometimes called a *clip*.

Brass — a commonly used term for an empty rifle or handgun cartridge case.

Break-Top — a firearm with a frame hinged to open at the top when unlocked, usually by pushing a lever to one side.

Breech — the rear end of the barrel, containing the firing chamber. Sometimes used to indicate the receiver and its working parts (action).

Breechblock — a solid block that rotates or slides into place to lock the cartridge into the breech.

Breechbolt — That part of a firearm's action which, when locked, sits firmly against the rear of the chamber. The breechbolt often houses such additional parts as the firing pin and extractor.

Breechloader — a firearm that is loaded from the rear, or breech end. (As distinguished from muzzleloader.)

Breech Plug — the plug that closes the breech of a muzzle-loading firearm.

Broach — a tool used to cut rifling grooves into the bore of a barrel.

Brush Load — a shotgun shell containing cardboard partitions to cause shot to spread rapidly.

Buckhorn Sight — an open metallic rear rifle sight once popular among hunters. It is distinguished by its high, curling sides that circle above the sighting notch.

Buckshot — large, round lead shot used to hunt deer and other large animals. It is also used

by military and police organizations.

Bull Barrel — An unusually thick or heavy rifle or handgun barrel, normally used for target shooting.

Bullet — an elongated projectile normally fired through a rifled barrel. There are many different bullet designs available for a variety of uses, including special bullets for shooting at paper targets, long-range marksmanship, or hunting both large and small game. Some bullets are cast from soft lead alloys, while others may feature complicated designs making use of several different kinds of metal in layers.

Bullet Energy — the energy, expressed in foot-pounds, a bullet in flight has available to deliver when it hits its target at a given range.

Bullet Jacket — the metal covering, or jacket, used in many rifle or pistol bullet designs.

Bullet Mould — a cavity-type mould used in casting simple bullets from melted lead.

Bullet Puller — a device for pulling or removing a bullet from a loaded cartridge without causing the cartridge to discharge.

Bullet Trap — a device used to capture bullets in flight.

Bullseye — the center of a target. Also the trade name for a popular gunpowder used by reloaders.

Butt — the bottom part of a handgun's grip. Also, the rear portion of the stock of a rifle or shotgun.

Buttplate — the protective metal or plastic plate attached to the end of a buttstock.

Buttstock — the rear portion of the stock on a rifle or shotgun that extends back from the action, or breech. That portion of the stock held against the shooter's shoulder when firing.

Button Rifling — a method of rifling a barrel in which a very hard "button" or die is forced through the bore. The outer edge of this button displaces the barrel steel to form a pattern of rounded, relatively shallow lands and grooves.

Caliber — the diameter of a rifle barrel bore, measured either in hundredths of an inch or in millimeters. Caliber is also sometimes used to designate a particular cartridge, such as 243 Winchester or 6mm (millimeter) Remington. Some caliber or cartridge designations can be confusing to the beginner, particularly those bearing two sets of identifying numbers. For instance, the 30-30 Winchester has a 30-caliber bullet (hence the first set of numbers), and when it was introduced back in 1895 it was loaded with 30 grains of powder (designated by the second number pair). In contrast, the 30-06 (also with a 30-caliber bullet) uses the year the cartridge was adopted by the military (1906) as its second set of numbers. Still other caliber (cartridge) designations, like the 25-06, indicate a 25-caliber projectile used in a necked-down 30-06 case. European calibers, like the 7x57, indicate the diameter of the projectile with the first number, while the set of numbers following the "x" gives the length of the cartridge case (both measurements in millimeters).

Calling a Shot — predicting where the bullet will hit at the moment the trigger is pulled.

Cam Lock — it is most commonly encountered on "trapdoor" Springfield type rifles and features a top-hinged breechblock locked by a "cam" piece at the rear.

Cannelure — a groove encircling the circumference of a bullet or cartridge case.

Cannon Lock — the earliest means used to produce ignition. Fire was applied directly to the powder charge through a touchhole located at the rear of the barrel.

Cant — tilting a rifle or handgun to the side, an act that usually causes the bullet to strike to that side of the aiming point and low.

Cap — percussion cap that serves as the primer on caplock firearms. Contains an easily detonated charge of fulminate.

Cap and Ball — indicates a firearm loaded with a charge of blackpowder, a lead ball, and ignited by a separate percussion cap.

Caplock — a gun lock, or firing mechanism used on blackpowder firearms, featuring a hollow nipple over which a percussion cap is placed. When the cap is crushed against the nipple by the hammer, the exploding fulminate mixture contained in the cap generates a flame that in turn ignites the main propellant charge.

Carbine — a short rifle with a barrel less than 22 inches long.

Cartridge — a self-contained firearm load that consists of a case, primer, propellant powder charge, and projectile. The term is usually reserved for metallic cartridges used in rifles or handguns, although a metal and plastic (or paper) shotgun shell is also technically considered a cartridge.

Case — a metallic cartridge container that holds the other necessary components — primer, powder and projectile. (The case of a shotgun "cartridge" is known as a "hull.")

Case Capacity — the volume available inside a cartridge case.

Case Harden — to give steel or iron a very hard, durable surface by heating, adding carbon, and then cooling the metal rapidly. Mottled blue, red or yellow coloring is characteristic of metals so treated.

Case Head Separation — the separation of the cartridge case head (base of the cartridge) from the main cartridge body.

Case Life — the number of times an individual cartridge case or hull can be reloaded and fired safely.

Case Mouth — the open end of a cartridge case.

CB Cap — 22 rimfire Conical Bullet Cap, a very low powered cartridge used in 22 rimfire rifles and handguns.

Centerfire — a firearm or cartridge that uses a primer located in the center of a cartridge case base to ignite the propellant charge.

Chamber — the rear portion of the bore in rifles, shotguns and auto pistols that is enlarged to admit a loaded cartridge or shotshell. Revolvers have several chambers (usually 5 or 6) located in a revolving cylinder located behind the barrel. Also called the firing chamber.

Chamber Pressure — the pressure generated inside the firing chamber by the expanding gases produced by the burning propellant powder. Modern centerfire rifles may support a chamber pressure of 50,000 pounds per square inch or more at the moment of firing, but this pressure lasts only a small fraction of a second.

Chamfer — a bevel scraped on the inside or outside edge of a cartridge case mouth during the reloading process to make it easier to seat a new bullet and crimp it in place.

Charge — a predetermined weight or measure of gunpowder.

Checkering — a pattern cut into the surface of a wooden firearm stock. Checkering can be both decorative and useful, as it provides a non-slip gripping surface for the shooter's hands.

Checkering Tool — a tool used to cut checkering into stock surfaces.

Cheekpiece — a raised portion of a rifle or shotgun buttstock against which the shooter rests his cheek while aiming the firearm.

Choke — an internal constriction (or flaring) at the muzzle of a shotgun barrel. By varying the bore size at or just behind the muzzle, the rate at which the shot pattern spreads can be partially controlled. The degree of choke is indicated by the percentage of shot pellets that can be contained by a 30-inch diameter circle at a range of 40 yards. A "Full choke" barrel should deliver a 65 to 75 percent pattern, while other chokes run as follows: Improved Modified — 55 to 65 percent; Modified — 45 to 55 percent; Improved Cylinder — 35 to 45 percent; Skeet — 30 to 35 percent; Cylinder — 25 to 35 percent. Choking is a relatively inexact science, however, and shotguns sometimes perform very differently than the choke markings on the barrel would lead you to expect. Changing ammunition or shot size will also affect the shot pattern and choke performance.

Choke Tube — a threaded insert which screws into the muzzle of a shotgun barrel. With a hand tool, the shooter may change these tubes and vary the degree of choke. Used in single and double-barrel shotguns.

Chronograph — an instrument that measures the velocity of a bullet as it passes between two measuring points, or screens.

Clay Bird or Clay Pigeon — a saucer-shaped disc of clay or some similar material used as a shotgun target. The disc is thrown in front of the shooter and shatters when hit by a charge of shot.

Cleaning — firearms maintenance usually consisting of scrubbing powder residue and possible lead fragments from the bore, wiping away dirt and any moisture that may cling to metal surfaces, and finally applying a very thin coat of oil to the bore and other steel parts.

Click-Adjustable — sight adjustment screws that emit an audible click when turned. Each click indicates one degree of adjustment.

Clip — a strip of metal (also called "stripper clip") used to hold a number of cartridges which may be more quickly loaded into a gun's magazine than by loading cartridges singly. Often used in error to mean "magazine."

Clock — a system using the face of a clock as a reference point for calling target hits. The shooter visualizes a clock face superimposed on his target, with the bullseye as the center.

A hit directly to the left of center becomes a "9 o'clock hit," while a hit directly over the bullseye is said to be at 12 o'clock. The system is also used to describe wind direction with the shooter at the center of a clock and facing toward noon. Thus a breeze coming from the right would be called coming from 3 o'clock.

Coarse Bead — the sight picture seen through open sights in which the entire front bead is visible, or even raised slightly above the rear sighting notch.

Cock — a term once used to describe the hammer of a flintlock or other early firearm. Now commonly used as a verb — the act of pulling back, or "cocking" the hammer or striking piece of a firearm.

Cocking Indicator — a projecting pin or some other device that indicates when the action is cocked and ready for firing.

Collimator — an instrument used for checking the optical alignment of a scope sight with the bore of the firearm it is mounted on.

Comb — the upper edge of a buttstock.

Combination Gun — a break-open firearm featuring both rifled and smooth bore (shotgun) barrels, to allow the use of both rifle and shotgun ammunition.

Combustion — the process of burning.

Compensator — a device attached to the muzzle of a firearm to reduce barrel movement caused by recoil.

Components — a term used to describe the separate materials that make up a cartridge or shotshell (including the primer, cartridge case or hull, powder, shot charge or bullet, and any wadding that may be used).

Crane — the swing-out yoke that holds the cylinder of a revolver.

Creep — the movement of the trigger, under pressure from the shooter's finger, prior to release of the hammer or cocking piece.

Crimp — tightening of the case mouth to hold the bullet or shot charge in place.

Cross-Bolt Safety — a safety button, usally located in the trigger guard, that is pushed to one side or the other to engage or release the safety mechanism.

Crosshairs — an aiming device consisting of two thin lines that intersect at right angles in the reticle of a telescopic sight.

Crown — the rounded edge found at the muzzle of most rifles and handguns.

Cylinder — a cylinder-shaped device that contains several (usually 5 or 6) firing chambers, used primarily in revolving handguns.

Cylinder Latch — a mechanical latch, usually located on the left side of a revolver, that allows the cylinder to be swung out on its crane from the frame of the gun.

Cylinder Gap — the gap or space that exists between the forward edge of the cylinder and the rear edge of the barrel in a revolver.

Damascus Barrel — shotgun barrels made by twisting strips of metal together, winding these strips around a form or mandrel, and then welding the seams. This form of barrel making was popular in the late 1800s, and some Damascus barreled shotguns are still in service. Such guns can be used with black-powder loads, but are much too weak for modern smokeless powder shotshells. Never attempt to shoot a smokeless powder shell in any gun that displays the intricate, coiled pattern characteristic of a Damascus barrel.

Derringer — a small pocket-sized handgun roughly patterned after the guns made by Henry Deringer.

Double-Action — the capability of a handgun to fire each time the trigger is pulled, without manually cocking the hammer between shots.

Double-Barreled Gun — sometimes referred to as simply a "double." Usually any gun with two barrels joined together and mounted side-by-side.

Double-Trigger — two triggers mounted one behind the other and enclosed by a single trigger guard, usually found only on double-barreled shotguns or — more rarely — double-barreled rifles. Each trigger fires one barrel (as distinguished from double-set triggers below).

Double-Set Triggers — a pair of triggers, mounted one behind the other and usually found only on target rifles and some European (or European-inspired) hunting rifles. In use, pulling on the front trigger alone will release the cocking piece and cause the rifle to fire, although the trigger used in this mode usually displays a fairly heavy pull and some creep. For more precise bullet placement, the shooter first pulls the rear trigger — this doesn't cause the rifle to fire, but only "sets" the front trigger so that a very light pressure is all that is then required on the forward trigger to activate the firing pin.

Doubling — the (usually accidental) discharge or firing of both barrels of a double-barreled gun simultaneously.

Dovetail — a slot cut into the barrel or action of a firearm in which either the front blade or the rear sight of a set of iron sights is mounted.

Drift — the tendency of a bullet to move to one side in flight. *Drift* alone usually refers to the sideward movement of the projectile caused by its rotational spin, while *Wind Drift* refers to any deflection caused by wind pressure.

Drift Pin — a small metal bar or pin used to hold parts of a gun's action together. These pins are held in place by friction or spring pressure and are removed by pushing on the end of the pin with a tool of similar diameter.

Drilling — a three-barreled combination gun, usually consisting of two shotgun barrels mounted in conjunction with a rifle barrel.

Drop — the distance between the upper portion of a gun's buttstock and an imaginary line extending backward from the top of the barrel. This is usually measured at two points: the comb (the front part of the buttstock); and at the heel (the rear, or butt end of the stock). Drop is also used to refer to the distance a bullet drops from a direct line of sight over a particular range.

Dud — a cartridge that fails to fire.

Ear Protectors — devices worn in or over the ears to protect them from damage caused by the high noise levels produced by gunfire.

Ejection — the removal of a cartridge or shotshell from the action of a gun, usually automatically by a mechanical device called the *ejector* when the action is opened. The ejector usually operates with enough force to throw an empty case completely clear of the gun.

Ejector — a mechanical device that ejects a cartridge or cartridge case from the breech or action of a firearm after the case has been pulled from the chamber by the extractor.

Ejector Rod — a rod that extends forward from a revolver's cylinder. In double-action revolvers with swing-out cylinders, the ejector rod swings away from the frame with the cylinder and empties all the chambers simultaneously when pushed. In single-action revolvers, the ejector rod is mounted permanently underneath the barrel and is used to push empty cases or unfired cartridges from their chambers one at a time.

Elevation — the up and down movement of an adjustable sight that raises or lowers the impact point of the bullet on target.

Engine Turned — the decorative polishing of a metal surface (often a rifle bolt) in a pattern of overlapping circular spots, or whorls. Sometimes called demascening.

English Stock — a shotgun stock featuring a straight grip and a high, straight comb.

Expanding Bullet — a bullet with an exposed tip of soft lead, or hollowpoint, designed to expand, or "mushroom" on impact.

Extraction — the removal of a cartridge, fired case or shotshell from the chamber of a gun.

Extractor — a device that grips the base of a cartridge or shell and withdraws the cartridge or shell from the firing chamber.

Extractor Groove — a groove just ahead of the base in rimless rifle and handgun cartridges, designed to provide a gripping surface for the extractor.

Extension Magazine — a magazine of greater than normal capacity, often used on police shotguns.

Extreme Range — the maximum distance a bullet or shot charge is likely to travel after being fired from a gun.

Eye Relief — the distance between the shooter's eye and the rear lens of a scope sight that allows the eye to see the full viewing field offered by the sight. This distance is critical in large-bore rifles with heavy recoil, as too little eye relief forces the shooter to move his eye so far forward that the scope may strike it under recoil.

Falling Block — single shot action in which the breechblock moves up and down at right angles to the bore. A very strong, solid type of action.

False Muzzle — a removable barrel extension used on some high-grade muzzle-loading rifles to protect the rifling near the muzzle and also fit the soft lead ball into the bore with a minimum of distortion. The false muzzle is used only during the loading process and must be removed before the rifle is fired.

Fanning — a method of firing fast repeat shots from a single-action revolver by holding the trigger back with the trigger finger and using the heel of the free hand to repeatedly strike the hammer back. Not recommended, as accuracy is impossible and fanning is likely to damage the mechanism.

Fast Draw — a shooting sport in which a revolver loaded with blank cartridges is drawn from a holster and fired in the fastest time possible. Not a sport for beginning shooters, as even blank cartridges can cause damage if carelessly used.

Feeding — the process of moving a cartridge from the magazine, through the action of a gun, and into the chamber. Single shot firearms have no magazine, and their chambers are fed manually.

Feed Guide or Ramp — a surface, projecting backwards from the chamber of a gun, that is designed to guide the bulleted end of a cartridge smoothly into the chamber during feeding.

Feed Mechanism — the assembly of action parts that accomplish feeding in a firearm.

Feet Per Second — abbreviated as fps, is the unit used in the United States to measure the velocity of a bullet or shot charge.

Field Gun — a shotgun designed primarily for bird hunting.

Field Load — a shotshell load designed for hunting upland game. Usually indicates a lighter-than-maximum loading.

Field-of-View — the area visible through a telescopic sight (or other magnifying optical aid, like binoculars), usually listed as the diameter (in feet or meters) of the viewing field at a range of 100 yards (or meters).

Fine Sight or Fine Bead — the sight picture seen through open iron sights in which the front bead is barely visible through the rear sighting notch.

Finish — a term used to describe the process used to protect the wood used in the stock (oil or polyurethane finish), or to protect the metal parts (blueing or case hardening, to name just two possibilities).

Firearm — a device that projects a bullet or shot charge through a barrel by the pressure of expanding gases produced by a burning powder charge.

Fireforming — expanding a cartridge case by firing it in a chamber slightly larger than the case was designed to fit. This is a procedure that should be used only by reloaders experienced enough to know what they're doing *and* how to do it safely.

Firing Line — the area, or line from which shooters at a target range fire their rifles, handguns or shotgun.

Firing Mechanism — those parts of a firearm that work together to cause a cartridge resting in the chamber to fire. The primary parts of the firing mechanism include the trigger, sear, hammer, mainspring and firing pin.

Firing Pin — the part of a gun's firing mechanism that strikes the primer of a cartridge to start ignition of the powder inside.

Floating Barrel — a rifle barrel that does not bear directly against the wood in a stock anywhere along its length. Many target rifles feature floating, or free-floating barrels.

Floorplate — a hinged or solid metal plate or covering fastened to the stock of a rifle underneath the magazine.

Flyer — a bullet that hits the target at some distance away from the holes made in the same target by a group of other bullets fired from the same gun or rifle.

Follower — the platform in a box-type magazine that supports the cartridges in the magazine and pushes them up into the action or breech by spring pressure. Or the solid section of rod that pushes cartridges through a tubular magazine, again by spring pressure.

Forcing Cone — the area in a shotgun bore located just ahead of the firing chamber. The forcing cone tapers from the diameter of the chamber to the diameter of the bore and serves to squeeze the shot column and following wads down to bore size. Revolvers, also, have forcing cones at the rear of their barrels which serve the same function.

Forend — the front part of a rifle or shotgun stock located underneath the barrel or barrels.

Forend Tension or Pressure — the pressure the stock places on the barrel. Except in the case of floating barrels, the wood in the barrel channel will bear directly against the barrel at some point or points along its length. Forend tension must be properly regulated to achieve best accuracy in a rifle.

Forend Tip — the extreme forward end, or tip of the forend, often made of wood or some other material that contrasts in color with the rest of the stock.

Fouling — metal or powder residue deposited in the barrel of a firearm as a result of firing.

Fouling Shot — a shot fired through a rifle or handgun (usually in competitive target shooting) before the gun is fired at a target. A bullet fired through a clean barrel may not strike the target at the same point that subsequent bullets from the now dirty, or fouled, bore will. Therefore a fouling shot is often fired by contestants at the beginning of a target match.

Fowling Piece — an obsolete term for a shotgun used for bird hunting.

Frame — the solid part of a revolver that encloses the cylinder, crane and the working mechanism of the gun. Or that part of an auto pistol that houses the trigger mechanism, magazine, and other working parts not contained in the slide.

Freebore — to remove a section of rifling in a rifled barrel just ahead of the chamber and throat.

Free Pistol — a specialized form of handgun used in certain kinds of target shooting.

Free Rifle — a specialized target rifle.

Front Sight — a metal bead, blade or post located near the muzzle of a firearm, usually used in conjunction with a metal rear sight mounted farther back along the barrel or receiver.

Full-Cock — a term used to describe the condition of a firearm with an exposed hammer when the hammer is locked all the way back against pressure from the mainspring. When a gun is at full-cock, all that is necessary to set the firing mechanism into motion is a pull on the trigger.

Gain Twist — a form of rifling used in some muzzle-loading firearms, in which the rate of twist is not uniform along the length of the barrel but increases toward the muzzle.

Gas-Check — a shallow metal (usually copper) cup attached to the base of a cast lead bullet to protect the soft lead from the hot gases produced by the burning gunpowder during firing.

Gas Cylinder — the mechanism used in gas-operated semi-automatic guns to convert gas tapped from the bore into mechanical energy to operate the gun's action. The gas cylinder contains a piston that moves back under gas pressure, and this movement is transmitted back to the action by an operating rod.

Gas Operated — a semi-automatic (or fully automatic) firearm that uses the expanding gases from the burning gunpowder to operate the action.

Gas Piston — the piston contained in the gas cylinder of a gas-operated firearm.

Gas Port — a small hole in the barrel of a gas-operated firearm through which expanding gas is tapped immediately after the gun is fired. The gas port usually feeds this gas directly to the gas piston.

Gauge — the term used to measure the size (diameter) of a shotgun bore. Traditionally, the gauge of a gun was determined by the number of lead balls sized to fit the bore that could be cast from a pound of lead. That is no longer true, and shotgun bore size has been pretty well standardized as follows: 10-gauge — bore size of .775-inch; 12-gauge — .729-inch; 16-gauge — .662-inch; 20-gauge — .615-inch; 28-gauge — .550-inch; .410 bore — .410-inch (the .410 is the one shotgun "gauge" that doesn't fit the orginal use of the term, as the .410 designation is the actual measurement — or caliber — of the bore size).

Glass Bedding — using a liquid fiberglass-epoxy compound inside a stock to provide a near-perfect fit between the wood surfaces and the steel rifle action or barrel. Glass bedding also reinforces the stock.

Grain — a unit of weight (437.5 grains = 1 ounce) used by handloaders in measuring bullets and powder charges.

Grip — the handle or butt section of a handgun, or the part of a shotgun, rifle or handgun stock that falls directly behind the action or breech to be gripped by the shooting hand.

Grip Adapter — a metal or plastic filler piece attached to the grip of a handgun, usually just behind the trigger guard, to offer a larger gripping surface to the shooter.

Grip Cap — a small cap of usually contrasting material fastened to the base of the pistol grip on rifles and shotguns.

Grip Safety — a mechanical safety device usually found on the backstrap of auto pistols that is automatically depressed or disengaged by the shooter's hand when the gun is gripped in firing position.

Grooves — the cuts in the bore of a firearm made to produce the rifling.

Group — a term used to describe the pattern made on a target by successive shots from a rifle or handgun.

Guard Screw — a screw that fastens a gun's trigger guard to the stock or action.

Gun — a device that propels a projectile of some kind through a barrel, causing the projectile to be thrown from the device, usually with considerable force.

Gun Case — a case used for carrying guns.

Gun Control — a term usually used in reference to various legislative acts and proposals designed to control the sale, ownership and use of guns.

Gun Powder — the powder burned in a firearm to propel a bullet or shot charge. Once used in reference to blackpowder only, this term is now commonly used to apply to smokeless powders as well.

Gunsafe — a heavy, securely lockable burglar and fire-resistant cabinet designed to hold and protect guns and ammunition.

Gun Stock — that part of a rifle or shotgun held in the shooter's hands, and rested against the shoulder when the gun is held in firing position. Traditionally made of wood, some modern gun stocks are formed of high-impact plastic and other space-age materials.

Hair Trigger — a trigger that requires very light pressure.

Half-Cock — a notch or detent that holds the hammer of an exposed hammer gun somewhere between the fully cocked and fully forward position. Serves as a safety mechanism in some firearms.

Hammer — the part of a gun's mechanism that pivots forward around an axis to strike the firing pin when released by trigger action.

Hammer Block — a safety mechanism that prevents the hammer from contacting the firing pin unless the trigger is pulled.

Hammerless — a term used to describe a gun without an exposed hammer.

Hammer Shroud — a cover placed over the hammer of a revolver to prevent the hammer from catching on clothing when carried in a holster or pocket.

Hammer Spur — the checkered or serrated upper surface of a hammer that provides leverage to the shooter's thumb when the gun is cocked.

Hand — a short, mechanical arm that turns the cylinder of a revolver each time the gun is cocked.

Handgun — a firearm designed to be fired with one hand. Handguns featuring a revolving cylinder with several individual firing chambers built into it are classified as revolvers, while handguns fed by a box magazine and featuring a self-loading action are called auto pistols, or simply pistols. While these firearms are designed for one-handed use, many modern handgun shooters use both hands to steady the gun for better control and accuracy.

Handloading — also known as *reloading*. The process of inserting new components — primer, powder, bullet, etc. — into a fired cartridge case to produce usable ammunition.

Hand Trap — A hand-held (sometimes spring-powered) device used to throw clay targets. It's commonly used by shotgunners when practicing wing shooting.

Hangfire — a momentary delay between the firing pin blow on the primer and the actual ignition of the powder charge.

Headspace — the distance from the breech (or bolt) face to that part of the chamber that supports the cartridge and prevents it from moving forward when struck by the firing pin. The method of headspacing varies with different types of cartridges. A rimmed cartridge is supported in the chamber by the rear edge of the chamber, and so headspace is measured from that point back to the face of the bolt. A rimless case is prevented from moving forward in the chamber by the chamber walls narrowing to conform to the case shoulder. A belted rifle cartridge "headspaces" on the front edge of the belt, and a rimless pistol cartridge seats the edge of its case mouth against a shoulder near the forward end of the chamber. Headspace is measured from those points back to the face of the breech. If a firearm chamber has too much headspace, the cartridge will not be properly supported and may rupture when fired.

Headspace Gauge — a gauge used to determine whether or not a gun has the proper amount of headspace.

Heel — the rear upper edge of a buttstock.

High Base or High Brass — a term used to describe shotshells featuring relatively high brass heads, or bases. Usually found on high-velocity hunting loads, although the height of the base has little to do with hull strength.

Hinged Frame — a firearm action hinged near the breech that is opened by "breaking" the gun at the hinge.

Holdover — aiming at a point above a distant target to compensate for long range without making mechanical adjustments to the sight. If a rifle is sighted-in to place the bullets exactly at the point-of-aim at 100 yards, and the shooter knows that the bullet will drop an additional 10 inches by the time it reaches the 200-yard mark, he will aim 10 inches above the target at the longer range.

Hollowpoint — a bullet with a cavity in its nose, designed to expand the bullet to a much larger diameter when it strikes. Used primarily for hunting large and small game.

Holster — a sheath or moulded pocket used to carry a handgun. Holsters are made to carry guns on the shooter's belt, under his armpit or concealed in a number of ingenious places like on the ankle, under the waistband of a pair of trousers, or even (for a policewoman) in the bra. Holsters are designed for a number of different uses, including fast draw competition, hunting, military or police service duty, horseback riding and concealment.

Hooded Sight — a front sight covered by a metal hood. The hood serves to both protect the sight as well as to give the sight greater optical definition by excluding light.

Hull — a fired (or unloaded) shotshell case.

Ignition — describes the process in which the priming compound is crushed and detonated by the firing pin, and the resulting flame causes the propellant powder to begin burning.

Incendiary Bullet — a bullet that contains incendiary material in its hollow base that ignites upon impact. For military use only — U.S. incendiary rounds are marked with blue paint on the bullet tip and should never be fired in a sporting rifle.

Inertia Firing Pin — a firing pin, contained within the breechblock, that is held away from the face of the breech by spring pressure until struck by the hammer.

Inletting — cutting a channel into the wood of a gunstock shaped to exactly fit (or as nearly so as possible) the barreled action of the firearm when the gun is assembled.

International Shooting Union (ISU) — an international association that organizes and oversees international shooting events throughout the world.

Iron Sights — any metal firearm sight, usually featuring separate front and rear units used together in aiming. Once referred mainly to the open rifle and handgun sights supplied by the factory, but the definition now includes aperture receiver sights. Most open iron sights are step-adjustable for elevation and may also be adjusted for windage, while some high-grade hunting rifles have two or even three sets of folding rear sight leaves, with each leaf regulated for a different range. Telescopic sights (and other sights using magnifying or non-magnifying lenses) are *not* iron sights, even though the tube, or body of these sights may be made of metal.

Jacket — a thin metal covering surrounding the lead core of a bullet.

Jag — a cleaning rod tip designed to hold cloth patches.

Jam — a condition caused by a loaded cartridge or empty case sticking in the action or chamber. When this happens, the gun cannot be operated.

Keeper — a leather loop or metal clamp used on a rifle sling to hold two or more layers of the sling together.

Kentucky Rifle — a term used in reference to the relatively long, narrow-stocked flintlock rifles used by settlers on the American frontier (Kentucky) in the late 18th century. Since most of these rifles were actually made in Pennsylvania, they are sometimes known as *Pennsylvania Rifles*.

Kentucky Windage — aiming to one side of a target to compensate for wind pressure. The term also refers to such adjustments made by rough estimation or trial and error.

Keyholing — a bullet tumbling in flight, or yawing to one side rather than flying perfectly straight-on.

Kick — a term used to decribe the sensation felt by a shooter, caused by recoil, when he fires a gun.

Lands — the raised surfaces between the grooves cut in a rifled bore.

Lead — (pronounced lēd), the distance a gun must be pointed ahead of a moving target when the trigger is pulled in order for the projectile or shot charge to hit the target.

Leading — a deposit left in the barrel of a gun by the passage of a lead bullet through the bore. These deposits of lead or lead alloy will usually affect accuracy unless removed by thorough cleaning.

Length of Pull — the distance between the buttplate of a shotgun (or rifle) and the trigger.

Lever-Action — a firearm action now used almost exclusively in rifles. (Except for a certain break-top single shot shotgun operated by an under lever, the production of lever-action shotguns ceased around the turn of the century). This term is applied to any firearm featuring an operating lever located under the receiver or breech area. While this definition includes some break-top guns and falling block single shots, a true lever-action rifle contains a reciprocating bolt, linked to the operating lever in such a manner that swinging the lever (which usually includes the trigger guard) down and away from the rifle opens the bolt; and pulling it back to the original position closes the bolt and locks it in place. These rifles are fed cartridges from tubular, box or spool-type magazines, and may be either hammerless or feature an exposed hammer at the rear of the action.

Load — used as a verb, it means to insert ammunition into a gun's magazine or chamber ("chamber loaded") or to reload an empty cartridge case with fresh components. As a noun, it refers to a single cartridge, or sometimes is used in reference to the weight of the bullet or shot charge contained by a cartridge or shotshell.

Loading Gate — a hinged gate that covers the loading port of a rifle or revolver.

Loading Press — a tool used for reloading metallic cartridge cases or paper or plastic-cased shotshells.

Lock — used as a noun to refer to the firing mechanism of a muzzle-loading firearm. Also used in reference to certain double-barreled shotgun (sidelock, boxlock) actions. As a verb it means closing the action of a firearm in preparation for firing.

Locked-Breech Action — Upon firing, the gun's breechblock and barrel remain locked together (and travel to the rear together) until the bullet leaves the muzzle. Residual rearward pressure continues to move the breechblock and barrel to the back of the gun's action, where these two units are mechanically separated and the fired shell casing ejected. The barrel now returns (under spring pressure) to its original position, and the breechblock follows behind, picking up a fresh round from the magazine.

Lockplate — the metal plate that holds the lock mechanism of blackpowder guns, or that covers the lock mechanism of double-barreled shotguns.

Lock Time — the time that elapses between completing the trigger pull and having the firing pin actually strike the primer.

Locking Lug — a short, usually heavy stud projecting from the side of a firearm's bolt, that cams into a mating slc cut into the receiver to firmly lock the bolt closed against the breech face.

Lockwork — the moving parts of a firing mechanism, or action.

Long Gun — a gun fired from the shoulder, usually with a longer barrel and stock than used in handguns. The term is used primarily for rifles and shotguns.

Long Rifle — a 22-caliber rimfire cartridge used for target shooting, plinking and hunting small game. This is the most popular rifle and handgun cartridge in existence.

Low Base or Low Brass — a term used to describe shotshells with relatively low brass heads, or bases. Commonly used in Skeet and trap target ammunition.

Machine Gun — a military weapon capable of sustained automatic fire. While the term is often applied to any automatic firing weapon, the military definition refers to a relatively heavy gun, usually mounted on a tripod, bipod or some other mechanical support.

Machine Rest — a mechanical device a firearm is fastened to for support during firing. Used in accuracy tests to eliminate human error.

Magazine — a container that carries the ammunition supply for a repeating rifle, handgun or shotgun. The cartridges or shotshells are pushed by spring pressure up into the action. Magazines may be permanently attached to the firearm (as in tubular or some rotary-spool magazines) or detachable. A detachable metal box magazine is sometimes called a *clip*.

Magazine Cut-Off — a mechanical device that, when engaged, prevents cartridges from feeding from a gun's magazine into the breech.

Magazine Follower — see *Follower*.

Magazine Plug — a plug used in the tubular magazines of semi-automatic, slide-, and bolt-actin shotguns to reduce the capacity to a total of three shells (including the one in the chamber) to comply with Federal Migratory Bird laws. Required in the United States for hunting game animals.

Magazine Safety — a device that prevents a gun from firing when the magazine has been removed.

Magnum — a cartridge or shotshell containing a particularly heavy or powerful powder charge. Also used to designate the firearms designed to fire these cartridges or shotshells.

Mainspring — the spring that powers the hammer or striker of a gun.

Malfunction — the failure of a firearm to operate, for whatever reason.

Mannlicher Stock — a rifle stock, originally designed by Ferdinand Ritter von Mannlicher in the 19th century, that runs full length along the barrel to end at the muzzle. Popular on some modern carbines.

Master Eye — the eye that is strongest of the pair.

Match — today used in reference to organized shooting competition to designate a particular shoot or event. Also refers to the burning fuse (match) used in early matchlock firearms.

Match Ammunition — special high-quality ammunition designed for target match competition.

Matchlock — a form of firearm ignition dating back to the 15th century in which a match, or slow-burning fuse, is mechanically held to the side of the gun's breech until time to fire. Then the burning end of the fuse is pulled down and back, either by direct pressure on the arm holding the fuse or by a trigger-operated lever, until it contacts the priming powder to ignite the propellant charge.

Matte Finish — a dull, glare-free metal finish.

Metallic Sights — see *Iron Sights*.

Minute of Angle — abbreviated MOA. An angular measure equal to 1/60 of one degree, used in referring to the accuracy of a firearm. One minute of angle is approximately equal to one inch of movement at 100 yards, and a rifle capable of "minute of angle accuracy" is capable of grouping three or more successive shots into a 1-inch cluster at that distance.

Misfire — the failure of a cartridge to fire.

Monte Carlo Comb or Stock — a raised comb found on some rifle and shotgun buttstocks.

Mushroom — the expansion of a bullet upon impact into the approximate shape of a mushroom.

Musket — a term used for early smoothbore firearms that fired a single ball.

Muzzle — the forward end of a barrel.

Muzzle Blast — the shock wave and resultant noise that occurs at the muzzle when the projectile leaves the bore, caused by the release of hot gases under pressure.

Muzzle Brake — a device attached to the muzzle of a firearm or a pattern of holes or slots cut into the barrel near the muzzle to reduce recoil.

Muzzle Energy — the kinetic energy, or force, carried by a projectile as it leaves the muzzle of a firearm.

Muzzleloader — any firearm that must be loaded from the muzzle end. The term is usually applied only to guns designed for blackpowder loads. While muzzle-loading guns became obsolete more than a century ago, some of these early firearms are still in use and modern replicas are being manufactured today, as shooting muzzleloaders has become a popular sport in recent years.

Muzzle Velocity — the velocity of a projectile fired from a gun at the moment it leaves the muzzle.

National Matches — rifle and handgun matches held annually at Camp Perry, Ohio, under the direction of the National Rifle Association.

National Rifle Association — an association of firearms shooters and enthusiasts, dating back to 1871, dedicated to promote shooting as a safe and wholesome sport. It is head-quartered in Washington, D.C.

National Shooting Sports Foundation — an industry organization founded in 1961 to promote the shooting sports.

Neck — the part of a cartridge case extending backward from the mouth of the case to the beginning of the case shoulder.

Neck Sizing — squeezing only the mouth and neck (the portion of case found directly behind the mouth) of a fired cartridge case back to its original size during the reloading operation.

Needle Gun — a breech-loading rifle developed in the 19th century that used a long, needle-like firing pin to reach a primer located in front of, instead of at the back of, the propellant powder charge in the cartridge.

Nipple — the hollow, protruding seat on which a percussion cap is placed. Found only on blackpowder muzzle-loading guns.

Nose — the forward tip of a bullet.

Oil Finish — a gunstock finish made by applying several coats of special oil and rubbing each coat in.

Open Sight — a rear iron sight with a flat-topped leaf featuring a V- or U-shaped aiming notch.

Operating Handle — a handle protruding from the bolt or operating rod of a semi-automatic rifle or shotgun, used to manually open the action.

Operating Rod — the arm or rod in a semi-automatic rifle or shotgun operated by gas pressure that transfers the energy from the gas piston to the action.

Over/Under — a double-barreled gun or action in which one barrel is fastened directly over the other. Also sometimes referred to as a "Superposed" gun, after the well-known Browning over/under shotgun.

Overtravel — movement of the trigger past the point where the sear of hammer is released.

Palm Rest — an adjustable, rounded support used in certain target rifles. The palm rest is attached to the lower part of the forend and is held by the non-shooting hand to support the rifle.

Pan — the small container that holds the priming powder in pre-percussion firearms.

Parkerize — a type of grey non-reflective finish generally applied to military guns.

Paster — a small square or patch of paper used to paste over bullet holes in a target.

Patch — a small piece of cloth used to clean the bore of a gun. The term also refers to a piece of cloth or paper partially wrapped around a lead ball when the ball is loaded into a blackpowder muzzleloader.

Patch Box — a box incorporated into the buttstock of a blackpowder muzzleloader to hold patches and other loading materials.

Patridge Sight — an open iron rifle and handgun sight that features a square-topped front sight blade in combination with a flat-topped rear sight using a square sighting notch.

Pattern — the pattern made by a charge of shot fired from a shotgun when it hits a large sheet of paper at a distance of 40 yards. One way of evaluating a shot pattern is to draw a 30-inch diameter circle around the heaviest concentration of shot holes, and then count the number of hits within the circle. This number is then divided by the number of pellets contained within the entire shot charge to give you the percentage of hits in the marked-off area. The choke of a gun is evaluated by its pattern percentages.

Peep Sight — see *aperture sight.*

Pellet — an individual shot, or pellet used in shotgun shells. Also a lead projectile used in airguns.

Pennsylvania Rifle — see *Kentucky rifle.*

Pepperbox — an early percussion-fired pistol featuring a number of parallel barrels arranged to form a single cylinder. This cylinder was rotated to line up each barrel in turn with the firing mechanism.

Percussion Cap — a small, cup-shaped metal container partially filled with fulminate priming compound. The container, or cap is designed to fit around a hollow nipple.

Percussion Firearm — any gun or rifle fired by striking a percussion cap, usually with a trigger-activated hammer.

Percussion Lock — the lockwork in a percussion firearm.

Piece — slang or jargon for rifle or handgun.

Pinfire Cartridge — an early self-contained metal cartridge that was fired by the hammer striking a small pin which projected from the side of the cartridge case which in turn ignited the priming compound inside the cartridge.

Pistol — technically, a handgun with a fixed chamber (magazine-fed selfloader or semi-automatic, or single shot), although some people mistakenly use the term in describing revolvers, as well. Small, full-automatic hand-held military weapons are sometimes called machine pistols.

Pistol Grip — the curved portion of a rifle or shotgun buttstock falling immediately behind the receiver and serving as a grip for the shooting hand.

Pistol Scope — a telescopic sight intended specifically for mounting on a handgun used for hunting. Differs from a rifle scope primarily in that eye relief is much longer.

Pitting — a term used to describe pits formed by rust or oxidation, usually in the bore of a firearm.

Plinker — a small caliber rifle or handgun used in plinking.

Plinking — informal shooting at printed paper targets, tin cans, water-filled ballons or other "fun" targets. Plinking is a popular pastime for owners of 22 rimfire firearms.

Plug — see *magazine plug.*.

Point-Blank — shooting done at very close range.

Point of Aim — the point on which the sights of a firearm are aligned when firing.

Point of Impact — where a projectile fired from a gun or rifle actually strikes. (Because a bullet's flight describes a parabola in relation to the sighting picture a shooter sees, the point-of-impact and the point-of-aim will exactly coincide at only two points along the flight path).

Powder — that component of a loaded cartridge or loose firearm charge that burns upon ignition, producing the expanding gases which force the projectile down the bore.

Powder Charge — a measured amount of gunpowder, usually that amount needed in a single cartridge or shotshell.

Powder Flask or Horn — a container for holding blackpowder.

Powder Measure — a device used to measure individual charges of powder by volume.

Press — a mechanical device used by reloaders to force a cartridge case through a series of dies during the loading process.

Primer — the fulminate-filled container used to ignite the propellant charge in a firearm cartridge or shotshell. The primer used in centerfire cartridges is a small metal cup-shaped device containing the priming mixture between the rear face of the primer and a solid anvil. This is held in a centrally located hole in the base of the cartridge by friction. The primer in rimfire cartridges consists of the entire cartridge rim, which is hollow and contains the priming compound.

Primer Pocket — the hole or depression located in the base of a cartridge, designed to hold the primer.

Projectile — anything fired from a gun, or thrown by any controlled means. Firearms projectiles are usually either solid bullets, or a shot charge containing many pellets.

Proof — to test the strength of a firearm by firing it with a cartridge loaded well above normal strength. Proofing is done under carefully regulated conditions, and once a gun passes a proof test it is stamped with a mark called a "proof mark" to show that it is safe to use with *standard* loads.

Proof Load — an overloaded cartridge used in proof tests.

PSI — common abbreviation for "pounds per square inch," — a measure of pressure.

Pump-Action — an action used in rifles and shotguns in which the bolt or breechblock is moved forward and back by the forward and back movement of the forend. Also called slide-action or trombone action.

Propellant — the powder charge burned to produce the gases that move or propel a projectile from a firearm.

Punt Gun — a large shotgun mounted on the front of a small boat and used by early market hunters to kill large numbers of ducks. The use of these guns is illegal in the United States.

QD Swivels (quick-detachable swivels) — steel loops used to fasten a sling or carrying strap to the stock of a rifle or shotgun. Plain swivels are permanently attached to the gunstock, while QD swivels can be removed in seconds to allow the sling to be transferred to another firearm.

Ramp Sight — a front sight that is mounted on a long, sloping ramp.

Ramrod — a rod used to push, or ram the various load components (wads and projectile) down the barrel of a muzzle-loading firearm.

Range — an area set aside for target shooting. Also the distance over which a gun is fired — the distance measured between the shooter and the target.

Rangefinder — a device used for determining range, or optically measuring the distance to a distant target.

Rate of Twist — the number of complete revolutions the rifling makes over a given length of barrel. Usually expressed as one turn (or twist) in so many inches.

Receiver — the assembly on a rifle or shotgun that *contains* the working parts of the action, including the bolt, trigger, and magazine housing. Usually located at the rear of the barrel.

Receiver Sight — see *aperture sight.*

Recoil — the rear-directed forces caused by firing a projectile in the other direction.

Recoil Lug — a piece of metal projecting from the action of a rifle or shotgun through which recoil force is transmitted to the gunstock.

Recoil Operated — a semi-automatic firearm action that is operated by the forces of recoil rather than by expanding gases.

Recoil Pad — a rubber or synthetic pad fastened to the butt end of the stock to help soften the effect of recoil on the shooter's shoulder.

Recoil Shield — the metal flange or projection extending from the sides of a revolver frame directly behind the cylinder.

Reloading — an empty cartridge case or shotshell hull with fresh components so that it can be fired again. Handloading.

Repeater — a gun capable of being fired more than once before reloading becomes necessary. Most repeating guns feature a magazine that feeds fresh ammunition to the firing chamber, although a revolver (which doesn't have a magazine, but several individual chambers) is classified as a repeater.

Resizing — squeezing a fired cartridge case back down to or near its original diameter by forcing it through a loading die designed for the purpose.

Rest — a support used for shooting.

Reticle — the arrangement of crosshairs, posts or dots used as an aiming device while looking through a telescopic sight.

Revolver — a handgun featuring a revolving cylinder containing several individual firing chambers evenly distributed around the radius of the cylinder. Cocking an exposed hammer (or simply pulling on the trigger in the case of double-action revolvers) rotates the cylinder to align each chamber with the bore, in turn.

Rib — a flat strip of metal fastened to the top of a rifle, shotgun or handgun barrel, usually running full length along it. A ventilated rib is raised above the barrel's surface and anchored with a series of metal posts.

Ricochet — the glancing of a bullet from any hard or unyielding surface.

Rifle — a firearm featuring both a rifled barrel and a stock that is held to the shoulder with both hands while the rifle is in firing position.

Rifled Slug — a single lead projectile designed to be fired in a shotgun for hunting deer and other large game. A rifled slug may be of all-lead construction and "finned" or grooved around its outside diameter, or it may have fiber or plastic wads attached to its base and lack "rifling" grooves entirely.

Rifling — the grooves cut in a spiral pattern on the inner surfaces of a gun's bore. These grooves form lands, and together they grip the bullet as it travels down the bore, imparting spin to stabilize the bullet in flight.

Rifling Marks — the marks made on a bullet's surface as it passes through a rifled bore.

Rim — the projecting edge around the base of some cartridges and all shotshells. The extractor grips this rim when removing the cartridge from the firing chamber.

Rimfire — a cartridge that contains its priming compound in the rim of its case. Rimfire cartridges cannot be reloaded after firing, as there's no practical way to replace the priming material. The most popular rimfire cartridges in use today are the 22 Long Rifle and 22 Short.

Rolled Crimp — a crimp, or roll at the mouth of a shotgun shell used to hold a cardboard wad over the shot.

Rolling Block — a firearm action featuring a "rolling" breechblock that rotates around a pin to open and close the breech. Used in single-shot rifles.

Rotary Magazine — a magazine that holds cartridges in a spool-like container, which rotates to feed each round into the chamber.

Round — refers to a single cartridge or shotshell and is also used in reference to a sequence of 25 clay targets thrown in either Skeet or trap.

Runover — a line of checkering that continues past the border of the checkering pattern.

Ruptured Case — a cartridge case that separates into two sections as a result of firing.

SA — abbreviation for single-action handguns.

Saddle Ring — a steel ring fastened to the left side of the receiver of a rifle or carbine.

Safety — a mechanical device designed to prevent accidental or unintentional firing of a gun. Most firearm safeties are operated by pushing a cross-bolted pin, a sliding button, or a lever. When engaged, these safeties block the trigger, hammer, or sear and prevent them from functioning.

Safety Notch — a notch at the base of the hammer in a single-action handgun that holds the hammer slightly to the rear of the firing pin or firing pin slot.

Scabbard — a gun sheath made of leather or some other material, designed to be fastened to a saddle, trail bike or some other means of transportation.

Scattergun — a shotgun.

Scatter Load — see *brush load*.

Schnabel Forend — a downward-curved tip carved at the end of the forend on some rifle stocks.

Scope — popular abbreviation for telescopic sight.

This optical aiming aid magnifies the target and superimposes a set of crosshairs or some other reticle over the target when a shooter looks through its lenses.

Scope Bases or Mounts — small blocks or ribs of steel fastened firmly to the receiver or action of a firearm to provide a suitable surface for the attachment of scope rings.

Scope Rings — metal rings fastened tightly around the tube of a scope sight, with clamps or projections extending downward to allow the rings to be fastened to a set of scope bases or mounts.

Sear — that part of the firing mechanism that holds the hammer or striker in the cocked position until the trigger is pulled.

Sectional Density — the ratio of the mass, or weight of a bullet to its cross-sectional area.

Selective Ejector — a mechanism, usually found in double-barreled shotguns, that throws fired hulls clear of the action while leaving unfired shotshells chambered when the action is opened.

Semi-automatic or Selfloader — a firearm that fires a cartridge, extracts and ejects the empty cartridge case, cocks the hammer or striker, and automatically feeds and chambers a fresh cartridge each time the trigger is pulled. The first round must be fed manually to get the sequence started, but after that the gun will fire repeatedly each time the trigger is pulled until the magazine is empty.

Setscrew — a screw used in adjusting the trigger pull in some firearms. Also refers to a screw used to anchor a larger screw in place.

Set Trigger — see *double set triggers*.

Shell — a metal and plastic or metal and paper cartridge loaded with primer, powder, wads and a charge of shot pellets, used in shotguns.

Shellholder — a slotted device that grips the rim or base of a cartridge case to hold it in a loading press.

Shoot — to fire a gun. The term also refers to an organized trap or Skeet target competition.

Shooting Glasses — eyeglasses designed for shooters. They feature large lenses in a variety of different colors for differing light conditions, and the lenses are tempered to protect the eyes from the possibility of hot gas leaking back through the action if a cartridge case ruptures.

Short — refers to the 22 rimfire Short rifle and handgun cartridge.

Shot — a tiny pellet or sphere of steel, lead alloy or other soft metal used in shotgun shells. Shot is available in many different sizes, with each size designated by a numbering scale that runs from 00 (double-ought) through 12. The larger the number, the smaller the size of the shot and vice versa. A particularly large variety of shot, known as buckshot, is used for hunting larger animals, while the smaller shot sizes are used for clay targets, hunting birds and other small game.

Shot Cartridge — a metallic cartridge loaded with small shot in place of a solid bullet for use in rifles and handguns.

Shot Collar or Protector — a thin-walled plastic tube surrounding the shot load to protect individual pellets from deforming as the load and shot collar travel down the bore.

Shot Column — the tubular-shaped column formed by a charge of shot as it travels down the bore.

Shotgun — a firearm featuring a smooth, unrifled bore and a shoulder stock, designed to fire shotshells.

Shotgun Plug — see *magazine plug.*

Shot Pattern — see *pattern.*

Shotshell — see *shell.*

Shot String — the elongated, strung-out arrangement of shot pellets after they leave the bore of a gun.

Shot Tower — a tall tower from which droplets of molten lead are dropped into a tank of water. As the droplets fall, they form a perfectly spherical shape and keep that shape as the water hardens them. These droplets are now called "shot," screened for size and used by the ounce (and/or fraction of) in the loading of shotshells.

Shoulder — the tapering radius of a bottleneck cartridge case where the case body narrows to form the neck.

Shoulder Holster — a holster designed to carry a handgun under or near the shooter's armpit.

Shoulder Pad — a pad of leather, corduroy or some other contrasting material sewn to the shoulder area of a jacket or shooting shirt.

Side Arm — usually refers to a handgun, although the term may be used in reference to other small, easily carried weapons.

Side-by-Side — a double-barreled shotgun with barrels aligned horizontally, one beside the other.

Sidehammer Gun — a firearm featuring a hammer offset to one side of the action.

Sideplate — a metal plate that covers the working parts of a gun's action.

Sight — any device used by a shooter to aim a rifle, handgun or shotgun toward a target.

Sight Disc — an insert used with aperture sights.

Sighting-In — the process of adjusting rifle or handgun sights so that the bullet strikes the point the sight intersects at a given range.

Sighting Shot — a shot fired at the beginning of a target match to check that the sights are in proper alignment.

Sight Radius — the distance between the front blade or post and the rear sight, used in reference to iron sights.

Silencer — an illegal (in this country) device that absorbs or muffles the sound made by a gun as it fires.

Single Action — a term used to describe a revolver that must be manually cocked *each* time it's fired, or an auto pistol that must be manually cocked by pulling back the hammer or slide before the *first* shot can be fired (subsequent shots can be fired by merely pulling on the trigger).

Single Shot — a firearm that must be manually loaded before each shot is fired. A gun without a magazine.

Single Stage Trigger — a trigger that displays little or no discernible movement before releasing the sear.

Single Selective Trigger — a single trigger used on a double-barreled gun, in which the firing sequence can be selected by the shooter.

Single Trigger — a single trigger used on a double-barreled gun, in which the firing sequence is pre-set at the factory. Shotguns with single (non-selective) triggers normally fire the more open-choked barrel first, followed by the tighter-choked tube.

Six O'clock Hold — a hold, or sight picture used in target shooting in which the front sight post is placed at the very bottom edge of the bullseye.

Sizing or Resizing — restoring an expanded cartridge case to or near its original size by forcing it into a metal die.

Skeet — a shotgun target game in which the shooter stands in one of 8 positions around a semicircle and on its radius and shoots at clay targets thrown from a high platform on the left, or a lower one on the right.

Slide — the forend and action bars in a slide- or

pump-operated rifle or shotgun. Also refers to the metal housing used to cover the barrel and striker mechanism in an auto pistol. In operation the pistol slide moves back and forth to effect extraction, ejection, cocking, and reloading.

Slide Action — a rifle or shotgun action operated by moving a reciprocating forend back and forth.

Sling — a strap attached to the buttstock and forend (or barrel) of a rifle or shotgun. Used for carrying the firearm from the shoulder when not in use, or (in the case of a rifle sling) wrapped around the supporting arm during firing to provide additional support.

Sling Swivel — the steel loop used to fasten a sling to the stock of a firearm.

Slug — see *rifled slug*.

Slug Gun — a shotgun specially intended for firing rifled slugs, usually equipped with rifle-style sights.

Small Bore — a shotgun with a smaller than 16-gauge bore, or a rifle of less than 25 caliber. Commonly used to describe rifles and handguns chambered for 22 rimfire cartridges.

Smokeless Powder — a nitrocellulose-based propellant burned in modern firearms to produce the gases needed to drive a projectile through the bore and out the barrel. Smokeless powder was invented in the 1880s.

Smoothbore — a gun with a smooth, unrifled bore. The term is commonly used in reference to a shotgun.

Snap Cap — a dummy cartridge or chamber insert with a cushioned dummy primer. Used to protect firing pins from breaking when practicing shooting with an unloaded gun.

Snap Shot — a shot taken quickly without careful aiming.

Solid Frame — a firearm that does not allow the barrel to be separated from the frame or action without a major disassembly effort.

Speed Loader — a device which enables a revolver cylinder to be fully loaded with fresh cartridges in one motion; often used by police.

Spin — the rotational movement imparted to a bullet by the spiral lands and grooves of a rifled barrel.

Spitzer — a bullet that gradually tapers to a point.

Splinter Forend — a very narrow, thin forend used on some double-barreled shotguns.

Sporterizing — any work done to a military rifle to make it more suitable for civilian sporting use. The term usually refers to reforming the stock or adding a new one (the simplest form of sporterizing), although many more refinements may be made.

Spotting Scope — a high-magnification telescope separate from a gun, used by target shooters to examine targets from the firing line.

Spreader Load — see *brush load*.

Sprue — a small cone of lead that is left on a cast bullet when it leaves the casting mould. This is normally removed before the bullet is used. The term also refers to the opening or port in a casting mould through which the molten lead is poured.

Sprue Cutter — a device incorporated into a bullet mould to remove all or part of the sprue when the mould is opened.

Stackbarrel — a term used in reference to double-barreled guns with the barrels aligned vertically; an over/under.

Stacking Swivel — a swivel or loop found near the muzzle on some military rifles, used to "stack" or prop the rifles off the ground in a tripod arrangement.

Standing Breech — the solid breech on break-top rifles and shotguns.

Star Crimp — a 5- or 6-pointed star-shaped crimp used to fold the mouth of the case back over the components in a shotgun shell.

Steel Shot — steel pellets or spheres loaded in shotshells intended for hunting. Developed to eliminate environmental hazards shown to be caused by lead shot.

Stock — the wooden or synthetic handle attached to a rifle or shotgun, and used to support the firearm and steady it against the shooter's shoulder.

Striker — a spring-activated rod or piece of metal released by pulling the trigger to move forward against the rear face of the firing pin. Serves the same function as a hammer. In some guns the striker serves as the firing pin.

Superposed — a Browning trade name for a European-made over/under double-barreled shotgun.

Sustained Lead — a method used to hit a moving target by picking a point in front of the target to aim the gun at. As the gun is swung with the moving target, the distance between this arbitrary aiming point and the target itself is maintained until the gun is fired.

Swing-Through Lead — a method used to hit a moving target, usually with a shotgun charge, by moving the gun muzzle faster than the target, from a point behind the target, until the muzzle passes in front of the target. The trigger is pulled just as the gun barrel is in alignment with the target, and if things are done correctly, the continuing movement or "swing" will provide the right amount of lead to allow a hit.

Swivel — see *sling swivel*.

Takedown — partially disassembling a firearm. Also refers to a rifle or shotgun designed to allow the barrel to be removed from the receiver and frame very easily.

Tamping — the act of firmly compressing (with a ramrod) blackpowder wads and projectile(s) in the rearmost area of a blackpowder firearm's bore.

Tang — the extended arm or arms at the rear of a receiver used in fastening the receiver to the buttstock.

Target — a mark used to shoot at. Usually refers to a sheet of cardboard or paper on which a series of concentric rings are printed.

Target Pit — an excavated pit found on some large target ranges. During a match, a crew of men work in this pit to record the scores on the targets mounted above them.

Telescopic Sight — see *scope sight*.

Throat — that part of a firearm's bore located immediately in front of the chamber. It tapers from the diameter of the firing chamber to the true diameter of the bore.

Thumbhole Stock — a rifle stock with comb extending straight back from the receiver, featuring a contoured hole to accommodate the thumb of the shooting hand.

Thumbrest — a ledge or indentation on the left side of the handgun grip (for right-handed shooters) designed to allow the shooter's thumb to rest against it.

Thumb Safety — a safety activated by the shooter's thumb.

Tip-Off Mount — a set of rings and mounts designed to allow easy removal of a scope sight from the receiver it's mounted on. Many 22 rimfire rifle receivers are grooved to accept a small-scale tipoff mount used with a number of inexpensive scope sights intended for rimfire use.

Toe — The extreme lower end of a buttstock.

Tolerance — the amount or degree of variation allowed in the dimensions or fit of parts used in firearms or ammunition.

Tong Tool — a hand-held handloading tool that performs the same functions of a bench-mounted loading press, but with less speed and convenience.

Top-Break — used to describe a gun action that is hinged at the bottom and opens or "breaks" at the top. See *break-top*.

Top Lever — a lever commonly found on double barreled rifles and shotguns. This lever is located at the rear of the action, usually over the tang, and pivots to the side to unlock the action.

Top Strap — the upper portion of a revolver's frame that extends from the rear sight notch to the rear of the barrel.

Touch Hole — the hole or vent used in early cannon lock firearms to ignite the powder charge in the firing chamber.

Tracer — a bullet or shot column treated to give off a bright light which can be easily seen as the projectile(s) travels toward the target. Generally used only by the military, but has been used as a training aid for beginning trap and Skeet shooters.

Trajectory — refers to the curved path of a bullet in flight.

Transfer Bar — part of the mechanism of some handguns that interposes between the hammer and firing pin to minimize the possibility of accidental discharge until the trigger is pulled.

Trap — a shotgun target game in which the shooter stands at a pre-determined distance behind the platform or station that the clay targets are thrown from. Also refers to the device used for throwing clay shotgun targets.

Trapdoor Action — a firearms action that features a breechblock that pivots up and forward when opened. This is an obsolete single shot action used today only in replicas of the military Springfield rifles dating back to the last century.

Trigger — that part of a gun's firing mechanism that is pressed by the shooter's finger to start the firing process. Usually a curved arm or bar projecting downward from the area of the action.

Trigger Guard — a loop of metal, plastic or other material that partially encircles the trigger to help prevent accidental firing.

Trigger Stop — a bar or other projection that limits the rearward travel of the trigger.

Trombone Action — pump-, or slide-action.

Try Gun — a shotgun with a fully adjustable stock, used to custom fit a made-to-order gunstock to a particular shooter. The try gun stock is adjusted to fit the customer, and the custom stock is made to conform to its dimensions.

Tube — used in reference to the body of a tubular magazine, or to the barrel of a firearm.

Tubular Magazine — a tube-shaped magazine in which the cartridges or shotshells are placed end to end.

Turnbolt Action — see *bolt-action*.

Turret — that part of a telescopic sight used to house the elevation and windage adjustment screws.

Twist — the spiral described by the lands and grooves in a rifled bore. See *rate of twist*.

Underhammer — a firearm action with a hammer located underneath the action or barrel. Used mainly in early percussion lock guns.

Underlever — see *lever-action*.

Upland Game — small game or birds usually hunted with a shotgun and shot at close range.

Variable Choke — a shotgun choke device that can be adjusted to throw varying shot patterns.

Variable Scope — a telescopic sight that provides an image of varying magnification. The magnification, or "power" of the scope is adjusted by turning a ring mounted just ahead of the rear eyepiece housing.

Varmint Rifle — a small-caliber rifle, usually equipped with a telescopic sight, designed for hunting small animals designated as "varmints," often at long range. Varmints include ground squirrels, rock chucks, marmots, foxes, bobcats, coyotes and other small animals that can usually be hunted year-around.

Velocity — the speed of a bullet or shot charge as it travels to the target. Because such projectiles are subject to air pressure, their velocities vary continuously until the projectiles end their flight.

Vent — an opening in a firearm to allow gas to escape without injuring the shooter in the event of a case rupture or failure. Vents are also found in the barrel and forend of gas-operated guns to direct spent gases from the operating cylinder. The hole connecting the firing chamber to the percussion cap or priming in a muzzle-loading firearm is also called a vent.

Ventilated Rib — see *rib*.

Wad — a short column of felt, plastic or paper used to separate the active components in a shotgun shell.

Wad Pressure — the pressure used by a handloader in seating a wad in a shotshell.

Wadcutter — a perfectly cylindrical bullet with a blunt, flat nose used in handguns primarily for target shooting.

Wax Bullet — a bullet made of wax, used for short-range target practice indoors.

WCF — abbreviation for Winchester Center Fire.

Wheellock — a form of lock, or an ignition system used in early muzzle-loading firearms, that featured a spring-powered wheel that rotated against a piece of flint to produce sparks.

Wildcat — a cartridge designed by a handloader, and not available from a commercial ammunition manufacturer.

Wind Deflection or Drift — the sideways movement of a bullet in flight, caused by wind pressure.

Windage — a sight adjustment or adjusting mechanism used to move the sight or reticle sideways. In blackpowder muzzle-loading guns, the terms refers to the difference in the diameter of the bore and the diameter of the ball or projectile to be loaded into the gun.

WMRF — Winchester Magnum Rimfire. Also WMR.

Worm — a claw-like device with corkscrew-shaped arms attached to the end of a ramrod to allow "pulling" or disassembling a loaded charge from a muzzle-loading firearm.

WRF —Winchester Rimfire

Wrist —the narrow diameter of a buttstock at the pistol grip or grip.

Wundhammer Swell — a slight bulge on the right side of a rifle's pistol grip, designed to fit the palm of the shooting hand.

X-Ring — a small inner ring within the bullseye of a rifle or handgun target.

Yoke — another name for revolver crane.

Zero — as a noun, refers to a particular sight setting that will put the bullet of a particular load on target at a specified range. As a verb, the term refers to the act of adjusting the sights on a rifle or handgun, and usually infers firing at a target to confirm the setting.

EP BM We hope you enjoyed this title
from Echo Point Books & Media

Before Closing this Book, Two Good Things to Know

1. Buy Direct & Save

Go to www.echopointbooks.com (click "Our Titles" at top or click "For Echo Point Publishing" in the middle) to see our complete list of titles. We publish books on a wide variety of topics—from spirituality to auto repair.

Buy direct and save 10% at www.echopointbooks.com

DISCOUNT CODE: EPBUYER

2. Make Literary History and Earn $100 Plus Other Goodies Simply for Your Book Recommendation!

At Echo Point Books & Media we specialize in republishing out-of-print books that are united by one essential ingredient: high quality. Do you know of any great books that are no longer actively published? If so, please let us know. If we end up publishing your recommendation, you'll be adding a wee bit to literary culture and a bunch to our publishing efforts.

Here is how we will thank you:

- A free copy of the new version of your beloved book that includes acknowledgement of your skill as a sharp book scout.

- A free copy of another Echo Point title you like from echopointbooks.com.

- And, oh yes, we'll also send you a check for $100.

Since we publish an eclectic list of titles, we're interested in a wide range of books. So please don't be shy if you have obscure tastes or like books with a practical focus. To get a sense of what kind of books we publish, visit us at www.echopointbooks.com.

If you have a book that you think will work for us, send us an email at editorial@echopointbooks.com

www.ingramcontent.com/pod-product-compliance
Lightning Source LLC
Chambersburg PA
CBHW050643150426

42813CB00054B/1167